Symmetry in Applied Continuous Mechanics

Editors

Marin Marin
Dumitru Baleanu
Sorin Vlase

MDPI • Basel • Beijing • Wuhan • Barcelona • Belgrade • Manchester • Tokyo • Cluj • Tianjin

Symmetry in Applied Continuous Mechanics

Editors
Marin Marin
University of Brasov
Romania

Dumitru Baleanu
Institute of Space Sciences
Romania
Cankaya University
Turkey

Sorin Vlase
"Transilvania" University of Brasov
Romania

Editorial Office
MDPI
St. Alban-Anlage 66
4052 Basel, Switzerland

This is a reprint of articles from the Special Issue published online in the open access journal *Symmetry* (ISSN 2073-8994) (available at: https://www.mdpi.com/journal/symmetry/special_issues/Symmetry_Applied_Continuous_Mechanics).

For citation purposes, cite each article independently as indicated on the article page online and as indicated below:

LastName, A.A.; LastName, B.B.; LastName, C.C. Article Title. *Journal Name* **Year**, *Article Number*, Page Range.

ISBN 978-3-03943-030-7 (Hbk)
ISBN 978-3-03943-031-4 (PDF)

© 2020 by the authors. Articles in this book are Open Access and distributed under the Creative Commons Attribution (CC BY) license, which allows users to download, copy and build upon published articles, as long as the author and publisher are properly credited, which ensures maximum dissemination and a wider impact of our publications.

The book as a whole is distributed by MDPI under the terms and conditions of the Creative Commons license CC BY-NC-ND.

Contents

About the Editors vii

Preface to "Symmetry in Applied Continuous Mechanics" ix

Marin Marin, Dumitru Băleanu and Sorin Vlase
Symmetry in Applied Continuous Mechanics
Reprinted from: *Symmetry* 2019, *11*, 1286, doi:10.3390/sym11101286 1

Yadong Zhou, Youchao Sun and Tianlin Huang
Impact-Damage Equivalency for Twisted Composite Blades with Symmetrical Configurations
Reprinted from: *Symmetry* 2019, *11*, 1292, doi:10.3390/sym11101292 7

Noelia Bazarra, José A. López-Campos, MarcosLópez, Abraham Segade
and José R. Fernández
Analysis of a Poro-Thermo-Viscoelastic Model of Type III
Reprinted from: *Symmetry* 2019, *11*, 1214, doi:10.3390/sym11101214 19

Iuliu Negrean and Adina-Veronica Crișan
Synthesis on the Acceleration Energies in the Advanced Mechanics of the Multibody Systems
Reprinted from: *Symmetry* 2019, *11*, 1077, doi:10.3390/sym11091077 37

M. Marin, S. Vlase, R. Ellahi and M.M. Bhatti
On the Partition of Energies for the Backward in Time Problem of Thermoelastic Materials with
a Dipolar Structure
Reprinted from: *Symmetry* 2019, *11*, 863, doi:10.3390/sym11070863 57

Eliza Chircan, Maria-Luminița Scutaru and Cătălin Iulian Pruncu
Two-Dimensional Finite Element in General Plane Motion Used in the Analysis of
Multi-Body Systems
Reprinted from: *Symmetry* 2019, *11*, 848, doi:10.3390/sym11070848 73

Zhang Yue and Dumitru Baleanu
Inference about the Ratio of the Coefficients of Variation of Two Independent Symmetric or
Asymmetric Populations
Reprinted from: *Symmetry* 2019, *11*, 824, doi:10.3390/sym11060824 81

Ji-Jun Pan, Mohammad Reza Mahmoudi, Dumitru Baleanu and Mohsen Maleki
On Comparing and Classifying Several Independent Linear and Non-Linear Regression Models
with Symmetric Errors
Reprinted from: *Symmetry* 2019, *11*, 820, doi:10.3390/sym11060820 93

Mariana D. Stanciu, Sorin Vlase and Marin Marin
Vibration Analysis of a Guitar considered as a Symmetrical Mechanical System
Reprinted from: *Symmetry* 2019, *11*, 727, doi:10.3390/sym11060727 103

Xinghan Xu and Weijie Ren
A Hybrid Model Based on a Two-Layer Decomposition Approach and an Optimized Neural
Network for Chaotic Time Series Prediction
Reprinted from: *Symmetry* 2019, *11*, 610, doi:10.3390/sym11050610 119

Krzysztof Kamil Żur and Piotr Jankowski
Multiparametric Analytical Solution for the Eigenvalue Problem of FGM Porous Circular Plates
Reprinted from: *Symmetry* **2019**, *11*, 429, doi:10.3390/sym11030429 . . . . **137**

Elsayed M. Abd-Elaziz, Marin Marin and Mohamed I. A. Othman
On the Effect of Thomson and Initial Stress in a Thermo-Porous Elastic Solid under G-N Electromagnetic Theory
Reprinted from: *Symmetry* **2019**, *11*, 413, doi:10.3390/sym11030413 . . . . **161**

Behzad Ghanbari, Dumitru Baleanu and Maysaa Al Qurashi
New Exact Solutions of the Generalized Benjamin–Bona–Mahony Equation
Reprinted from: *Symmetry* **2019**, *11*, 20, doi:10.3390/sym11010020 . . . . **179**

Silviu Nastac, Carmen Debeleac and Sorin Vlase
Hysteretically Symmetrical Evolution of Elastomers-Based Vibration Isolators within α-Fractional Nonlinear Computational Dynamics
Reprinted from: *Symmetry* **2019**, *11*, 924, doi:10.3390/sym11070924 . . . . **191**

Marin Marin , Mohamed I. A. Othman , Sorin Vlase and Lavinia Codarcea-Munteanu
Thermoelasticity of Initially Stressed Bodies with Voids: A Domain of Influence
Reprinted from: *Symmetry* **2019**, *11*, 573, doi:10.3390/sym11040573 . . . . **209**

About the Editors

Marin Marin Marin MARIN is a Professor in Mathematics at the Department of Mathematics and Computer Science at Transilvania University of Brasov, Romania. He is an eminent scientist in this field and has contributed various books as well as research papers in many SCI indexed journals. His H-index is 37.

Dumitru Baleanu is a well-known and highly regarded specialist in Physics and Applied Mathematics. He has published an impressive number of scientific articles in the most prestigious journals and books in the most valuable publishing houses in the world. He is a Professor of Applied Mathematics at the Cankaya University in Ankara, Turkey.

Sorin Vlase is a Professor at the Department of Engineering Mechanics at Transilvania University of Brasov, the head of department, and a member of the Romanian Academy of Technical Sciences. He is a specialist in the field of Mechanics, with many contributions in this domain presented in papers published in prestigious journal.

Preface to "Symmetry in Applied Continuous Mechanics"

This Special Issue is dedicated to applications in the field of continuous mechanics in which the studied structures have certain symmetries. From the very beginning of the development of the Theory of Elasticity, it was found that the existence of certain types of symmetries in the geometry of a body or a structure leads to some advantages in the calculation of that structure. As a result, these observations have been used in industrial applications. At present, in all fields of engineering, there are many situations in which the systems used have different identical parts or symmetries. As a result, identifying the properties that these symmetries can bring in certain stages of design and calculation becomes an important operation. The present works show this tendency to identify and use, in order to save time and costs, these properties in the design and practical realization of such structures.

Marin Marin, Dumitru Baleanu, Sorin Vlase
Editors

Editorial

Symmetry in Applied Continuous Mechanics

Marin Marin [1],*, Dumitru Bǎleanu [2,3] and Sorin Vlase [4]

[1] Department of Mathematics and Computer Science, Transilvania University of Brașov, B-dul Eroilor 29, 500036 Brașov, Romania
[2] Department of Mathematics, Faculty of Art and Sciences, Cankaya University, 0630 Ankara, Turkey; dumitru@cankaya.edu.tr
[3] Institute of Space Sciences, Magurele-Bucharest, R 76900, Romania
[4] Department of Mechanical Engineering, Faculty of Mechanical Engineering, Transilvania University of Brașov, B-dul Eroilor 29, 500036 Brașov, Romania; svlase@unitbv.ro
* Correspondence: m.marin@unitbv.ro; Tel.: +34-950-015791; Fax: +34-950-015491

Received: 10 October 2019; Accepted: 10 October 2019; Published: 14 October 2019

Abstract: Engineering practice requires the use of structures containing identical components or parts, which are useful from several points of view: less information is needed to describe the system, design is made quicker and easier, components are made faster than a complex assembly, and finally the time to achieve the structure and the cost of manufacturing decreases. Additionally, the subsequent maintenance of the system becomes easier and cheaper. This Special Issue is dedicated to this kind of mechanical structure, describing the properties and methods of analysis of these structures. Discrete or continuous structures in static and dynamic cases are considered. Theoretical models, mathematical methods, and numerical analysis of the systems, such as the finite element method and experimental methods, are expected to be used in the research. Such applications can be used in most engineering fields including machine building, automotive, aerospace, and civil engineering.

Keywords: symmetry; topology; mechanical structures; robots; vibration; mechanical engineering; applied mechanics

---

1. Introduction

In engineering, including civil engineering, machinery construction industry, automotive industry, and the aerospace industry, there are products, elements, machines, and components that contain identical, repetitive parts, which have different types of symmetries. In the constructions, most buildings, works of art, halls etc have identical parts and have symmetries, because symmetry is beneficial for an easy, fast, cheaper and aesthetic design. These properties can be successfully used to facilitate static and dynamic analysis of some structures. The symmetries of different types that offer structure-specific properties have long been observed and used especially in the static case. They are presented in the classical courses of Strength of Materials or Structural Analysis. Symmetries in mechanics have been studied mainly from the point of view of mathematicians [1,2]. In January 2018, a Special Issue of Symmetry Magazine dedicated to applications in structural mechanics was launched [3]. A European project was also funded to study this type of problem [4] and courses were held at the Center for Solid Mechanics CISM from UDINE (Similarity, Symmetry and Group Theoretical Methods in Mechanics, 7 September 2015. Lectures were delivered at the International Center for Mechanical Sciences). Symmetry in Applied Continuous Mechanics was developed in the last decades [5,6].

2. Statistics of the Special Issue

The statistics of papers called for this Special Issue related to published or rejected items were [7–20]: total submissions (21), published (13; 62%), and rejected (8; 38%). The authors' geographical distribution by countries of authors in published papers is shown in Table 1, and it can be seen that 35 authors are from 11 different countries. Note that it is usual for a paper to be signed by more than one author and for authors to collaborate with authors with different affiliations.

Table 1. Geographic distribution by countries of authors.

Country	Number of Authors
Romania	10
Saudi Arabia	2
Pakistan	1
China	7
England	1
Turkey	1
Iran	3
Japan	1
Egypt	2
Poland	2
Spain	5
Total	35

3. Authors of the Special Issue

The authors of this Special Issue and their main affiliations are summarized in Table 2, and it can be seen that there are four authors on average per manuscript.

Table 2. Affiliations and bibliometric indicators for authors.

Author	Affiliation	Reference
Iuliu Negrean	Technical University of Cluj-Napoca, Romania	[9]
Adina Veronica Crisan	Technical University of Cluj-Napoca, Romania	[9]
Silviu Năstac	"Dunarea de Jos" University of Galati, Romania	[10]
Carmen Debeleac	"Dunarea de Jos" University of Galati, Romania	[10]
Sorin Vlase	Transilvania University of Brașov, Romania	[10,11,15,17]
Marin Marin	Transilvania University of Brașov, Romania	[11,15,17,19]
R. Ellahi	King Fahd University of Petroleum & Minerals, Saudi Arabia International Islamic University (IIUI), Pakistan	[11]
M.M. Bhatti	Shandong University of Science and Technology, China Shanghai University, China	[11]
Eliza Chircan	Transilvania University of Brașov, Romania	[12]
Maria Luminița Scutaru	Transilvania University of Brașov, Romania	[12]

Table 2. *Cont.*

Author	Affiliation	Reference
Cătălin Iulian Pruncu	Imperial College, London, UK; University of Birmingham, UK	[12]
Zhang Yue	Hainan Radio and TV University, China	[13]
Dumitru Băleanu	Cankaya University, Ankara, Turkey; Institute of Space Sciences, Bucharest-Magurele, Romania	[13,14,20]
Ji-Jun Pan	Dianxi Science and Technology, Normal University, China	[14]
Mohammad Reza Mahmoudi	Fasa University, Iran	[14]
Mohsen Maleki	Shiraz University, Iran	[14]
Mariana Stanciu	Transilvania University of Brașov, Romania	[15]
Xinghan Xu	Kyoto University, Japan	[16]
Weijie Ren	Dalian University of Technology, China	[16]
Mohamed I. A. Othman	Zagazig University, P.O. Box 44519 Zagazig, Egypt	[17,19]
Lavinia Codarcea-Munteanu	Transilvania University of Brașov, Romania	[17]
Krzysztof Kamil Żur	Bialystok University of Technology, Poland	[18]
Piotr Jankowski	Bialystok University of Technology, Poland	[18]
Elsayed M. Abd-Elaziz	Zagazig Higher Institute of Engineering & Technology, Egypt	[19]
Behzad Ghanbari	Kermanshah University of Technology, Iran	[20]
Maysaa Al Qurashi	King Saud University, Riyadh, Saudi Arabia	[20]
Noelia Bazarra	Universidade de Vigo, Spain	[7]
José A. López-Campos	Departamento de Ingeniería Mecánica, Escola de Enxeñaría Industrial, Vigo, Spain	[7]
Marcos López	Departamento de Ingeniería Mecánica, Escola de Enxeñaría Industrial, Vigo, Spain	[7]
Abraham Segade	Departamento de Ingeniería Mecánica, Escola de Enxeñaría Industrial, Vigo, Spain	[7]
José R. Fernández	Universidade de Vigo, Spain	[7]
Yadong Zhou	Nanjing University of Aeronautics and Astronautics, Nanjing, China	[8]
Youchao Sun	Nanjing University of Aeronautics and Astronautics, Nanjing, China	[8]
Tianlin Huang	Commercial Aircraft Engine Co., LTD, Shanghai, China	[8]

4. Brief Overview of the Contributions to the Special Issue

The analysis of the topics identifies the research undertaken. This section classifies the manuscripts according to the topics proposed in the Special Issue. It was observed that there are three topics that have dominated the others: symmetry in mechanical engineering; symmetry in applied mathematics and symmetry in civil engineering.

Author Contributions: Conceptualization, M.M. and S.V.; methodology, D.B., software M.M.; validation, M.M., D.B. and S.V.; formal analysis, D.B.; investigation, S.V.; resources, S.V.; data curation, M.M.; writing-original draft preparation M.M. and S.V.; writing-review and editing, M.M.; visualization, D.B.; supervision, S.V., project administration, M.M.

Funding: This research received no external funding.

Conflicts of Interest: The authors declare no conflicts of interest.

References

1. Marsden, J.E.; Ratiu, T.S. *Introduction to Mechanics and Symmetry: A Basic Exposition of Classical Mechanical Systems*; Springer: Berlin/Heidelberg, Germany, 2003; p. 586. ISBN 978-0387986432.
2. Holm, D.D.; Stoica, C.; Ellis, D.C.P. *Geometric Mechanics and Symmetry*; Oxford University Press: Oxford, UK, 2009.
3. Zavadskas, E.K.; Bausys, R.; Antuchevičienė, J. Civil Engineering and Symmetry. Available online: https://www.mdpi.com/journal/symmetry/special_issues/Civil_Engineering_Symmetry (accessed on 4 October 2019).
4. Marin, M.; Vlase, S. Effect of internal state variables in thermoelasticity of microstretch bodies. *An. Sti. Univ. Ovidius Constanta* **2016**, *24*, 241. [CrossRef]
5. Marin, M.; Baleanu, D.; Vlase, S. Effect of microtemperatures for micropolar thermoelastic bodies. *Struct. Eng. Mech.* **2017**, *61*, 381–387. [CrossRef]
6. Othman, M.I.A.; Marin, M. Effect of thermal loading due to laser pulse on thermoelastic porous medium under G-N theory. *Results Physics* **2017**, *7*, 3863–3872. [CrossRef]
7. Bazarra, N.; López-Campos, J.A.; López, M.; Segade, A.; Fernández, J.R. Analysis of a Poro-Thermo-Viscoelastic Model of Type III. *Symmetry* **2019**, *11*, 1214. [CrossRef]
8. Zhou, Y.; Sun, Y.; Huang, T. Impact damage equivalency for twisted composite blades 2 with symmetrical configurations. *Symmetry* **2019**, 1292, in press. [CrossRef]
9. Negrean, I.; Crișan, A.D. Synthesis on the Acceleration Energies in the Advanced Mechanics of the Multibody Systems. *Symmetry* **2019**, *11*, 1077. [CrossRef]
10. Nastac, S.; Debeleac, C.; Vlase, S. Hysteretically Symmetrical Evolution of Elastomers-Based Vibration Isolators within α-Fractional Nonlinear Computational Dynamics. *Symmetry* **2019**, *11*, 924. [CrossRef]
11. Marin, M.; Vlase, S.; Ellahi, R.; Bhatti, M.M. On the Partition of Energies for the Backward in Time Problem of Thermoelastic Materials with a Dipolar Structure. *Symmetry* **2019**, *11*, 863. [CrossRef]
12. Chircan, E.; Scutaru, M.; Pruncu, C.I. Two-Dimensional Finite Element in General Plane Motion Used in the Analysis of Multi-Body Systems. *Symmetry* **2019**, *11*, 848. [CrossRef]
13. Zang, Y.; Baleanu, D. Inference about the Ratio of the Coefficients of Variation of Two Independent Symmetric or Asymmetric Populations. *Symmetry* **2019**, *11*, 824. [CrossRef]
14. Pan, J.; Mahmoudi, M.R.; Baleanu, D.; Maleki, M. On Comparing and Classifying Several Independent Linear and Non-Linear Regression Models with Symmetric Errors. *Symmetry* **2019**, *11*, 820.
15. Stanciu, M.D.; Vlase, S.; Marin, M. Vibration Analysis of a Guitar considered as a Symmetrical Mechanical System. *Symmetry* **2019**, *11*, 727. [CrossRef]
16. Xu, X.; Ren, W. A Hybrid Model Based on a Two-Layer Decomposition Approach and an Optimized Neural Network for Chaotic Time Series Prediction. *Symmetry* **2019**, *11*, 610. [CrossRef]
17. Marin, M.; Othman, M.I.A.; Vlase, S.; Codarcea-Munteanu, L. Thermoelasticity of Initially Stressed Bodies with Voids: A Domain of Influence. *Symmetry* **2019**, *11*, 573. [CrossRef]
18. Żur, K.K.; Jankowski, P. Multiparametric Analytical Solution for the Eigenvalue Problem of FGM Porous Circular Plates. *Symmetry* **2019**, *11*, 429. [CrossRef]

19. Abd-Elaziz, E.M.; Marin, M.; Othman, M.I.A. On the Effect of Thomson and Initial Stress in a Thermo-Porous Elastic Solid under G-N Electromagnetic Theory. *Symmetry* **2019**, *11*, 413. [CrossRef]
20. Ghanbari, B.; Baleanu, D.; al Qurashi, M. New Exact Solutions of the Generalized Benjamin–Bona–Mahony Equation. *Symmetry* **2019**, *11*, 20. [CrossRef]

© 2019 by the authors. Licensee MDPI, Basel, Switzerland. This article is an open access article distributed under the terms and conditions of the Creative Commons Attribution (CC BY) license (http://creativecommons.org/licenses/by/4.0/).

Article

Impact-Damage Equivalency for Twisted Composite Blades with Symmetrical Configurations

Yadong Zhou [1,*], Youchao Sun [1,*] and Tianlin Huang [2]

1. College of Civil Aviation, Nanjing University of Aeronautics and Astronautics, Nanjing 211100, China
2. Commercial Aircraft Engine Co., LTD, Aero Engine Cooperation of China, Shanghai 201100, China
* Correspondence: yzhou@nuaa.edu.cn (Y.Z.); sunyc@nuaa.edu.cn (Y.S.)

Received: 8 August 2019; Accepted: 8 October 2019; Published: 15 October 2019

Abstract: In spite of potential advantages for aircraft structures, composite laminates can be subjected to bird-strike hazard in civil aviation. For purpose of future surrogate experiments, in this study, impact-damage equivalency for twisted composite blades is numerically investigated by Smoothed Particle Hydrodynamics (SPH) and finite element method (FEM). Cantilever slender flat plates are usually used for basic impact tests, the impact-damage equivalency is being considered by comparing damage modes and energies of three impact configurations: (1) twisted blade; (2) flat blade (axisymmetric); and (3) inclined flat blade (centrosymmetric). The damage maps and energy variations were comparatively investigated. Results indicate that both symmetrical flat and inclined flat blades can be, to a certain extent, regarded as alternatives for real twisted blades under bird impact; however, both types of blade have their own merits and drawbacks, and hence should be used carefully. These results aim to serve as tentative design guideline for future prototype or model experimental study of laminated blades in real aeronautical structures.

Keywords: bird-strike impact; composite laminate; damage equivalency; fan blade; symmetrical configurations

1. Introduction

Since the beginning of aviation history, bird-strike events have become an increasingly serious and catastrophic issue for aircraft structures [1], particularly for aircraft safety below 3000 feet from the ground. For aeronautical structures in civil aviation industry, the crashworthiness ability to withstand bird strike has been regulated by airworthiness certification requirements. The front-facing components of aircraft susceptible to bird strikes, mainly refer to the engine fan blades [2–4], windshield [5–8], wing leading edge section [9–11], wing flaps [12,13], and other auxiliary units, etc. Among these, the rotating fan blades have relatively high probability of bird strike, due to the air ingestion of aero-engines. Under bird impacts, many vital parameters, e.g., geometric effects of birds [2,4,14], material models of birds [15,16], velocities and angles of bird projectile [17–19], etc., determine the final deformation and damage mechanisms of such forward-facing structures.

Thanks to good performance and high efficiency, composite laminates have been more and more exploited as key materials for the engine primary structures, e.g., the fan blades of high-bypass ratio engines in the modern commercial aircraft, which can offer the potential for reducing the weight of civil aircrafts. Additionally, some civil aircraft manufacturers of models such as ARJ21 and C919 in China, are seeking further use of composite laminates in key components. However, bird-strike certification compliance against impact-induced damage should be considered [20], e.g., according to FAR Part 25. As an important form of foreign object damage (FOD), when subject to bird impact, the laminated configurations will become critically vital [21,22]. The damage and failure modes in composite materials largely depend on the structural geometries [7,23], the impact velocity/locations [10,12,24], the shape and mass of the projectile [12], number and placement of layers [7,22,25], the fiber and matrix properties [26], etc.

In recent years, evaluating the crashworthiness characteristics of composite laminates has gained much attention [27,28]. Heimbs et al. [29] presented an experimental and numerical analysis of the T800S/M21 composite laminate plate under bird impact, with two opposite ends of the plate clamped. Hu et al. [23] conducted a composite cockpit prototype test under bird strike. Recently, Mohagheghian et al. [21] employed three different polymer interlayer materials for laminated aircraft windshields, where square plates were tested under bird strike. Orlando et al. [13] presented the prototype test of a composite flap configuration compliant with EASA and FAA bird strike requirements. According to the airworthiness requirements, Guida et al. [30] presented the simplified square plate for leading edge design, subjected to the bird-strike phenomenon at 180 m/s and with an impacting mass of 1.8 kg. For composite fan blades under bird strike, Friedrich [31], Hou et al. [32], and Liu et al. [33] conducted surrogate experiments by using cantilever slender plate-type structures. In numerical aspect, Nishikawa et al. [34] also used the cantilever narrow-and-flat plate (1000 mm × 500 mm × 10 mm) to discuss the bird-strike problem of composite fan blade. The authors [35] recently studied the effect of rotational speeds on bird-strike damage modes in rotating composite laminates, but only rotating flat configurations were considered.

How to design a rotating structure that meets the requirements of bird impact resistance? The most straightforward method is "structural design–test–redesign–retest"; but the production cycle and cost should be absolutely considered in civil aviation industry. As typical forward-facing structures, composite fan blades constitute a rotationally symmetric system, and each blade has complex twisted surfaces due to the angle of attack, such as an example shown in Figure 1. In primary design phase, basic experiments should be conducted to obtain blades' dynamic behaviours and damage mechanisms under bird impact; however, cantilever slender flat plates are usually used for these basic tests for simplicity and repeatability. Prototype experiments of blades will be high cost and time-consuming. It can be promising to do laboratory-scale model experiments at early stage, using bird substitute materials (with better reproducibility [6]) and simplified blade model (less cost for basic data and mechanisms research [32,33]). However, its validity remains unclear, and to what degree it can replace the prototype experiments also remains questioned. In brief, the impact-damage equivalency has been considered inadequately in alternative testing. For these considerations, the main objective of present work is to numerically investigate the configuration effect on the impact-induced damage of laminated blade-like plates, which tentatively gain design guideline for future prototype or surrogate model experimental study of laminated blades in real aeronautical structures.

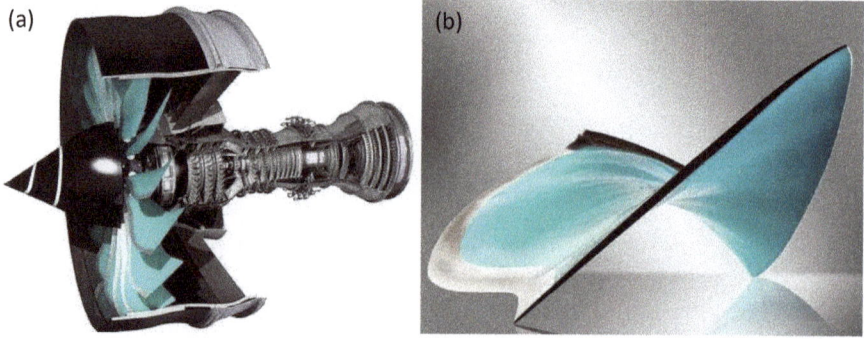

Figure 1. An example of aero-engine and one of its fan blades (**a**) the turbofan engine; (**b**) the twisted fan blade [36].

2. Numerical Method

Generally speaking, characterized by soft-body impact, bird-strike phenomenon is a short duration (millisecond range) and high-intensity loads event. The bird's mechanical property actually changes from the low-velocity to the high-velocity regimes [4]. There are three most suggested methods [9,37], named as Lagrangian, Arbitrary Lagrangian Eulerian (ALE) and Smoothed Particle Hydrodynamics (SPH), to simulate the bird behaviour. Among these methods, SPH could better describe bird fragmentation into discrete particles with high efficiency, yet required fewer elements and normally had a shorter solution time compared to the Lagrangian and Eulerian model. In the SPH approach, a continuous field is represented by a set of discrete but interacting particles. Therefore, SPH has recently experienced high popularity [2,7,9,13,21,23,38,39] in bird-strike and high-velocity dynamics studies. In the present study, an SPH method has been employed, which is a meshless particle-based method. In this investigation, SPH modelling was similar to previous work [35], and with the same parameters in the equation of state (EOS). Different substitute bird impactor geometries were proposed, e.g., the cylinder, the cylinder with hemispherical ends, the ellipsoid and the sphere [37], but there is no standardized artificial bird shape. Here the bird body is modelled as a cylinder (aspect ratio = 2:1) projectile, with 200 mm length and 100 mm diameter. The bird-strike velocity was fixed to 180 m/s. The projectile has an initial density of ρ_0 = 950 kg/m^3, and the weight of projectile model is ≈1.5 kg, which is to verify the compliance of airworthiness regulation. The linear Mie–Grüneisen EOS (also called U_s - U_p equation) was used for the bird, with parameters: the intercept of U_s - U_p curve c_0 = 1483 m/s; Grüneisen gamma Γ_0 = 0; and s = 0.

For the fiber breakage and matrix cracking simulation, Continuum Damage Mechanics (CDM) enables to predict with good accuracy the onset and growth of the intra-laminar damage by introducing a degradation factor for material mechanical properties. The laminate edges' length and width are 500 mm and 150 mm, respectively. The initial impact radius is 50 mm and initial distance between the projectile's front and the target is 100 mm, which can define the initial position of the projectile's centre of mass. The total thicknesses of composite laminated blades is 48 mm, with a basic laminate layup [0/45/0/-45/-45/0/45/0]$_s$. Namely, the 16-ply blades were considered, and the geometric details are as shown in Table 1. The real laminated blades may consist of hundreds of layups, e.g., 400 layups in GE90-115B. Herein, the number of layups was simplified by using the virtual lamina thickness.

Table 1. The considered plies and dimensions for blades.

Ply Angles	Ply Number	Total Thickness (mm)	In-Plane Dimensions (mm × mm)
[0/45/0/-45/-45/0/45/0]$_s$	16	48	500 × 150

The blade was clamped supported on the root edge. The final numerical model of the impacted blade consists of 1340 conventional shell finite elements (linear quadrilateral elements S4R in ABAQUS). The failure modes are based on Hashin's criteria formulation [40]. The material mechanical properties and fracture energies are needed in CDM modelling, of which the values are taken from References [41,42] as given in Table 2. Configuration designs of composite blades under bird impact are shown as Figure 2 (view from the tip). Figure 2a represents the twisted blade under impact; while (b) and (c) represent the vertical and inclined impact on flat blades respectively. It can be seen that (b) and (c) are axisymmetric and centrosymmetric, respectively. Therefore, the question is to investigate the damage equivalency of symmetrical configurations for the twisted configuration. In present simulations, the bird projectile impacts the half span of the blade, therefore the inclined angle (Figure 2c) is determined to the half of the twisted angle (θ in Figure 2a). The considered twisted angle was 36° and hence the inclined angle was 18°.

Table 2. Material mechanical properties of the unidirectional lamina.

Properties	Value
Mass density	1600 kg/m^3
Orthotropic elastic properties	E_{11} = 181,000 MPa; E_{22} = 10,300 MPa; ν_{12} = 0.28; G_{12} = 7170 MPa; G_{13} = 1000 MPa; G_{23} = 500 MPa
Strength properties	$X_T = X_C$ = 1500 MPa; Y_T = 40 MPa; Y_C = 246 MPa; S_{12} = 200 MPa; S_{13} = 123 MPa
Fracture energies	G_{1c}^T = 11.5 kJ/m^2; G_{2c}^T = 4.1 kJ/m^2; G_{1c}^C = 0.35 kJ/m^2; G_{2c}^C = 3.2 kJ/m^2

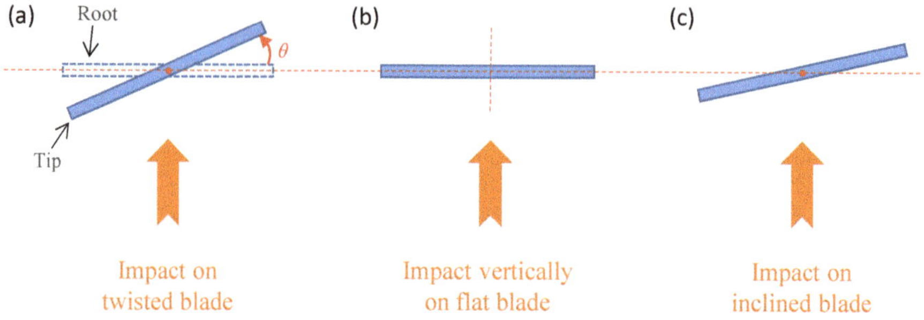

Figure 2. Configuration designs for composite blades under bird impact (view from the tip).

3. Results and Discussion

For the extremely discontinuous processes, an explicit dynamic analysis is appropriate to carry out using ABAQUS/Explicit, which can be computationally efficient for complex mechanics models with relatively short dynamic duration. In the present simulation, the projectile begins to initially contact the impacted plate at approximately 0.5 ms, and the total simulation duration is 6 ms.

Bird impacts with high kinetic energy can produce severe intra-laminar damages, e.g., fiber breakage and matrix cracking. In the failure modes of composite laminates, the fiber breakage could probably lead to catastrophic failure of overall laminated structures and then threaten the blade containment of aero-engines. Therefore, in the beginning of this section, attention is focused on the fiber-tension damage mode.

Figure 3 illustrates the damage maps of fiber-tension mode of the three configurations: (a) the twisted blade; (b) the flat blade; (c) the inclined-impacted flat blade. First, the maximum damages are close to each other: 0.3916, 0.3817, and 0.3719 respectively. Then, it can be observed that blade roots are all the highly-damaged areas for the three configurations, while damage distributions appear slightly different. At the roots, the vertically-impacted flat case (b) shows that high damage almost runs through along the root; however, in the twisted blade, high damage distributes at two corners of the root, with the left corner larger damaged area. This fact can be due to the twisted surface, with the left side is more front-facing than the right. From Figure 3c, it can be seen that the damage of inclined-impacted flat blade can reflect the corner-distributed feature. Thus, the inclined-impacted flat blade is more equivalent to the twisted blade in terms of fiber-tension damage.

For the fiber-compression damage mode, Figure 4 shows that both the axisymmetric and centrosymmetric configurations can describe the damage distribution of the twisted case relatively well. However, one disadvantage exists in that the damage distribution at the right corner cannot be well reflected. The fiber-compression damage in the twisted case indicates that the rear-forwarding area (right corner at the root) can be more easily subjected to fiber compression, which can lead to local buckling of fibers.

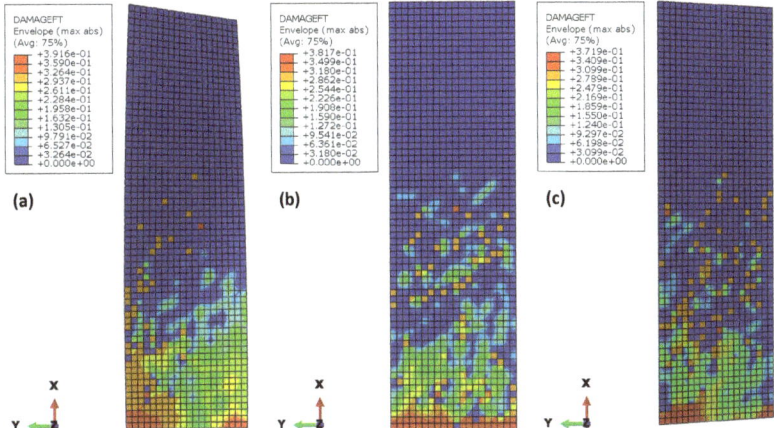

Figure 3. Fiber-tension damage maps of the three impacted configurations. (**a**) the twisted blade; (**b**) the flat blade; (**c**) the inclined-impacted flat blade.

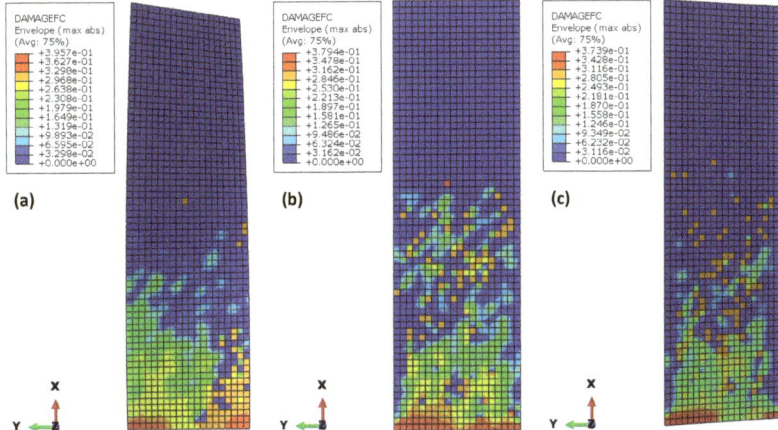

Figure 4. Fiber-compression damage maps of the three impacted configurations. (**a**) the twisted blade; (**b**) the flat blade; (**c**) the inclined-impacted flat blade.

Figure 5 illustrates the damage maps of matrix-compression mode of the three configurations: (a) the twisted blade; (b) the vertically-impacted flat blade; (c) the inclined-impacted flat blade. First, the maximum damages are also close to each other: 0.3426, 0.3548, and 0.3654 respectively. Then, it can be observed that blade roots are all the highly-damaged area for the three configurations, while damage distributions appear quite different. At the roots, the result in vertically-impacted flat case (b) shows that high damage distributes at the root corners; however, in the twisted blade, although high damage also distributes at two corners of the root, but with the much larger damaged area distributed in both corners. This could also be due to the twisted surface, aggravating the matrix-compression damage. From Figure 5c, it can be seen that the damage of inclined-impacted flat blade cannot reflect the corner-distributed feature for the matrix-compression mode. Thus, in terms of matrix-compression damage, the two flat blades both failed to reflect the damage distribution of the twisted blade, to certain degree. Such phenomenon indicates that the flat laminates cannot perfectly reproduce the compressive deformations that the twisted laminate experiences, which is somewhat similar with the local distribution in fiber-compression case.

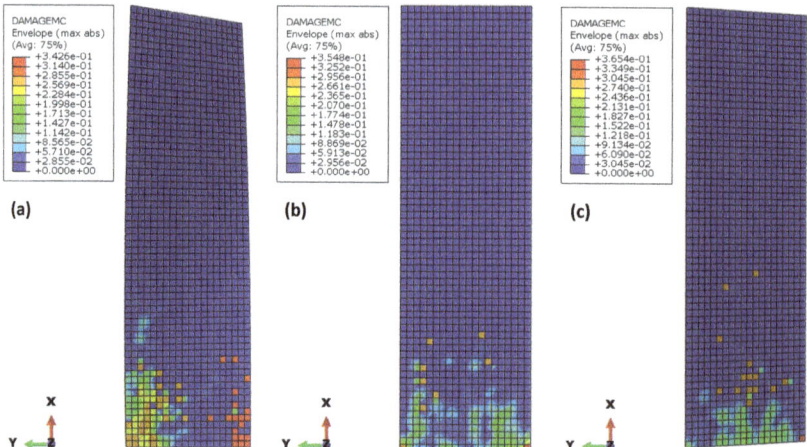

Figure 5. Matrix-compression damage maps of the three impacted configurations. (**a**) the twisted blade; (**b**) the flat blade; (**c**) the inclined-impacted flat blade.

The matrix-tension damage can largely induce the stiffness degradation of laminated blades. Figure 6 shows that the matrix-tension damage distributions of the three cases are in good agreement on the whole, where extensive damages appeared except around the blade tip.

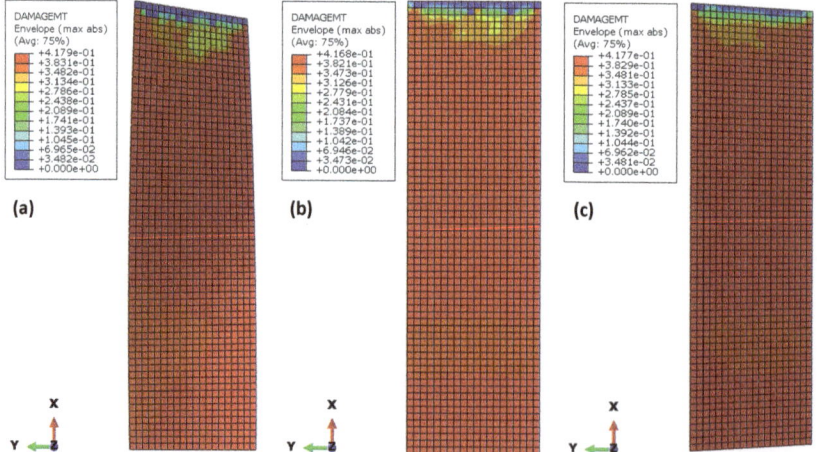

Figure 6. Matrix-tension damage maps of the three impacted configurations. (**a**) the twisted blade; (**b**) the flat blade; (**c**) the inclined-impacted flat blade.

As an evolution process of inconvertible energy dissipation, damage dissipation energy denotes the energy dissipated by the laminate's failure modes during the impact event. Figure 7 compares the time histories of the damage dissipation energies of the laminates in three configurations. Results indicate that vertically-impacted flat blade is quite adequate to reflect the real damage history in twisted blade case, while the inclined flat case performs inadequately, especially as the time increases. It is necessary to mention that it is difficult for the damage dissipation energy to be experimentally measured, due to the lack of straightforward measurement index.

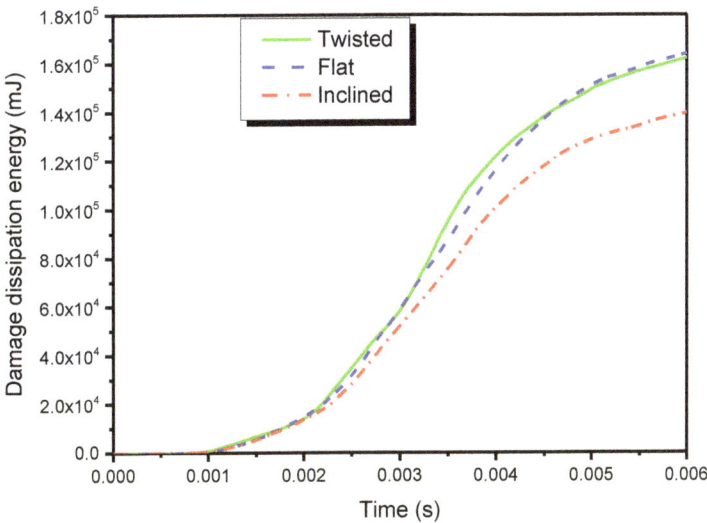

Figure 7. Time histories of damage dissipation energies in the three configurations.

To further explain the damage results, Figure 8 demonstrates the deformation series of the three blades. In the twisted blade, the deformations had an oblique because of the asymmetry. However, the deformations were completely symmetric in the vertically-impacted flat plate, due to the strict symmetry of the system (the impacted plate and the bird projectile). It can be observed that the deformations of the inclined case also indicated an oblique impact, just as with the twisted blade. Therefore, it can be deduced that both the bending and torsion deformations happen in the twisted and inclined blades, while only bending deflection happens in the vertically-impacted flat blade.

The total contact forces generated in the three models along the impact direction are compared in Figure 9. Because the projectile began to initially contact the impacted plate at approximately 0.5 ms, the contact forces achieved the highest peak value also at approximately 0.5 ms in the three blades, which is due to the shock of Hugoniot pressure. It can be observed that the inclined blade is more equivalent to the twisted one, by comparing the highest peak values of the contact forces. However, the vertically-impacted flat blade is prone to overestimate the peak value of contact force. Such a result may be mainly due to the fact that the impact energy is mostly absorbed by the plate in such axisymmetric cases. On the contrary, the initial twist/inclination configuration and the induced torsion deformations can partially disperse the kinetic energy of the bird projectile.

Hereto, simulation results revealed some important laws for the damage distributions, damage energies, and contact forces in the different blade-like configurations. It can be seen that the replaceability and applicability of the simplified blade-like configurations (axisymmetric and centrosymmetric) can be different from different perspectives, which depend on the actual requirements to examine the crashworthiness of real fan blades. For instance, when the fatal fiber fracture or the peak force is the primary concern, the inclined case is the relative adequate choice. However, the vertically-impact flat case can well reflect the damage dissipation energy. These numerical results can provide a useful reference for aerospace engineers dealing with simplified and surrogate bird-strike experiments of complex fan blades. Finally, it is necessary to mention that the present study only considered the case in which the bird projectile impacts the central axis of the blade. It is worthwhile to investigate the replaceability when the impact location is off the central axis, where more complicated deformations can happen.

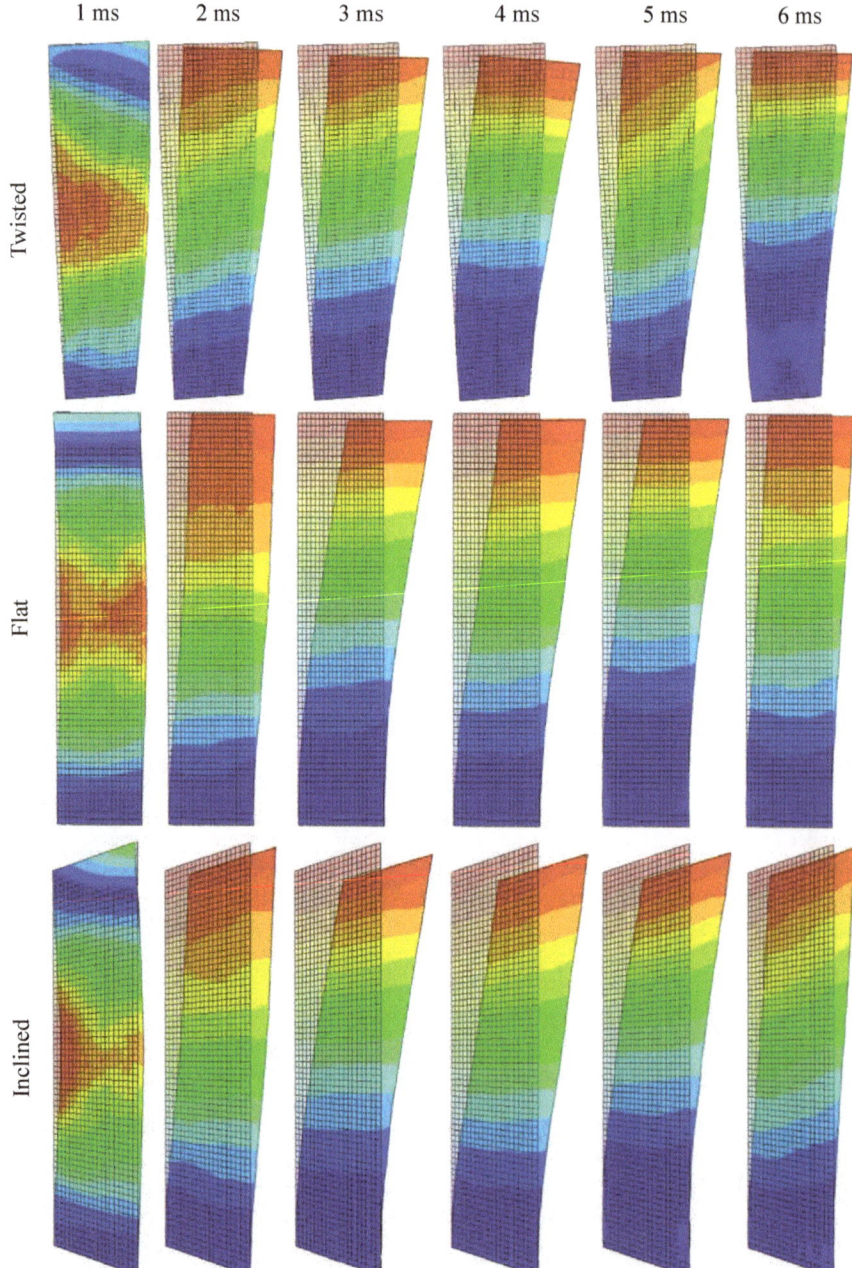

Figure 8. Deformation series of the three blades.

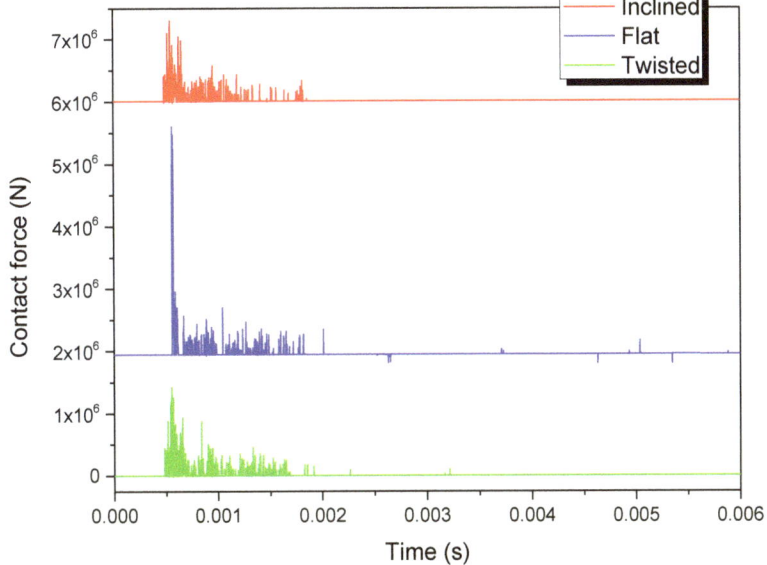

Figure 9. Contact forces of the three models.

4. Conclusions

Considering the fact that cantilever slender flat plates are usually used for basic impact tests, for the purpose of a future simplified and surrogate experimental method, this article numerically investigated the damage equivalency for twisted composite blades under bird strike, by using SPH-FEM simulations. The impact-damage equivalency was considered by comparing damage modes and energy variations of three impact configurations: (1) the twisted blade to represent the real aero-engine fan blades; (2) the vertically-impacted flat blade (axisymmetric); and (3) the inclined flat blade (centrosymmetric). The damage maps, energy variations and deformations were comparatively investigated. It was found that:

- The inclined-impacted flat blade is more equivalent to the twisted blade in terms of the fiber-tension damage and the contact force, which benefits from the similarity between the twisted and inclined configurations, such as the coupling bending and torsion deformations. On the whole, the inclined flat blade exhibits advantages over the vertically-impacted flat one in reproducing the fatal damage mode, the corner-distributed feature, and the oblique deformation.
- For the fiber-compression damage mode, both the axisymmetric and centrosymmetric configurations can relatively well describe the damage distribution of the twisted case. However, in terms of the matrix-compression damage, the two flat blades both failed to adequately reflect the damage distribution of the twisted blade. It can be indicated that the flat laminates cannot perfectly reproduce the local compressive deformations that the twisted laminate experiences.
- The three configurations showed similar matrix-tension damage maps, which were globally distributed. The vertically-impacted flat blade is quite adequate to reflect the real damage history in the twisted blade case, while the inclined flat case performs inadequately.

In brief, results indicate that both the vertically-impacted flat and inclined flat blades can be, to a certain extent, regarded as alternatives for real twisted blades under bird impact; however, both vertically-impacted flat and inclined flat blades have their own merits and drawbacks, and hence their replaceability and applicability should be carefully exploited for real engine blades in experimental evaluation, for the purpose of bridging the gap between the mechanism testing and the structural damage prediction.

Author Contributions: Conceptualization, Y.Z.; Data curation, Y.Z.; Formal analysis, Y.Z.; Funding acquisition, Y.S.; Investigation, Y.Z.; Methodology, Y.Z.; Project administration, Y.S. and T.H.; Resources, Y.S.; Software, Y.Z.; Supervision, Y.S.; Validation, T.H.; Visualization, Y.Z.; Writing—original draft, Y.Z.; Writing—review & editing, Y.Z.

Funding: This research was funded by the National Natural Science Foundation and the Civil Aviation Administration of China (U1333119); Shanghai Engineering Research Center of Civil Aircraft Health Monitoring (GCZX-2015-05); the Defense Industrial Technology Development Program (JCKY2013605B002); Fundamental Research Funds for the Central Universities (56XBC18206, 56XBA18201, IEA180017A19); the National Science and Technology Major Project (2017-VIII-0003-0114).

Conflicts of Interest: The authors declare no conflict of interest.

References

1. Abrate, S. Soft impacts on aerospace structures. *Prog. Aerosp. Sci.* **2016**, *81*, 1–17. [CrossRef]
2. Zhang, D.; Fei, Q. Effect of bird geometry and impact orientation in bird striking on a rotary Jet-Engine fan analysis using SPH method. *Aerosp. Sci. Technol.* **2016**, *54*, 320–329. [CrossRef]
3. Zhang, Z.; Li, L.; Zhang, D. Effect of arbitrary yaw/pitch angle in bird strike numerical simulation using SPH method. *Aerosp. Sci. Technol.* **2018**, *81*, 284–293. [CrossRef]
4. Meguid, S.A.; Mao, R.H.; Ng, T.Y. FE analysis of geometry effects of an artificial bird striking an aeroengine fan blade. *Int. J. Impact Eng.* **2008**, *35*, 487–498. [CrossRef]
5. Mohagheghian, I.; Wang, Y.; Zhou, J.; Yu, L.; Guo, X.; Yan, Y.; Charalambides, M.N.; Dear, J.P. Deformation and damage mechanisms of laminated glass windows subjected to high velocity soft impact. *Int. J. Solids Struct.* **2017**, *109*, 46–62. [CrossRef]
6. Zhou, J.; Liu, J.; Zhang, X.; Yan, Y.; Jiang, L.; Mohagheghian, I.; Dear, J.P.; Charalambides, M.N. Experimental and numerical investigation of high velocity soft impact loading on aircraft materials. *Aerosp. Sci. Technol.* **2019**, *90*, 44–58. [CrossRef]
7. Grimaldi, A.; Sollo, A.; Guida, M.; Marulo, F. Parametric study of a SPH high velocity impact analysis—A birdstrike windshield application. *Compos. Struct.* **2013**, *96*, 616–630. [CrossRef]
8. Hedayati, R.; Ziaei-Rad, S.; Eyvazian, A.; Hamouda, A.M. Bird strike analysis on a typical helicopter windshield with different lay-ups. *J. Mech. Sci. Technol.* **2014**, *28*, 1381–1392. [CrossRef]
9. Riccio, A.; Cristiano, R.; Saputo, S.; Sellitto, A. Numerical methodologies for simulating Bird-Strike on composite wings. *Compos. Struct.* **2018**, *202*, 590–602. [CrossRef]
10. Yu, Z.; Xue, P.; Yao, P.; Zahran, M.S. Analytical determination of the critical impact location for wing leading edge under birdstrike. *Lat. Am. J. Solids Struct.* **2019**, *16*, 1–17. [CrossRef]
11. Di Caprio, F.; Cristillo, D.; Saputo, S.; Guida, M.; Riccio, A. Crashworthiness of wing leading edges under bird impact event. *Compos. Struct.* **2019**, *216*, 39–52. [CrossRef]
12. Smojver, I.; Ivančević, D. Numerical simulation of bird strike damage prediction in airplane flap structure. *Compos. Struct.* **2010**, *92*, 2016–2026. [CrossRef]
13. Orlando, S.; Marulo, F.; Guida, M.; Timbrato, F. Bird strike assessment for a composite wing flap. *Int. J. Crashworthiness* **2018**, *23*, 219–235. [CrossRef]
14. Vignjevic, R.; Orłowski, M.; De Vuyst, T.; Campbell, J.C. A parametric study of bird strike on engine blades. *Int. J. Impact Eng.* **2013**, *60*, 44–57. [CrossRef]
15. Liu, J.; Li, Y.; Gao, X. Bird strike on a flat plate: Experiments and numerical simulations. *Int. J. Impact Eng.* **2014**, *70*, 21–37. [CrossRef]
16. Allaeys, F.; Luyckx, G.; Van Paepegem, W.; Degrieck, J. Characterization of real and substitute birds through experimental and numerical analysis of momentum, average impact force and residual energy in bird strike on three rigid targets: A flat plate, a wedge and a splitter. *Int. J. Impact Eng.* **2017**, *99*, 1–13. [CrossRef]
17. Dar, U.A.; Zhang, W.; Xu, Y. FE Analysis of Dynamic Response of Aircraft Windshield against Bird Impact. *Int. J. Aerosp. Eng.* **2013**, *2013*, 1–12. [CrossRef]
18. Mao, R.H.; Meguid, S.A.; Ng, T.Y. Transient three dimensional finite element analysis of a bird striking a fan blade. *Int. J. Mech. Mater. Des.* **2008**, *4*, 79–96. [CrossRef]

19. Dar, U.A.; Awais, M.; Mian, H.H.; Sheikh, M.Z. The effect of representative bird model and its impact direction on crashworthiness of aircraft windshield and canopy structure. *Proc. Inst. Mech. Eng. Part G J. Aerosp. Eng.* **2019**, *233*, 5150–5163. [CrossRef]
20. Georgiadis, S.; Gunnion, A.J.; Thomson, R.S.; Cartwright, B.K. Bird-Strike simulation for certification of the Boeing 787 composite moveable trailing edge. *Compos. Struct.* **2008**, *86*, 258–268. [CrossRef]
21. Mohagheghian, I.; Charalambides, M.N.; Wang, Y.; Jiang, L.; Zhang, X.; Yan, Y.; Kinloch, A.J.; Dear, J.P. Effect of the polymer interlayer on the High-Velocity soft impact response of laminated glass plates. *Int. J. Impact Eng.* **2018**, *120*, 150–170. [CrossRef]
22. Hedayati, R.; Sadighi, M. Effect of Using an Inner Plate between Two Faces of a Sandwich Structure in Resistance to Bird-Strike Impact. *J. Aerosp. Eng.* **2016**, *29*, 04015020. [CrossRef]
23. Hu, D.; Song, B.; Wang, D.; Chen, Z. Experiment and numerical simulation of a Full-Scale helicopter composite cockpit structure subject to a bird strike. *Compos. Struct.* **2016**, *149*, 385–397. [CrossRef]
24. Wang, J.; Xu, Y.; Zhang, W. Finite element simulation of PMMA aircraft windshield against bird strike by using a rate and temperature dependent nonlinear viscoelastic constitutive model. *Compos. Struct.* **2014**, *108*, 21–30. [CrossRef]
25. Roberts, G.D.; Revilock, D.M.; Binienda, W.K.; Nie, W.Z.; Mackenzie, S.B.; Todd, K.B. Impact Testing and Analysis of Composites for Aircraft Engine Fan Cases. *J. Aerosp. Eng.* **2002**, *15*, 104–110. [CrossRef]
26. Higuchi, R.; Okabe, T.; Yoshimura, A.; Tay, T.E. Progressive failure under High-Velocity impact on composite laminates: Experiment and phenomenological mesomodeling. *Eng. Fract. Mech.* **2017**, *178*, 346–361. [CrossRef]
27. Minak, G.; Abrate, S.; Ghelli, D.; Panciroli, R.; Zucchelli, A. Low-Velocity impact on carbon/epoxy tubes subjected to torque–Experimental results, analytical models and FEM analysis. *Compos. Struct.* **2010**, *92*, 623–632. [CrossRef]
28. Saghafi, H.; Minak, G.; Zucchelli, A. Effect of preload on the impact response of curved composite panels. *Compos. Part B Eng.* **2014**, *60*, 74–81. [CrossRef]
29. Heimbs, S.; Bergmann, T. High-Velocity Impact Behaviour of Prestressed Composite Plates under Bird Strike Loading. *Int. J. Aerosp. Eng.* **2012**, *2012*, 1–11. [CrossRef]
30. Guida, M.; Sellitto, A.; Marulo, F.; Riccio, A. Analysis of the Impact Dynamics of Shape Memory Alloy Hybrid Composites for Advanced Applications. *Materials* **2019**, *12*, 153. [CrossRef] [PubMed]
31. Friedrich, L.A. *Impact Resistance of Hybrid Composite Fan Blade Materials*; NASA: Washington, DC, USA, 1974.
32. Hou, J.P.; Ruiz, C. Soft body impact on laminated composite materials. *Compos. Part A Appl. Sci. Manuf.* **2007**, *38*, 505–515. [CrossRef]
33. Liu, L.; Luo, G.; Chen, W.; Zhao, Z.; Huang, X. Dynamic Behavior and Damage Mechanism of 3D Braided Composite Fan Blade under Bird Impact. *Int. J. Aerosp. Eng.* **2018**, *2018*, 1–16. [CrossRef]
34. Nishikawa, M.; Hemmi, K.; Takeda, N. Finite-Element simulation for modeling composite plates subjected to soft-body, High-Velocity impact for application to bird-strike problem of composite fan blades. *Compos. Struct.* **2011**, *93*, 1416–1423. [CrossRef]
35. Zhou, Y.; Sun, Y.; Cai, W. Bird-Striking damage of rotating laminates using SPH-CDM method. *Aerosp. Sci. Technol.* **2019**, *84*, 265–272. [CrossRef]
36. Ten Years of "Twisting" a Fan Blade, This Innovative Technology Is about to Affect Your Airplane. 2018. Available online: http://www.sohu.com/a/234993331_660773 (accessed on 27 July 2019).
37. Heimbs, S. Computational methods for bird strike simulations: A review. *Comput. Struct.* **2011**, *89*, 2093–2112. [CrossRef]
38. Liu, J.; Li, Y.; Yu, X.; Tang, Z.; Gao, X.; Lv, J.; Zhang, Z. A novel design for reinforcing the aircraft tail leading edge structure against bird strike. *Int. J. Impact Eng.* **2017**, *105*, 89–101. [CrossRef]
39. Taddei, L.; Awoukeng Goumtcha, A.; Roth, S. Smoothed particle hydrodynamics formulation for penetrating impacts on ballistic gelatine. *Mech. Res. Commun.* **2015**, *70*, 94–101. [CrossRef]
40. Hashin, Z.; Rotem, A. A Fatigue Failure Criterion for Fiber Reinforced Materials. *J. Compos. Mater.* **1973**, *7*, 448–464. [CrossRef]

41. Riccio, A.; Ricchiuto, R.; Saputo, S.; Raimondo, A.; Caputo, F.; Antonucci, V.; Lopresto, V. Impact behaviour of omega stiffened composite panels. *Prog. Aerosp. Sci.* **2016**, *81*, 41–48. [CrossRef]
42. Smojver, I.; Ivančević, D. Bird strike damage analysis in aircraft structures using Abaqus/Explicit and coupled Eulerian Lagrangian approach. *Compos. Sci. Technol.* **2011**, *71*, 489–498. [CrossRef]

© 2019 by the authors. Licensee MDPI, Basel, Switzerland. This article is an open access article distributed under the terms and conditions of the Creative Commons Attribution (CC BY) license (http://creativecommons.org/licenses/by/4.0/).

Article

Analysis of a Poro-Thermo-Viscoelastic Model of Type III

Noelia Bazarra [1], José A. López-Campos [2], Marcos López [2], Abraham Segade [2] and José R. Fernández [1,*]

[1] Departamento de Matemática Aplicada I, Universidade de Vigo, ETSI Telecomunicación, Campus As Lagoas Marcosende s/n, 36310 Vigo, Spain; noabaza@hotmail.com
[2] Departamento de Ingeniería Mecánica, Máquinas y Motores Térmicos y Fluídos, Escola de Enxeñaría Industrial, Campus As Lagoas Marcosende s/n, 36310 Vigo, Spain; joseangellopezcampos@uvigo.es (J.A.L.-C.); mllago@uvigo.es (M.L.); asegade@uvigo.es (A.S.)
* Correspondence: jose.fernandez@uvigo.es; Tel.: +34-986-818-746

Received: 26 August 2019; Accepted: 26 September 2019; Published: 29 September 2019

Abstract: In this work, we numerically study a thermo-mechanical problem arising in poro-viscoelasticity with the type III thermal law. The thermomechanical model leads to a linear system of three coupled hyperbolic partial differential equations, and its weak formulation as three coupled parabolic linear variational equations. Then, using the finite element method and the implicit Euler scheme, for the spatial approximation and the discretization of the time derivatives, respectively, a fully discrete algorithm is introduced. A priori error estimates are proved, and the linear convergence is obtained under some suitable regularity conditions. Finally, some numerical results, involving one- and two-dimensional examples, are described, showing the accuracy of the algorithm and the dependence of the solution with respect to some constitutive parameters.

Keywords: viscoelasticity; type III thermal law; finite elements; error estimates; numerical results

1. Introduction

Since the first works by Nunziato and Cowin [1,2], many research papers have been published involving the theory of elasticity of voids (see, for instance, [3–20] and also the monographs [21,22]). The main idea of such a theory is the assumption that the mass at each material point is found as the product of the mass density by the volume fraction, adding a new variable in the constitutive equations. This theory has become very interesting because it has been useful in applications appearing in solid mechanics (rocks, woods or even bones).

In this paper, we also consider an improvement of the classical Fourier heat law to remove the well-known paradox of infinite wave speed. It is based on the work by Green and Nagdhi [23], where the thermal displacement was included as a new independent variable into the model (even of the elastic displacement), and it is usually called type III thermal law. Since then, many authors worked on this theory (see, e.g., [4,24–29]). Here, our aim is to extend the results provided in [30], where a general thermoelastic law was studied from both mathematical (existence and uniqueness) and numerical (stability, a priori error estimates) points of view. We will consider the viscoelastic case and we will also include the porosity into the model. Therefore, using the well-known finite element method and the implicit Euler scheme, for the spatial approximation and the discretization of the time derivatives, we present a fully discrete algorithm, we perform an a priori error analysis, which leads to the linear convergence of the algorithm under some regularity conditions on the continuous solution, and we perform some numerical simulations, in one and two dimensions, to show the accuracy of the algorithm and the dependence of the solution on some constitutive parameters.

The paper is structured as follows. The thermomechanical model is described in Section 2 following [26,27,31]. Its variational formulation is also obtained and an existence and uniqueness result is stated. Then, in Section 3 a fully discrete algorithm is studied and analyzed. Finally, some numerical simulations are shown in Section 4 and some conclusions are presented in Section 5.

2. The Thermo-Mechanical Problem and Its Variational Formulation

In this section, we describe the model, the constitutive assumptions, the variational formulation of the thermomechanical problem, and we recall an existence and uniqueness result (see [26,27,31] for further details).

Denote by $\Omega \subset \mathbb{R}^d$, $d = 1, 2, 3$, and $[0, T]$, $T > 0$, the domain occupied by the thermo-viscoelastic body and the time interval, respectively. Let also be $x \in \Omega$ and $t \in [0, T]$ the spatial and time variables.

According to [26,27,31,32], using the linear theory of centrosymmetric isotropic and homogeneous materials, a thermo-viscoelastic body is considered. Thus, let $u, v \in \mathbb{R}^d$ and $\varphi, e, \theta, \psi \in \mathbb{R}$ be the displacement and velocity fields, the volume fraction, the volume fraction speed, the temperature and the thermal displacement, respectively. As usual, we have the following relations among them:

$$u(t) = \int_0^t v(s)\,ds + u_0, \quad \varphi(t) = \int_0^t e(s)\,ds + \varphi_0, \quad \psi(t) = \int_0^t \theta(s)\,ds + \psi_0, \tag{1}$$

where u_0, φ_0 and ψ_0 represent initial conditions for the displacement, the volume fraction and the thermal displacement.

Therefore, the thermomechanical problem of a centrosymmetric, isotropic and homogeneous thermo-viscoelastic body within the type III thermal theory is written as follows (see [26,27,32]).

Problem P. Find the displacement $u : \overline{\Omega} \times [0, T] \to \mathbb{R}^d$, the volume fraction $\varphi : \overline{\Omega} \times [0, T] \to \mathbb{R}$ and the thermal displacement $\psi : \overline{\Omega} \times [0, T] \to \mathbb{R}$ such that,

$$\rho \ddot{u}_i = \mu^* \Delta \dot{u}_i + (\mu^* + \lambda^*) \dot{u}_{j,ji} + \mu \Delta u_i + (\mu + \lambda) u_{j,ji} + \gamma \varphi_{,i} - \beta \dot{\psi}_{,i} \quad \text{in} \quad \Omega \times (0, T) \quad \text{for } i = 1, \ldots, d, \tag{2}$$

$$J \ddot{\varphi} = a_0 \Delta \varphi + \gamma u_{i,i} - \xi \varphi + d \dot{\psi} + m \Delta \psi \quad \text{in} \quad \Omega \times (0, T), \tag{3}$$

$$a \ddot{\psi} = \kappa \Delta \psi + m \Delta \varphi - d \dot{\varphi} - \beta \dot{u}_{i,i} + \kappa^* \Delta \dot{\psi} \quad \text{in} \quad \Omega \times (0, T), \tag{4}$$

$$u = 0, \quad \psi = \varphi = 0 \quad \text{on} \quad \partial\Omega \times (0, T), \tag{5}$$

$$u(x, 0) = u_0(x), \quad \dot{u}(x, 0) = v_0(x), \quad \varphi(x, 0) = \varphi_0(x) \quad \text{for a.e. } x \in \Omega, \tag{6}$$

$$\dot{\varphi}(x, 0) = e_0(x), \quad \psi(x, 0) = \psi_0(x), \quad \dot{\psi}(x, 0) = \theta_0(x) \quad \text{for a.e. } x \in \Omega. \tag{7}$$

Here, v_0, e_0 and θ_0 are given initial conditions for the velocity, the volume fraction speed and the temperature, respectively. As usual, the summation over repeated indexes is assumed, the partial derivative with respect to a variable is represented by a subscript preceded by a comma and one superposed dot denotes the first-order partial time derivative (two superposed dots represent the second-order partial time derivative).

In Equations (2)–(4), ρ is the mass density, J is the product of the mass density by the equilibrated inertia, λ and μ denote Lame's coefficients, λ^* and μ^* represent viscosity coefficients, a denotes the heat capacity, a_0 is the volume fraction diffusion, κ is the thermal diffusion and κ^* is the viscous thermal diffusion. The remaining coefficients γ, β, ξ, m, d are model parameters. We also note that, for the sake of simplicity, we have neglected in Equations (2) and (4) the respective terms corresponding to the volume forces and the heat flux.

The following assumptions are imposed on the above constitutive coefficients:

$$\lambda > 0, \quad \mu > 0, \quad (\lambda + \mu)\xi > \gamma^2, \quad a > 0, \quad \rho > 0, \quad J > 0, \quad a_0 > 0, \quad \kappa^* > 0, \\ a_0 \kappa > m^2, \quad \mu^* > 0, \quad \lambda^* + \mu^* > 0. \tag{8}$$

Now, we will derive the weak form of Problem P. Thus, denote by $Y = L^2(\Omega)$, $H = [L^2(\Omega)]^d$ and $Q = [L^2(\Omega)]^{d\times d}$, and their respective scalar products (resp. norms) by $(\cdot,\cdot)_Y$, $(\cdot,\cdot)_H$ and $(\cdot,\cdot)_Q$ (resp. $\|\cdot\|_Y$, $\|\cdot\|_H$ and $\|\cdot\|_Q$). Moreover, to simplify the writing let $E = H_0^1(\Omega)$ and $V = [H_0^1(\Omega)]^d$.

Using boundary conditions (5) and applying classical Green's formula and relations (1), we obtain the weak form of Problem P as follows.

Problem VP. *Find the velocity* $v : [0, T] \to V$, *the volume fraction speed* $e : [0, T] \to E$ *and the temperature* $\theta : [0, T] \to E$ *such that* $v(0) = v_0$, $e(0) = e_0$, $\theta(0) = \theta_0$ *and, for a.e.* $t \in (0, T)$,

$$\rho(\dot{v}(t), w)_H + \mu^*(\nabla v(t), \nabla w)_Q + (\lambda^* + \mu^*)(\operatorname{div} v(t), \operatorname{div} w)_Y + (\lambda + \mu)(\operatorname{div} u(t), \operatorname{div} w)_Y$$
$$+ \mu(\nabla u(t), \nabla w)_Q = \gamma(\nabla \varphi(t), w)_H - \beta(\nabla \theta(t), w)_H \quad \forall w \in V, \tag{9}$$

$$J(\dot{e}(t), r)_Y + a_0(\nabla \varphi(t), \nabla r)_H + \xi(\varphi(t), r)_Y + m(\nabla \psi(t), \nabla r)_H = d(\theta(t), r)_Y$$
$$- \gamma(\operatorname{div} u(t), r)_Y \quad \forall r \in E, \tag{10}$$

$$a(\dot{\theta}(t), z)_Y + \kappa^*(\nabla \theta(t), \nabla z)_H + \kappa(\nabla \psi(t), \nabla z)_H + m(\nabla \varphi(t), \nabla z)_H = -d(e(t), z)_Y$$
$$- \beta(\operatorname{div} v(t), z)_Y \quad \forall z \in E, \tag{11}$$

where the displacement, the volume fraction and the thermal displacement are then recovered from relations (1).

Proceeding as in [27,31], we can state the existence of a unique solution to Problem VP. Details are omitted for the sake of reading.

Theorem 1. *Assume that the coefficients satisfy conditions (8) and the following regularity on the initial data:*

$$u_0, v_0 \in [H^2(\Omega)]^d, \quad \varphi_0, e_0, \psi_0, \theta_0 \in H^2(\Omega). \tag{12}$$

Then, Problem VP admits a unique solution with the regularity:

$$v \in C([0, T]; V) \cap C^1([0, T]; H), \quad e, \theta \in C([0, T]; E) \cap C^1([0, T]; Y).$$

3. Fully Discrete Approximations: An a Priori Error Analysis

In this section, a finite element algorithm is shown for approximating solutions to Problem VP. This is done in two steps. First, to approximate the variational spaces V and E we define the finite element spaces V^h and E^h as follows,

$$V^h = \{w^h \in [C(\overline{\Omega})]^d \; ; \; w^h_{|Tr} \in [P_1(Tr)]^d \quad \forall Tr \in \mathcal{T}^h, \quad w^h = 0 \text{ on } \partial\Omega\}, \tag{13}$$

$$E^h = \{r^h \in C(\overline{\Omega}) \; ; \; r^h_{|Tr} \in P_1(Tr) \quad \forall Tr \in \mathcal{T}^h, \quad r^h = 0 \text{ on } \partial\Omega\}, \tag{14}$$

where we assume that Ω is a polyhedral domain and we denote by $\mathcal{T}^h$ a regular triangulation of $\overline{\Omega}$. Moreover, the space of polynomials of global degree less or equal to 1 in element Tr is represented by $P_1(Tr)$. Here, $h > 0$ denotes the spatial discretization parameter.

Secondly, in order to discretize the time derivatives we use a uniform partition of the time interval $[0, T]$, that we denote by $0 = t_0 < t_1 < \ldots < t_N = T$ ($k = T/N$ is the time step size). Moreover, for a continuous function $f(t)$ we denote $f_n = f(t_n)$ and, for the sequence $\{z_n\}_{n=0}^N$, let $\delta z_n = (z_n - z_{n-1})/k$ be its corresponding divided differences.

Using the classical implicit Euler scheme, the fully discrete approximation of Problem VP is the following.

Problem VP^{hk}. Find the discrete velocity $v^{hk} = \{v_n^{hk}\}_{n=0}^N \subset V^h$, the discrete volume fraction speed $e^{hk} = \{e_n^{hk}\}_{n=0}^N \subset E^h$ and the discrete temperature $\theta^{hk} = \{\theta_n^{hk}\}_{n=0}^N \subset E^h$ such that $v_0^{hk} = v_0^h$, $e_0^{hk} = e_0^h$, $\theta_0^{hk} = \theta_0^h$, and, for $n = 1, \ldots, N$,

$$\rho(\delta v_n^{hk}, w^h)_H + \mu^*(\nabla v_n^{hk}, \nabla w^h)_Q + (\lambda^* + \mu^*)(\operatorname{div} v_n^{hk}, \operatorname{div} w^h)_Y + (\lambda + \mu)(\operatorname{div} u_n^{hk}, \operatorname{div} w^h)_Y$$
$$+ \mu(\nabla u_n^{hk}, \nabla w^h)_Q = \gamma(\nabla \varphi_n^{hk}, w^h)_H - \beta(\nabla \theta_n^{hk}, w^h)_H \quad \forall w^h \in V^h, \tag{15}$$

$$J(\delta e_n^{hk}, r^h)_Y + a_0(\nabla \varphi_n^{hk}, \nabla r^h)_H + \xi(\varphi_n^{hk}, r^h)_Y + m(\nabla \psi_n^{hk}, \nabla r^h)_H = d(\theta_n^{hk}, r^h)_Y$$
$$- \gamma(\operatorname{div} u_n^{hk}, r^h)_Y \quad \forall r^h \in E^h, \tag{16}$$

$$a(\delta \theta_n^{hk}, z^h)_Y + \kappa^*(\nabla \theta_n^{hk}, \nabla z^h)_H + \kappa(\nabla \psi_n^{hk}, \nabla z^h)_H + m(\nabla \varphi_n^{hk}, \nabla z^h)_H = -d(e_n^{hk}, z^h)_Y$$
$$- \beta(\operatorname{div} v_n^{hk}, z^h)_Y \quad \forall z^h \in E^h, \tag{17}$$

where the discrete displacement, the discrete volume fraction and the discrete thermal displacement are then recovered from the relations:

$$u_n^{hk} = k \sum_{j=1}^n v_j^{hk} + u_0^h, \quad \varphi_n^{hk} = k \sum_{j=1}^n e_j^{hk} + \varphi_0^h, \quad \psi_n^{hk} = k \sum_{j=1}^n \theta_j^{hk} + \psi_0^h, \tag{18}$$

and the discrete initial conditions, denoted by u_0^h, v_0^h, φ_0^h, e_0^h, ψ_0^h and θ_0^h are given by

$$u_0^h = \mathcal{P}^{1h} u_0, \quad v_0^h = \mathcal{P}^{1h} v_0, \quad \varphi_0^h = \mathcal{P}^{2h} \varphi_0, \quad e_0^h = \mathcal{P}^{2h} e_0, \quad \psi_0^h = \mathcal{P}^{2h} \psi_0, \quad \theta_0^h = \mathcal{P}^{2h} \theta_0. \tag{19}$$

Here, $\mathcal{P}^{1h}$ and $\mathcal{P}^{2h}$ are the respective projection operators over the finite element spaces V^h and E^h (see [33]).

It is easy to show that discrete problem VP^{hk} admits a unique solution using well-known results on linear variational equations and conditions (8), so we omit the details.

The aim of this section is to obtain some a priori error estimates on the numerical errors $u_n - u_n^{hk}$, $v_n - v_n^{hk}$, $\varphi_n - \varphi_n^{hk}$, $e_n - e_n^{hk}$, $\psi_n - \psi_n^{hk}$ and $\theta_n - \theta_n^{hk}$.

First, we obtain the estimates on the velocity field. Subtracting Equation (9) at time $t = t_n$ for a test function $w = w^h \in V^h \subset V$ and discrete Equation (15), we have, for all $w^h \in V^h$,

$$\rho(\dot{v}_n - \delta v_n^{hk}, w^h)_H + \mu^*(\nabla(v_n - v_n^{hk}), \nabla w^h)_Q + (\lambda^* + \mu^*)(\operatorname{div}(v_n - v_n^{hk}), \operatorname{div} w^h)_Y$$
$$+ (\lambda + \mu)(\operatorname{div}(u_n - u_n^{hk}), \operatorname{div} w^h)_Y + \mu(\nabla(u_n - u_n^{hk}), \nabla w^h)_Q - \gamma(\nabla(\varphi_n - \varphi_n^{hk}), w^h)_H$$
$$+ \beta(\nabla(\theta_n - \theta_n^{hk}), w^h)_H = 0,$$

and so,

$$\rho(\dot{v}_n - \delta v_n^{hk}, v_n - v_n^{hk})_H + \mu^*(\nabla(v_n - v_n^{hk}), \nabla(v_n - v_n^{hk}))_Q + \mu(\nabla(u_n - u_n^{hk}), \nabla(v_n - v_n^{hk}))_Q$$
$$+ (\lambda^* + \mu^*)(\operatorname{div}(v_n - v_n^{hk}), \operatorname{div}(v_n - v_n^{hk}))_Y + (\lambda + \mu)(\operatorname{div}(u_n - u_n^{hk}), \operatorname{div}(v_n - v_n^{hk}))_Y$$
$$- \gamma(\nabla(\varphi_n - \varphi_n^{hk}), v_n - v_n^{hk})_H + \beta(\nabla(\theta_n - \theta_n^{hk}), v_n - v_n^{hk})_H$$
$$= \rho(\dot{v}_n - \delta v_n^{hk}, v_n - w^h)_H + \mu^*(\nabla(v_n - v_n^{hk}), \nabla(v_n - w^h))_Q + \mu(\nabla(u_n - u_n^{hk}), \nabla(v_n - w^h))_Q$$
$$+ (\lambda^* + \mu^*)(\operatorname{div}(v_n - v_n^{hk}), \operatorname{div}(v_n - w^h))_Y + (\lambda + \mu)(\operatorname{div}(u_n - u_n^{hk}), \operatorname{div}(v_n - w^h))_Y$$
$$- \gamma(\nabla(\varphi_n - \varphi_n^{hk}), v_n - w^h)_H + \beta(\nabla(\theta_n - \theta_n^{hk}), v_n - w^h)_H \quad \forall w^h \in V^h.$$

It is straightforward to show that

$$(\dot{v}_n - \delta v_n^{hk}, v_n - v_n^{hk})_H \geq (\dot{v}_n - \delta v_n, v_n - v_n^{hk})_H$$
$$+ \frac{1}{2k}\left[\|v_n - v_n^{hk}\|_H^2 - \|v_{n-1} - v_{n-1}^{hk}\|_H^2\right],$$
$$(\nabla(u_n - u_n^{hk}), \nabla(v_n - v_n^{hk}))_Q \geq (\nabla(u_n - u_n^{hk}), \nabla(\dot{u}_n - \delta u_n))_Q$$
$$+ \frac{1}{2k}\left[\|\nabla(u_n - u_n^{hk})\|_Q^2 - \|\nabla(u_{n-1} - u_{n-1}^{hk})\|_Q^2\right],$$
$$(\text{div}(u_n - u_n^{hk}), \text{div}(v_n - v_n^{hk}))_Y = (\text{div}(u_n - u_n^{hk}), \text{div}(\dot{u}_n - \delta u_n))_Y$$
$$+ \frac{1}{2k}\left[\|\text{div}(u_n - u_n^{hk})\|_Y^2 - \|\text{div}(u_{n-1} - u_{n-1}^{hk})\|_Y^2\right.$$
$$+ \|\text{div}(u_n - u_n^{hk} - (u_{n-1} - u_{n-1}^{hk}))\|_Y^2\Big],$$
$$(\nabla(\theta_n - \theta_n^{hk}), v_n - w^h)_H = -(\theta_n - \theta_n^{hk}, \text{div}(v_n - w^h))_Y,$$
$$(\nabla(\varphi_n - \varphi_n^{hk}), v_n - w^h)_H = -(\varphi_n - \varphi_n^{hk}, \text{div}(v_n - w^h))_Y,$$
$$(\nabla(\theta_n - \theta_n^{hk}), v_n - v_n^{hk})_H = -(\theta_n - \theta_n^{hk}, \text{div}(v_n - v_n^{hk}))_Y,$$
$$(\nabla(\varphi_n - \varphi_n^{hk}), v_n - v_n^{hk})_H = -(\varphi_n - \varphi_n^{hk}, \text{div}(v_n - v_n^{hk}))_Y,$$

where we have used that $v_n^{hk} = \delta u_n^{hk} = (u_n^{hk} - u_{n-1}^{hk})/k$ and notations $\delta v_n = (v_n - v_{n-1})/k$ and $\delta u_n = (u_n - u_{n-1})/k$. Therefore, we find that, for all $w^h \in V^h$,

$$\frac{\rho}{2k}\left[\|v_n - v_n^{hk}\|_H^2 - \|v_{n-1} - v_{n-1}^{hk}\|_H^2\right] + C\|\nabla(v_n - v_n^{hk})\|_Q^2 + C\|\text{div}(v_n - v_n^{hk})\|_Y^2$$
$$+ \frac{\mu}{2k}\left[\|\nabla(u_n - u_n^{hk})\|_Q^2 - \|\nabla(u_{n-1} - u_{n-1}^{hk})\|_Q^2\right] + \beta(\theta_n - \theta_n^{hk}, \text{div}(v_n - v_n^{hk}))_Y$$
$$+ \frac{\lambda + \mu}{2k}\Big[\|\text{div}(u_n - u_n^{hk})\|_Y^2 - \|\text{div}(u_{n-1} - u_{n-1}^{hk})\|_Y^2$$
$$+ \|\text{div}(u_n - u_n^{hk} - (u_{n-1} - u_{n-1}^{hk}))\|_Y^2\Big] + \gamma(\varphi_n - \varphi_n^{hk}, \text{div}(v_n - v_n^{hk}))_Y$$
$$\leq C\Big(\|v_n - v_n^{hk}\|_H^2 + \|\dot{v}_n - \delta v_n\|_H^2 + \|\nabla(u_n - u_n^{hk})\|_Q^2 + \|\text{div}(u_n - u_n^{hk})\|_Y^2$$
$$+ \|v_n - w^h\|_H^2 + \|\nabla(v_n - w^h)\|_Q^2 + \|\text{div}(v_n - w^h)\|_Y^2 + \|\varphi_n - \varphi_n^{hk}\|_Y^2$$
$$+ \|\nabla(\dot{u}_n - \delta u_n)\|_Q^2 + \|\text{div}(\dot{u}_n - \delta u_n)\|_Y^2 + (\delta v_n - \delta v_n^{hk}, v_n - w^h)_H$$
$$+ \|\theta_n - \theta_n^{hk}\|_Y^2\Big). \tag{20}$$

Here and in what follows, $C > 0$ denotes a positive constant whose value may change for each expression, which depends on the continuous solution but it is independent of the discretization parameters h and k.

Secondly, we will find the error estimates on the volume fraction speed. Then, we subtract Equation (10) at time $t = t_n$ for a test function $r = r^h \in E^h \subset E$ and discrete Equation (16) to obtain, for all $r^h \in E^h$,

$$J(\dot{e}_n - \delta e_n^{hk}, r^h)_Y + a_0(\nabla(\varphi_n - \varphi_n^{hk}), \nabla r^h)_H + \xi(\varphi_n - \varphi_n^{hk}, r^h)_Y + m(\nabla(\psi_n - \psi_n^{hk}), \nabla r^h)_H$$
$$= d(\theta_n - \theta_n^{hk}, r^h)_Y - \gamma(\text{div}(u_n - u_n^{hk}), r^h)_Y.$$

Thus, we find, for all $r^h \in E^h$,

$$J(\dot{e}_n - \delta e_n^{hk}, e_n - e_n^{hk})_Y + a_0(\nabla(\varphi_n - \varphi_n^{hk}), \nabla(e_n - e_n^{hk}))_H + \xi(\varphi_n - \varphi_n^{hk}, e_n - e_n^{hk})_Y$$
$$+ m(\nabla(\psi_n - \psi_n^{hk}), \nabla(e_n - e_n^{hk}))_H - d(\theta_n - \theta_n^{hk}, e_n - e_n^{hk})_Y + \gamma(\text{div}(u_n - u_n^{hk}), e_n - e_n^{hk})_Y$$
$$= J(\dot{e}_n - \delta e_n^{hk}, e_n - r^h)_Y + a_0(\nabla(\varphi_n - \varphi_n^{hk}), \nabla(e_n - r^h))_H + \xi(\varphi_n - \varphi_n^{hk}, e_n - r^h)_Y$$
$$+ m(\nabla(\psi_n - \psi_n^{hk}), \nabla(e_n - r^h))_H - d(\theta_n - \theta_n^{hk}, e_n - r^h)_Y + \gamma(\text{div}(u_n - u_n^{hk}), e_n - r^h)_Y.$$

Since

$$(\dot{e}_n - \delta e_n^{hk}, e_n - e_n^{hk})_Y \geq (\dot{e}_n - \delta e_n, e_n - e_n^{hk})_Y + \frac{1}{2k}\left[\|e_n - e_n^{hk}\|_Y^2 - \|e_{n-1} - e_{n-1}^{hk}\|_Y^2\right],$$

$$(\nabla(\varphi_n - \varphi_n^{hk}), \nabla(e_n - e_n^{hk}))_H = (\nabla(\varphi_n - \varphi_n^{hk}), \nabla(\dot{\varphi}_n - \delta\varphi_n))_H$$
$$+ \frac{1}{2k}\left[\|\nabla(\varphi_n - \varphi_n^{hk})\|_H^2 - \|\nabla(\varphi_{n-1} - \varphi_{n-1}^{hk})\|_H^2 + \nabla(\varphi_n - \varphi_n^{hk} - (\varphi_{n-1} - \varphi_{n-1}^{hk}))\|_H^2\right],$$

$$(\varphi_n - \varphi_n^{hk}, e_n - e_n^{hk})_Y = (\varphi_n - \varphi_n^{hk}, \dot{\varphi}_n - \delta\varphi_n)_Y + \frac{1}{2k}\left[\|\varphi_n - \varphi_n^{hk}\|_Y^2 - \|\varphi_{n-1} - \varphi_{n-1}^{hk}\|_Y^2\right.$$
$$\left.+ \|\varphi_n - \varphi_n^{hk} - (\varphi_{n-1} - \varphi_{n-1}^{hk})\|_Y^2\right],$$

where we have used the fact that $e_n^{hk} = \delta\varphi_n^{hk} = (\varphi_n^{hk} - \varphi_{n-1}^{hk})/k$ and the notations $\delta e_n = (e_n - e_{n-1})/k$ and $\delta\varphi_n = (\varphi_n - \varphi_{n-1})/k$, it follows that, for all $r^h \in E^h$,

$$\frac{J}{2k}\left[\|e_n - e_n^{hk}\|_Y^2 - \|e_{n-1} - e_{n-1}^{hk}\|_Y^2\right]$$
$$+ \frac{a_0}{2k}\left[\|\nabla(\varphi_n - \varphi_n^{hk})\|_H^2 - \|\nabla(\varphi_{n-1} - \varphi_{n-1}^{hk})\|_H^2 + \nabla(\varphi_n - \varphi_n^{hk} - (\varphi_{n-1} - \varphi_{n-1}^{hk}))\|_H^2\right]$$
$$+ \frac{\zeta}{2k}\left[\|\varphi_n - \varphi_n^{hk}\|_Y^2 - \|\varphi_{n-1} - \varphi_{n-1}^{hk}\|_Y^2 + \|\varphi_n - \varphi_n^{hk} - (\varphi_{n-1} - \varphi_{n-1}^{hk})\|_Y^2\right]$$
$$+ m(\nabla(\psi_n - \psi_n^{hk}), \nabla(e_n - e_n^{hk}))_H - d(\theta_n - \theta_n^{hk}, e_n - e_n^{hk})_Y + \gamma(\operatorname{div}(u_n - u_n^{hk}), e_n - e_n^{hk})_Y$$
$$\leq C\Big(\|e_n - e_n^{hk}\|_Y^2 + \|\dot{e}_n - \delta e_n\|_Y^2 + \|\nabla(\varphi_n - \varphi_n^{hk})\|_H^2 + \|\varphi_n - \varphi_n^{hk}\|_Y^2 + \|e_n - r^h\|_Y^2$$
$$+ \|\nabla(e_n - r^h)\|_H^2 + \|\nabla(\dot{\varphi}_n - \delta\varphi_n)\|_H^2 + \|\dot{\varphi}_n - \delta\varphi_n\|_Y^2 + (\delta e_n - \delta e_n^{hk}, e_n - r^h)_Y$$
$$+ \|\theta_n - \theta_n^{hk}\|_Y^2 + \|\operatorname{div}(u_n - u_n^{hk})\|_Y^2\Big). \tag{21}$$

Finally, we get the estimates on the temperature field. Then, we subtract Equation (11) at time $t = t_n$ for a test function $z = z^h \in E^h \subset E$ and discrete Equation (17) to have, for all $z^h \in E^h$,

$$a(\dot{\theta}_n - \delta\theta_n^{hk}, z^h)_Y + \kappa^*(\nabla(\theta_n - \theta_n^{hk}), \nabla z^h)_H + \kappa(\nabla(\psi_n - \psi_n^{hk}), \nabla z^h)_H + m(\nabla(\varphi_n - \varphi_n^{hk}), \nabla z^h)_H$$
$$+ d(e_n - e_n^{hk}, z^h)_Y + \beta(\operatorname{div}(v_n - v_n^{hk}), z^h)_Y = 0,$$

and so, it follows that, for all $z^h \in E^h$,

$$a(\dot{\theta}_n - \delta\theta_n^{hk}, \theta_n - \theta_n^{hk})_Y + \kappa^*(\nabla(\theta_n - \theta_n^{hk}), \nabla(\theta_n - \theta_n^{hk}))_H + \kappa(\nabla(\psi_n - \psi_n^{hk}), \nabla(\theta_n - \theta_n^{hk}))_H$$
$$+ m(\nabla(\varphi_n - \varphi_n^{hk}), \nabla(\theta_n - \theta_n^{hk}))_H + d(e_n - e_n^{hk}, \theta_n - \theta_n^{hk})_Y + \beta(\operatorname{div}(v_n - v_n^{hk}), \theta_n - \theta_n^{hk})_Y$$
$$= a(\dot{\theta}_n - \delta\theta_n^{hk}, \theta_n - z^h)_Y + \kappa^*(\nabla(\theta_n - \theta_n^{hk}), \nabla(\theta_n - z^h))_H + \kappa(\nabla(\psi_n - \psi_n^{hk}), \nabla(\theta_n - z^h))_H$$
$$+ m(\nabla(\varphi_n - \varphi_n^{hk}), \nabla(\theta_n - z^h))_H + d(e_n - e_n^{hk}, \theta_n - z^h)_Y + \beta(\operatorname{div}(v_n - v_n^{hk}), \theta_n - z^h)_Y.$$

Keeping in mind that

$$(\dot{\theta}_n - \delta\theta_n^{hk}, \theta_n - \theta_n^{hk})_Y \geq (\dot{\theta}_n - \delta\theta_n, \theta_n - \theta_n^{hk})_Y + \frac{1}{2k}\left[\|\theta_n - \theta_n^{hk}\|_Y^2 - \|\theta_{n-1} - \theta_{n-1}^{hk}\|_Y^2\right],$$

$$(\nabla(\psi_n - \psi_n^{hk}), \nabla(\theta_n - \theta_n^{hk}))_H = (\nabla(\psi_n - \psi_n^{hk}), \nabla(\dot{\psi}_n - \delta\psi_n))_H$$
$$+ \frac{1}{2k}\left[\|\nabla(\psi_n - \psi_n^{hk})\|_H^2 - \|\nabla(\psi_{n-1} - \psi_{n-1}^{hk})\|_H^2 + \|\nabla(\psi_n - \psi_n^{hk} - (\psi_{n-1} - \psi_{n-1}^{hk}))\|_H^2\right],$$

$$(\operatorname{div}(v_n - v_n^{hk}), \theta_n - z^h)_Y = -(v_n - v_n^{hk}, \nabla(\theta_n - z^h))_H,$$

where we have used the fact that $\theta_n^{hk} = \delta \psi_n^{hk} = (\psi_n^{hk} - \psi_{n-1}^{hk})/k$ and the notations $\delta \theta_n = (\theta_n - \theta_{n-1})/k$ and $\delta \psi_n = (\psi_n - \psi_{n-1})/k$, we obtain, for all $z^h \in E^h$,

$$\frac{a}{2k}\left[\|\theta_n - \theta_n^{hk}\|_Y^2 - \|\theta_{n-1} - \theta_{n-1}^{hk}\|_Y^2\right] + C\|\nabla(\theta_n - \theta_n^{hk})\|_H^2$$
$$+ \frac{\kappa}{2k}\left[\|\nabla(\psi_n - \psi_n^{hk})\|_H^2 - \|\nabla(\psi_{n-1} - \psi_{n-1}^{hk})\|_H^2 + \|\nabla(\psi_n - \psi_n^{hk} - (\psi_{n-1} - \psi_{n-1}^{hk}))\|_H^2\right]$$
$$+ m(\nabla(\varphi_n - \varphi_n^{hk}), \nabla(\theta_n - \theta_n^{hk}))_H + d(e_n - e_n^{hk}, \theta_n - \theta_n^{hk})_Y + \beta(\text{div}\,(v_n - v_n^{hk}), \theta_n - \theta_n^{hk})_Y$$
$$\leq C\Big(\|\theta_n - \theta_n^{hk}\|_Y^2 + \|\dot\theta_n - \delta\theta_n\|_Y^2 + \|\nabla(\psi_n - \psi_n^{hk})\|_H^2 + \|\theta_n - z^h\|_Y^2 + \|\nabla(\theta_n - z^h)\|_H^2$$
$$+ \|v_n - v_n^{hk}\|_H^2 + \|e_n - e_n^{hk}\|_Y^2 + (\delta\theta_n - \delta\theta_n^{hk}, \theta_n - z^h)_Y + \|\nabla(\dot\psi_n - \delta\psi_n)\|_H^2$$
$$+ \|\nabla(\varphi_n - \varphi_n^{hk})\|_H^2\Big). \tag{22}$$

Combining now estimates (20)–(22) we find that

$$\frac{\rho}{2k}\left[\|v_n - v_n^{hk}\|_H^2 - \|v_{n-1} - v_{n-1}^{hk}\|_H^2\right] + C\|\nabla(v_n - v_n^{hk})\|_Q^2 + C\|\text{div}\,(v_n - v_n^{hk})\|_Y^2$$
$$+ \frac{\mu}{2k}\left[\|\nabla(u_n - u_n^{hk})\|_Q^2 - \|\nabla(u_{n-1} - u_{n-1}^{hk})\|_Q^2\right]$$
$$+ \frac{\lambda + \mu}{2k}\left[\|\text{div}\,(u_n - u_n^{hk})\|_Y^2 - \|\text{div}\,(u_{n-1} - u_{n-1}^{hk})\|_Y^2\right.$$
$$+ \|\text{div}\,(u_n - u_n^{hk} - (u_{n-1} - u_{n-1}^{hk}))\|_Y^2\Big] + \gamma(\varphi_n - \varphi_n^{hk}, \text{div}\,(v_n - v_n^{hk}))_Y$$
$$+ \frac{J}{2k}\left[\|e_n - e_n^{hk}\|_Y^2 - \|e_{n-1} - e_{n-1}^{hk}\|_Y^2\right] + m(\nabla(\varphi_n - \varphi_n^{hk}), \nabla(\theta_n - \theta_n^{hk}))_H$$
$$+ \frac{\xi}{2k}\left[\|\varphi_n - \varphi_n^{hk}\|_Y^2 - \|\varphi_{n-1} - \varphi_{n-1}^{hk}\|_Y^2 + \|\varphi_n - \varphi_n^{hk} - (\varphi_{n-1} - \varphi_{n-1}^{hk})\|_Y^2\right]$$
$$+ m(\nabla(\psi_n - \psi_n^{hk}), \nabla(e_n - e_n^{hk}))_H + \gamma(\text{div}\,(u_n - u_n^{hk}), e_n - e_n^{hk})_Y$$
$$+ \frac{a}{2k}\left[\|\theta_n - \theta_n^{hk}\|_Y^2 - \|\theta_{n-1} - \theta_{n-1}^{hk}\|_Y^2\right] + C\|\nabla(\theta_n - \theta_n^{hk})\|_H^2$$
$$+ \frac{\kappa}{2k}\left[\|\nabla(\psi_n - \psi_n^{hk})\|_H^2 - \|\nabla(\psi_{n-1} - \psi_{n-1}^{hk})\|_H^2 + \|\nabla(\psi_n - \psi_n^{hk} - (\psi_{n-1} - \psi_{n-1}^{hk}))\|_H^2\right]$$
$$+ \frac{a_0}{2k}\left[\|\nabla(\varphi_n - \varphi_n^{hk})\|_H^2 - \|\nabla(\varphi_{n-1} - \varphi_{n-1}^{hk})\|_H^2 + \|\nabla(\varphi_n - \varphi_n^{hk} - (\varphi_{n-1} - \varphi_{n-1}^{hk}))\|_H^2\right]$$
$$\leq C\Big(\|v_n - v_n^{hk}\|_H^2 + \|\dot v_n - \delta v_n\|_H^2 + \|\nabla(u_n - u_n^{hk})\|_Q^2 + \|\text{div}\,(u_n - u_n^{hk})\|_Y^2$$
$$+ \|v_n - w^h\|_H^2 + \|\nabla(v_n - w^h)\|_Q^2 + \|\text{div}\,(v_n - w^h)\|_Y^2 + \|\varphi_n - \varphi_n^{hk}\|_Y^2$$
$$+ \|\nabla(\dot u_n - \delta u_n)\|_Q^2 + \|\text{div}\,(\dot u_n - \delta u_n)\|_Y^2 + (\delta v_n - \delta v_n^{hk}, v_n - w^h)_H$$
$$+ \|\theta_n - \theta_n^{hk}\|_Y^2 + \|\dot\theta_n - \delta\theta_n\|_Y^2 + \|\nabla(\psi_n - \psi_n^{hk})\|_H^2 + \|\theta_n - z^h\|_Y^2 + \|\nabla(\theta_n - z^h)\|_H^2$$
$$+ \|e_n - e_n^{hk}\|_Y^2 + (\delta\theta_n - \delta\theta_n^{hk}, \theta_n - z^h)_Y + \|\nabla(\dot\psi_n - \delta\psi_n)\|_H^2 + \|\nabla(\varphi_n - \varphi_n^{hk})\|_H^2$$
$$+ \|\dot e_n - \delta e_n\|_Y^2 + \|e_n - r^h\|_Y^2 + \|\nabla(e_n - r^h)\|_H^2 + \|\nabla(\dot\varphi_n - \delta\varphi_n)\|_H^2 + \|\dot\varphi_n - \delta\varphi_n\|_Y^2$$
$$+ (\delta e_n - \delta e_n^{hk}, e_n - r^h)_Y\Big).$$

Taking into account that

$$\gamma(\varphi_n - \varphi_n^{hk}, \text{div}\,(v_n - v_n^{hk}))_Y + \gamma(e_n - e_n^{hk}, \text{div}\,(u_n - u_n^{hk}))_Y$$
$$= \gamma(\varphi_n - \varphi_n^{hk}, \text{div}\,(\dot u_n - \delta u_n))_Y + \gamma(\dot\varphi_n - \delta\varphi_n, \text{div}\,(u_n - u_n^{hk}))_Y$$
$$+ \gamma(\varphi_n - \varphi_n^{hk}, \text{div}\,(\delta u_n - \delta u_n^{hk}))_Y + \gamma(\delta\varphi_n - \delta\varphi_n^{hk}, \text{div}\,(u_n - u_n^{hk}))_Y,$$
$$m(\nabla(\psi_n - \psi_n^{hk}), \nabla(e_n - e_n^{hk}))_Y + m(\nabla(\theta_n - \theta_n^{hk}), \nabla(\varphi_n - \varphi_n^{hk}))_Y$$
$$= m(\nabla(\psi_n - \psi_n^{hk}), \nabla(\dot\varphi_n - \delta\varphi_n))_Y + m(\nabla(\dot\psi_n - \delta\psi_n), \nabla(\varphi_n - \varphi_n^{hk}))_Y$$
$$+ m(\nabla(\psi_n - \psi_n^{hk}), \nabla(\delta\varphi_n - \delta\varphi_n^{hk}))_Y + m(\nabla(\delta\psi_n - \delta\psi_n^{hk}), \nabla(\varphi_n - \varphi_n^{hk}))_Y,$$

$$\gamma(\varphi_n - \varphi_n^{hk}, \text{div}\,(\delta u_n - \delta u_n^{hk}))_Y + \gamma(\delta\varphi_n - \delta\varphi_n^{hk}, \text{div}\,(u_n - u_n^{hk}))_Y$$
$$= \frac{\gamma}{k}\Big[(\varphi_n - \varphi_n^{hk}, \text{div}\,(u_n - u_n^{hk}))_Y - (\varphi_{n-1} - \varphi_{n-1}^{hk}, \text{div}\,(u_{n-1} - u_{n-1}^{hk}))_Y$$
$$+ (\varphi_n - \varphi_n^{hk} - (\varphi_{n-1} - \varphi_{n-1}^{hk}), \text{div}\,(u_n - u_n^{hk} - (u_{n-1} - u_{n-1}^{hk})))_Y\Big],$$
$$m(\nabla(\psi_n - \psi_n^{hk}), \nabla(\delta\varphi_n - \delta\varphi_n^{hk}))_Y + m(\nabla(\delta\psi_n - \delta\psi_n^{hk}), \nabla(\varphi_n - \varphi_n^{hk}))_Y$$
$$= \frac{m}{k}\Big[(\nabla(\psi - \psi_n^{hk}), \nabla(\varphi_n - \varphi_n^{hk}))_Y - (\nabla(\psi_{n-1} - \psi_{n-1}^{hk}), \nabla(\varphi_{n-1} - \varphi_{n-1}^{hk}))_Y$$
$$+ (\nabla(\psi - \psi_n^{hk} - (\psi_{n-1} - \psi_{n-1}^{hk})), \nabla(\varphi_n - \varphi_n^{hk} - (\varphi_{n-1} - \varphi_{n-1}^{hk})))_Y,$$
$$\frac{\lambda + \mu}{2k}\|\text{div}\,(u_n - u_n^{hk} - (u_{n-1} - u_{n-1}^{hk}))\|_Y^2 + \frac{\xi}{2k}\|\varphi_n - \varphi_n^{hk} - (\varphi_{n-1} - \varphi_{n-1}^{hk})\|_Y^2$$
$$+ \frac{\gamma}{k}(\varphi_n - \varphi_n^{hk} - (\varphi_{n-1} - \varphi_{n-1}^{hk}), \text{div}\,(u_n - u_n^{hk} - (u_{n-1} - u_{n-1}^{hk})))_Y \geq 0,$$
$$\frac{a_0}{2k}\|\nabla(\varphi_n - \varphi_n^{hk} - (\varphi_{n-1} - \varphi_{n-1}^{hk}))\|_H^2 + \frac{\kappa}{2k}\|\nabla(\psi_n - \psi_n^{hk} - (\psi_{n-1} - \psi_{n-1}^{hk}))\|_H^2$$
$$+ \frac{m}{k}(\nabla(\varphi_n - \varphi_n^{hk} - (\varphi_{n-1} - \varphi_{n-1}^{hk})), \nabla(\psi_n - \psi_n^{hk} - (\psi_{n-1} - \psi_{n-1}^{hk})))_H \geq 0,$$

where we have employed conditions (8), multiplying the previous estimates by k and summing up to n, it follows that, for all $w^h = \{w_j^h\}_{j=1}^n \subset V^h$, $r^h = \{r_j^h\}_{j=1}^n \subset E^h$ and $z^h = \{z_j^h\}_{j=1}^n \subset E^h$,

$$\rho\|v_n - v_n^{hk}\|_H^2 + Ck\sum_{j=1}^n \|\nabla(v_j - v_j^{hk})\|_Q^2 + Ck\sum_{j=1}^n \|\text{div}\,(v_j - v_j^{hk})\|_Y^2$$
$$+ \mu\|\nabla(u_n - u_n^{hk})\|_Q^2 + (\lambda + \mu)\|\text{div}\,(u_n - u_n^{hk})\|_Y^2 + 2\gamma(\varphi_n - \varphi_n^{hk}, \text{div}\,(u_n - u_n^{hk}))_Y$$
$$+ J\|e_n - e_n^{hk}\|_Y^2 + a_0\|\nabla(\varphi_n - \varphi_n^{hk})\|_H^2 + \xi\|\varphi_n - \varphi_n^{hk}\|_Y^2 + a\|\theta_n - \theta_n^{hk}\|_Y^2$$
$$+ 2m(\nabla(\psi_n - \psi_n^{hk}), \nabla(\varphi_n - \varphi_n^{hk}))_H + C\sum_{j=1}^n k\|\nabla(\theta_j - \theta_j^{hk})\|_H^2 + \kappa\|\nabla(\psi_n - \psi_n^{hk})\|_H^2$$
$$\leq Ck\sum_{j=1}^n \Big(\|v_j - v_j^{hk}\|_H^2 + \|\dot{v}_j - \delta v_j\|_H^2 + \|\nabla(u_j - u_j^{hk})\|_Q^2 + \|\text{div}\,(u_j - u_j^{hk})\|_Y^2$$
$$+ \|v_j - w_j^h\|_H^2 + \|\nabla(v_j - w_j^h)\|_Q^2 + \|\text{div}\,(v_j - w_j^h)\|_Y^2 + \|\varphi_j - \varphi_j^{hk}\|_Y^2$$
$$+ \|\nabla(\dot{u}_j - \delta u_j)\|_Q^2 + \|\text{div}\,(\dot{u}_j - \delta u_j)\|_Y^2 + (\delta v_j - \delta v_j^{hk}, v_j - w_j^h)_H$$
$$+ \|\theta_j - \theta_j^{hk}\|_Y^2 + \|\dot{\theta}_j - \delta\theta_j\|_Y^2 + \|\nabla(\psi_j - \psi_j^{hk})\|_H^2 + \|\theta_j - z_j^h\|_Y^2 + \|\nabla(\theta_j - z_j^h)\|_H^2$$
$$+ \|e_j - e_j^{hk}\|_Y^2 + (\delta\theta_j - \delta\theta_j^{hk}, \theta_j - z_j^h)_Y + \|\nabla(\dot{\psi}_j - \delta\psi_j)\|_H^2 + \|\nabla(\varphi_j - \varphi_j^{hk})\|_H^2$$
$$+ \|\dot{e}_j - \delta e_j\|_Y^2 + \|e_j - r_j^h\|_Y^2 + \|\nabla(e_j - r_j^h)\|_H^2 + \|\nabla(\dot{\varphi}_j - \delta\varphi_j)\|_H^2 + \|\dot{\varphi}_j - \delta\varphi_j\|_Y^2$$
$$+ (\delta e_j - \delta e_j^{hk}, e_j - r_j^h)_Y\Big) + C\Big(\|v_0 - v_0^h\|_H^2 + \|\nabla(v_0 - v_0^h)\|_Q^2 + \|\text{div}\,(v_0 - v_0^h)\|_Y^2$$
$$+ \|\nabla(u_0 - u_0^h)\|_Q^2 + \|\text{div}\,(u_0 - u_0^h)\|_Y^2 + \|e_0 - e_0^h\|_Y^2 + \|\nabla(\varphi_0 - \varphi_0^h)\|_H^2$$
$$+ \|\varphi_0 - \varphi_0^h\|_Y^2 + \|\theta_0 - \theta_0^h\|_Y^2 + \|\nabla(\theta_0 - \theta_0^h)\|_H^2\Big). \tag{23}$$

Keeping in mind again assumptions (8), we find that there exist two constants $\zeta_1, \zeta_2 > 0$ such that $\gamma/\xi < \zeta_1 < (\lambda + \mu)/\gamma$ and $m/\kappa < \zeta_2 < a_0/m$. Then, we have

$$(\lambda + \mu)\|\text{div}\,(u_n - u_n^{hk})\|_Y^2 + \xi\|\varphi_n - \varphi_n^{hk}\|_Y^2 + 2m(\text{div}\,(u_n - u_n^{hk}), \varphi_n - \varphi_n^{hk})_Y$$
$$\geq (\lambda + \mu - \gamma\zeta_1)\|\text{div}\,(u_n - u_n^{hk})\|_Y^2 + \left(\xi - \frac{\gamma}{\zeta_1}\right)\|\varphi_n - \varphi_n^{hk}\|_Y^2,$$
$$a_0\|\nabla(\varphi_n - \varphi_n^{hk})\|_H^2 + \kappa\|\nabla(\psi_n - \psi_n^{hk})\|_H^2 + 2m(\nabla(\varphi_n - \varphi_n^{hk}), \nabla(\psi_n - \psi_n^{hk}))_H$$
$$\geq (a_0 - m\zeta_2)\|\nabla(\varphi_n - \varphi_n^{hk})\|_H^2 + \left(\kappa - \frac{m}{\zeta_2}\right)\|\nabla(\psi_n - \psi_n^{hk})\|_H^2.$$

We will need the following discrete version of Gronwall's lemma (see, for instance, [34,35]).

Lemma 1. Let $\{E_n\}_{n=0}^N$ and $\{G_n\}_{n=0}^N$ be two sequences of nonnegative real numbers satisfying, for a positive constant $C_1 > 0$ independent of G_n and E_n,

$$E_0 \leq C_1 G_0,$$
$$E_n \leq C_1 G_n + C_1 \sum_{j=1}^n k E_j, \quad n = 1, \ldots, N,$$

where k is a positive constant. If we denote $C = C_1(1 + C_1 T e^{2C_1 T})$ and $T = Nk$, then we have

$$\max_{0 \leq n \leq N} E_n \leq C \max_{0 \leq n \leq N} G_n.$$

Now, taking into account that

$$k \sum_{j=1}^n (\delta v_j - \delta v_j^{hk}, v_j - w_j^h)_H = \sum_{j=1}^n (v_j - v_j^{hk} - (v_{j-1} - v_{j-1}^{hk}), v_j - w_j^h)_H$$
$$= (v_n - v_n^{hk}, v_n - w_n^h)_H + (v_0^h - v_0, v_1 - w_1^h)_H$$
$$+ \sum_{j=1}^{n-1} (v_j - v_j^{hk}, v_j - w_j^h - (v_{j+1} - w_{j+1}^h))_H,$$

$$k \sum_{j=1}^n (\delta e_j - \delta e_j^{hk}, e_n - r_j^h)_Y = \sum_{j=1}^n (e_j - e_j^{hk} - (e_{j-1} - e_{j-1}^{hk}), e_n - r_j^h)_Y$$
$$= (e_n - e_n^{hk}, e_n - r_n^h)_Y + (e_0^h - e_0, e_1 - e_1^h)_Y$$
$$+ \sum_{j=1}^{n-1} (e_j - e_j^{hk}, e_j - r_j^h - (e_{j+1} - r_{j+1}^h))_Y,$$

$$k \sum_{j=1}^n (\delta \theta_j - \delta \theta_j^{hk}, \theta_n - z_j^h)_Y = \sum_{j=1}^n (\theta_j - \theta_j^{hk} - (\theta_{j-1} - \theta_{j-1}^{hk}), \theta_n - z_j^h)_Y$$
$$= (\theta_n - \theta_n^{hk}, \theta_n - z_n^h)_Y + (\theta_0^h - \theta_0, \theta_1 - z_1^h)_Y$$
$$+ \sum_{j=1}^{n-1} (\theta_j - \theta_j^{hk}, \theta_j - z_j^h - (\theta_{j+1} - z_{j+1}^h))_Y,$$

applying Lemma 1 to estimates (23), we conclude the following a priori error estimates result.

Theorem 2. Under the assumptions of Theorem 1, let us denote by (v, e, θ) the solution to Problem VP and by $(v^{hk}, e^{hk}, \theta^{hk})$ the solution to Problem VP^{hk}, then we obtain the following a priori error estimates, for all $w^h = \{w_j^h\}_{j=0}^N \subset V^h$, $r^h = \{r_j^h\}_{j=0}^N \subset E^h$ and $z^h = \{z_j^h\}_{j=0}^N \subset E^h$,

$$\max_{0 \leq n \leq N} \Big\{ \|v_n - v_n^{hk}\|_H^2 + \|\nabla(u_n - u_n^{hk})\|_Q^2 + \|\mathrm{div}\,(u_n - u_n^{hk})\|_Y^2 + \|e_n - e_n^{hk}\|_Y^2 + \|\nabla(\varphi_n - \varphi_n^{hk})\|_H^2$$
$$+ \|\varphi_n - \varphi_n^{hk}\|_Y^2 + \|\theta_n - \theta_n^{hk}\|_Y^2 + \|\nabla(\psi_n - \psi_n^{hk})\|_H^2 \Big\}$$
$$+ Ck \sum_{n=1}^N \Big[\|\nabla(v_n - v_n^{hk})\|_Q^2 + \|\mathrm{div}\,(v_n - v_n^{hk})\|_Y^2 + \|\nabla(\theta_n - \theta_n^{hk})\|_H^2 \Big]$$
$$\leq Ck \sum_{j=1}^N \Big(\|\dot{v}_j - \delta v_j\|_H^2 + \|v_j - w_j^h\|_H^2 + \|\nabla(v_j - w_j^h)\|_Q^2 + \|\mathrm{div}\,(v_j - w_j^h)\|_Y^2$$
$$+ \|\nabla(\dot{u}_j - \delta u_j)\|_Q^2 + \|\mathrm{div}\,(\dot{u}_j - \delta u_j)\|_Y^2 + \|\dot{\theta}_j - \delta \theta_j\|_Y^2 + \|\theta_j - z_j^h\|_Y^2 + \|\nabla(\theta_j - z_j^h)\|_H^2$$
$$+ \|\nabla(\dot{\psi}_j - \delta \psi_j)\|_H^2 + \|\dot{e}_j - \delta e_j\|_Y^2 + \|e_j - r_j^h\|_Y^2 + \|\nabla(e_j - r_j^h)\|_H^2 + \|\nabla(\dot{\varphi}_j - \delta \varphi_j)\|_H^2 + \|\dot{\varphi}_j - \delta \varphi_j\|_Y^2 \Big)$$

$$
\begin{aligned}
&+ C \max_{0 \leq n \leq N} \|v_n - w_n^h\|_H^2 + C \max_{0 \leq n \leq N} \|e_n - r_n^h\|_Y^2 + C \max_{0 \leq n \leq N} \|\theta_n - z_n^h\|_Y^2 \\
&+ \frac{C}{k} \sum_{j=1}^{N-1} \|v_j - w_j^h - (v_{j+1} - w_{j+1}^h)\|_H^2 + \frac{C}{k} \sum_{j=1}^{N-1} \|e_j - r_j^h - (e_{j+1} - r_{j+1}^h)\|_Y^2 \\
&+ \frac{C}{k} \sum_{j=1}^{N-1} \|\theta_j - z_j^h - (\theta_{j+1} - z_{j+1}^h)\|_Y^2 + C \Big(\|v_0 - v_0^h\|_H^2 + \|\nabla(v_0 - v_0^h)\|_Q^2 + \|\mathrm{div}\,(v_0 - v_0^h)\|_Y^2 \\
&+ \|\nabla(u_0 - u_0^h)\|_Q^2 + \|\mathrm{div}\,(u_0 - u_0^h)\|_Y^2 + \|e_0 - e_0^h\|_Y^2 + \|\nabla(\varphi_0 - \varphi_0^h)\|_H^2 \\
&+ \|\varphi_0 - \varphi_0^h\|_Y^2 + \|\theta_0 - \theta_0^h\|_Y^2 + \|\nabla(\theta_0 - \theta_0^h)\|_H^2 \Big).
\end{aligned}
\tag{24}
$$

The previous error estimates can be used to derive the convergence order. As a particular case, under suitable additional regularity, the linear convergence is derived.

Corollary 1. *Let the assumptions of Theorem 2 still hold. If we assume that the solution to Problem VP has the additional regularity:*

$$
\begin{aligned}
u &\in H^2(0,T;V) \cap H^3(0,T;H) \cap C^1([0,T];[H^2(\Omega)]^d), \\
e, \theta &\in H^2(0,T;E) \cap H^3(0,T;Y) \cap C^1([0,T];H^2(\Omega)),
\end{aligned}
\tag{25}
$$

and we use the finite element spaces V^h and E^h defined in (13) and (14), respectively, and the discrete initial conditions u_0^h, v_0^h, φ_0^h, e_0^h, ψ_0^h and θ_0^h given in (19), there exists a positive constant $C > 0$, independent of the discretization parameters h and k but depending on the continuous solution, such that

$$
\begin{aligned}
\max_{0 \leq n \leq N} \Big\{ &\|v_n - v_n^{hk}\|_H + \|\nabla(u_n - u_n^{hk})\|_Q + \|\mathrm{div}\,(u_n - u_n^{hk})\|_Y + \|e_n - e_n^{hk}\|_Y + \|\nabla(\varphi_n - \varphi_n^{hk})\|_H \\
&+ \|\varphi_n - \varphi_n^{hk}\|_Y + \|\theta_n - \theta_n^{hk}\|_Y + \|\nabla(\psi_n - \psi_n^{hk})\|_H \Big\} \leq C(h+k).
\end{aligned}
$$

4. Numerical Simulations

In this final section, we will describe the algorithm and some numerical results involving one- and two-dimensional examples.

4.1. Numerical Algorithm

As a first step, given the solution u_{n-1}^{hk}, v_{n-1}^{hk}, φ_{n-1}^{hk}, e_{n-1}^{hk}, ψ_{n-1}^{hk} and θ_{n-1}^{hk} at time t_{n-1}, the velocity v_n^{hk}, the volume fraction speed e_n^{hk} and the temperature θ_n^{hk} are obtained from the coupled Equations (15)–(17). The corresponding linear system can be expressed in terms of a product variable x_n and the resulting symmetric linear system is solved by using the well-known Cholesky method. Finally, the displacement field u_n^{hk}, the volume fraction φ_n^{hk} and the thermal displacement ψ_n^{hk} are updated using (18).

The numerical scheme was implemented on a Intel Core 3.2 GHz PC using MATLAB, and a typical one-dimensional run ($h = k = 10^{-3}$) took about 0.957 seconds of CPU time, meanwhile a two-dimensional run took about 2.74 s of CPU time.

4.2. A One-Dimensional Example: Numerical Convergence

To show the accuracy of the finite element approximations studied in the previous section, we will consider the following simpler one-dimensional problem:

Problem $\mathbf{P}^{1D}$. *Find the displacement $u : [0,1] \times [0,1] \to \mathbb{R}$, the volume fraction $\varphi : [0,1] \times [0,1] \to \mathbb{R}$ and the thermal displacement $\psi : [0,1] \times [0,1] \to \mathbb{R}$ such that*

$$\ddot{u} = 5\dot{u}_{xx} + 5u_{xx} + 2\varphi_x - \dot{\psi}_x + F_1 \quad \text{in} \quad (0,1) \times (0,1),$$
$$\ddot{\varphi} = 2\varphi_{xx} + 2u_x - 3\varphi + \psi + \psi_{xx} + F_2 \quad \text{in} \quad (0,1) \times (0,1),$$
$$\ddot{\psi} = \psi_{xx} + \varphi_{xx} - \dot{\varphi} - \dot{u}_x + \dot{\psi}_{xx} + F_3 \quad \text{in} \quad (0,1) \times (0,1),$$
$$u(0,t) = u(1,t) = \psi(0,t) = \psi(0,t) = \varphi(0,t) = \varphi(1,t) = 0 \quad \text{for} \quad t \in (0,1),$$
$$u(x,0) = \dot{u}(x,0) = \varphi(x,0) = x(x-1) \quad \text{for a.e.} \quad x \in (0,1),$$
$$\dot{\varphi}(x,0) = \psi(x,0) = \dot{\psi}(x,0) = x(x-1) \quad \text{for a.e.} \quad x \in (0,1),$$

where functions F_i, $i = 1, 2, 3$, are given by, for all $(x, t) \in (0,1) \times (0,1)$,

$$F_1(x,t) = e^t \left(x(x-1) - 19 - 2x \right),$$
$$F_2(x,t) = e^t \left(3x(x-1) - 4 - 4x \right),$$
$$F_3(x,t) = e^t \left(2x(x-1) - 7 + 2x \right).$$

We point out that Problem P^{1D} is the one-dimensional version of Problem P using the following data:

$$\Omega = (0,1), \quad T = 1, \quad \rho = 1, \quad \beta = 1, \quad J = 1, \quad a = 1, \quad \kappa^* = 1, \quad d = 1, \quad \mu = 2, \quad \lambda = 1,$$
$$\mu^* = 1, \quad \lambda^* = 1, \quad \gamma = 2, \quad \xi = 3, \quad m = 1, \quad a_0 = 2, \quad \kappa = 1,$$

and all the initial conditions equal to $x(x-1)$.

We note that, in this case, the exact solution to Problem P^{1D} can be easily found and it has the following form:

$$u(x,t) = \varphi(x,t) = \psi(x,t) = e^t x(x-1) \quad \text{for all} \quad (x,t) \in (0,1) \times (0,1).$$

The numerical errors, for several values of the discretization parameters h and k, and given by

$$\max_{0 \leq n \leq N} \left\{ \|v_n - v_n^{hk}\|_H + \|\nabla(u_n - u_n^{hk})\|_Q + \|\text{div}(u_n - u_n^{hk})\|_Y + \|e_n - e_n^{hk}\|_Y + \|\nabla(\varphi_n - \varphi_n^{hk})\|_H \right.$$
$$\left. + \|\varphi_n - \varphi_n^{hk}\|_Y + \|\theta_n - \theta_n^{hk}\|_Y + \|\nabla(\psi_n - \psi_n^{hk})\|_H \right\},$$

are shown in Table 1. Moreover, the evolution of the error depending on the parameter $h + k$ is depicted in Figure 1. The convergence of the numerical approximations is clearly observed, and its linear convergence, stated in Corollary 1, seems to be achieved.

Table 1. Example 1: Numerical errors for some h and k.

$h \downarrow k \rightarrow$	0.01	0.005	0.002	0.001	0.0005	0.0002	0.0001
$1/2^3$	0.434612	0.432825	0.431788	0.431452	0.431287	0.431188	0.431155
$1/2^4$	0.210312	0.207822	0.206588	0.206216	0.206038	0.205935	0.205902
$1/2^5$	0.107268	0.103366	0.101371	0.100897	0.100692	0.100580	0.100545
$1/2^6$	0.057972	0.053438	0.050939	0.050182	0.049865	0.049725	0.049685
$1/2^7$	0.034160	0.029016	0.026254	0.025397	0.024994	0.024772	0.024714
$1/2^8$	0.023231	0.017132	0.014037	0.013122	0.012688	0.012438	0.012359
$1/2^9$	0.018869	0.011671	0.008032	0.007020	0.006560	0.006299	0.006215
$1/2^{10}$	0.017445	0.009490	0.005197	0.004018	0.003510	0.003236	0.003149
$1/2^{11}$	0.017049	0.008778	0.004000	0.002601	0.002010	0.001708	0.001618
$1/2^{12}$	0.016947	0.008580	0.003583	0.002003	0.001301	0.00052	0.000854
$1/2^{13}$	0.016921	0.008529	0.003462	0.001794	0.001002	0.000589	0.000476

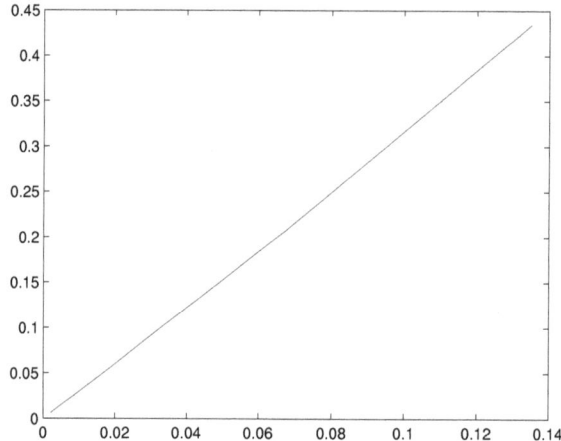

Figure 1. Example 1: Asymptotic behavior of the numerical scheme.

If we assume now that there are not volume forces, and we use the following data:

$$\Omega = (0,1), \quad T = 10, \quad \rho = 1, \quad \beta = 1, \quad J = 4, \quad a = 1, \quad \kappa^* = 1, \quad d = 1, \quad \mu = 2, \quad \lambda = 1,$$
$$\mu^* = 3, \quad \lambda^* = 4, \quad \gamma = 1, \quad \xi = 3, \quad m = 1, \quad a_0 = 2, \quad \kappa = 1,$$

and the initial conditions, for all $x \in (0,1)$,

$$u_0(x) = x(x-1), \quad v_0(x) = \varphi_0(x) = e_0(x) = \psi_0(x) = \theta_0(x) = 0,$$

taking the discretization parameters $h = k = 0.001$, the evolution in time of the discrete energy E_n^{hk} defined by

$$E_n^{hk} = \frac{1}{2}\Big\{\rho\|v_n^{hk}\|_Y^2 + (\lambda + 2\mu)\|(u_n^{hk})_x\|_Y^2 + J\|e_n^{hk}\|_Y^2 + \gamma(\varphi_n^{hk}, (u_n^{hk})_x)_Y + a_0\|(\varphi_n^{hk})_x\|_Y^2 + \xi\|\varphi_n^{hk}\|_Y^2$$
$$+ m((\psi_n^{hk})_x, (\varphi_n^{hk})_x)_Y + a\|\theta_n^{hk}\|_Y^2 + \kappa\|(\psi_n^{hk})_x\|_Y^2\Big\},$$

is shown in Figure 2 using both natural and semi-log scales. We can see that the discrete energy converges to zero and an exponential decay seems to be achieved.

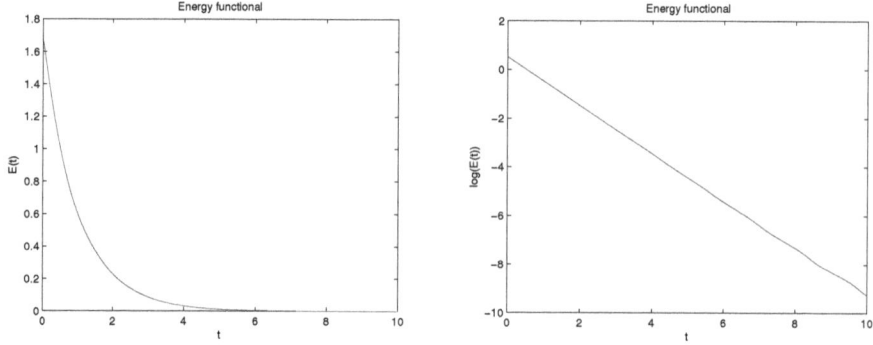

Figure 2. Example 1: Discrete energy evolution in natural and semi-log scales.

4.3. First Two-Dimensional Example: Application of a Surface Force

As a first two-dimensional example, we consider the square domain $[0,1] \times [0,1]$, which is assumed to be clamped on its left part $\{0\} \times [0,1]$. We also suppose that the thermal displacement and porosity vanish on the whole boundary, and we use the following expression for the mechanical surface force:

$$f_F(x,y,t) = (-ty^2, 0) \quad \text{for} \quad x=1, y \in (0,1),$$

which is assumed to be applied on the boundary $\Gamma_F = \{1\} \times (0,1)$, a part of the whole boundary $\partial\Omega$. Even if this case was not studied in the previous sections, it is straightforward to extend the analysis to this more general case.

The following data have been employed in this simulation:

$$T=1, \quad \rho=1, \quad \beta=1, \quad J=1, \quad a=1, \quad \kappa^*=1, \quad d=1, \quad \mu=2, \quad \lambda=1,$$
$$\mu^*=2, \quad \lambda^*=1, \quad \gamma=2, \quad \zeta=3, \quad m=1, \quad a_0=2, \quad \kappa=1,$$

and null initial conditions for all the variables.

Taking the time discretization parameter $k = 0.01$, the deformation generated by this mechanical force is shown in Figure 3. In Figure 4 we plot the displacement in vector arrows (on the left-hand side) and its norm (on the right-hand side) at final time. As can be seen, there is a deformation along the horizontal axis. Moreover, the porosity (left) and the thermal displacement (right) are plotted in Figure 5 at final time. We note that both neglect on the boundary due to the null boundary conditions and they are generated by the deformation of the body.

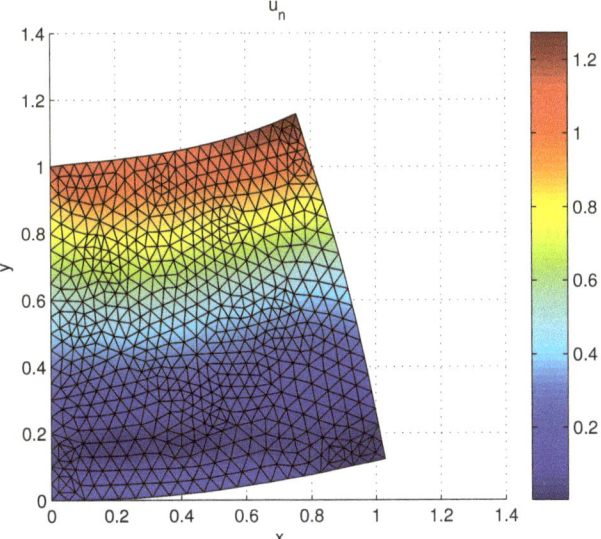

Figure 3. Example 2: Deformation and von Mises stress norm at final time.

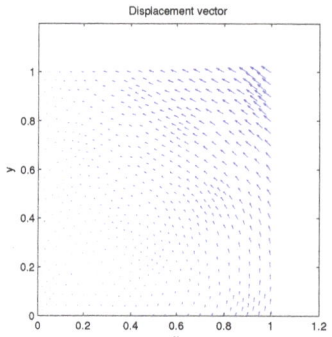 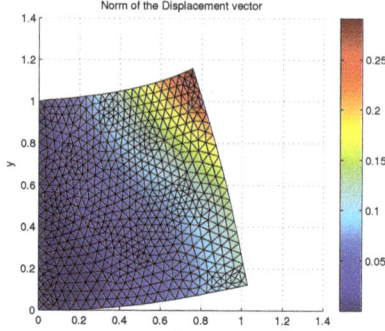

Figure 4. Example 2: Representation of the displacement with arrows (**left**) and its norm (**right**) at final time.

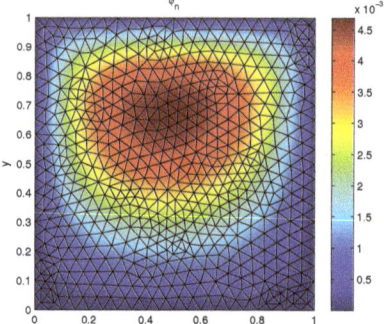 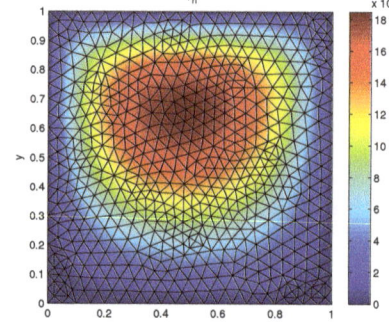

Figure 5. Example 2: Porosity (**left**) and thermal displacement (**right**) at final time.

4.4. Second Two-Dimensional Example: Dependence on the Type III Thermal Coefficient

In this second two-dimensional example, we will investigate the dependence on the type III thermal coefficient κ^*.

In this simulation, we have used the following data:

$$T = 1, \quad \rho = 100, \quad \beta = 1, \quad J = 4, \quad a = 3, \quad d = 1, \quad \mu = 2, \quad \lambda = 1, \quad \mu^* = 2,$$
$$\lambda^* = 1, \quad \gamma = 2, \quad \xi = 3, \quad m = 1, \quad a_0 = 2, \quad \kappa = 1,$$

with parameter κ^* varying between 0.001 and 100. No mechanical forces are now applied, and we also impose an initial condition $\varphi_0(x, t) = e_0(x, y) = x(x - 1) * y(y - 1)$ for $(x, y) \in (0, 1) \times (0, 1)$, being the remaining initial conditions assumed to be zero.

Taking the time discretization parameter $k = 0.01$, the evolution in time of the norm of the displacement at middle point $x = (0.5, 0.5)$ is shown in Figure 6. It seems that an increasing quadratic behavior is found. Moreover, in Figure 7 we plot the evolution in time of the porosity (left) and the thermal displacement (right). The porosity has an oscillating behavior and, as expected, the thermal displacement has a quadratic form, converging both variables to zero.

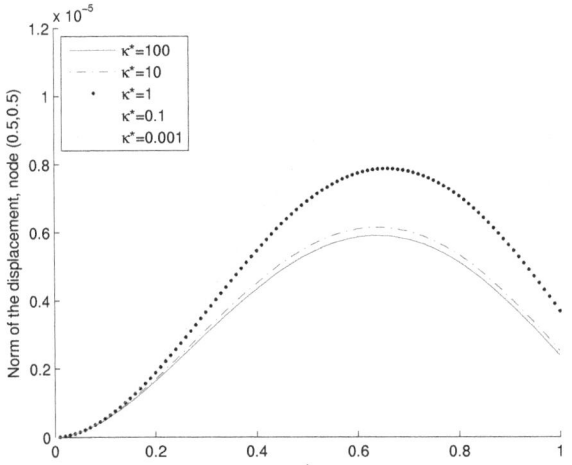

Figure 6. Example3 2: Evolution in time of the norm of the displacement at point $x = (0.5, 0.5)$.

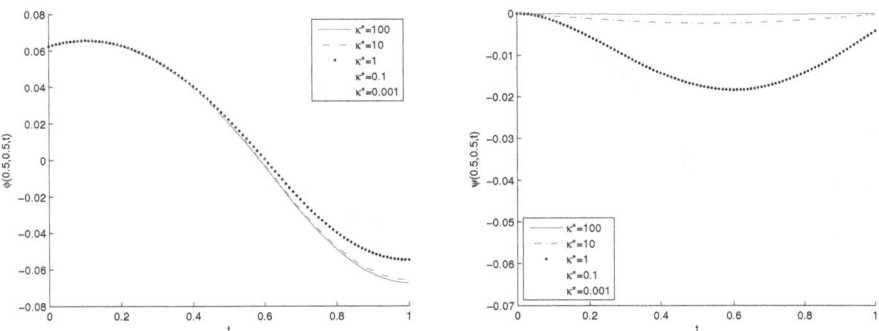

Figure 7. Example 3: Evolution in time of the porosity (**left**) and thermal displacement (**right**) at point $x = (0.5, 0.5)$.

5. Conclusions

In this paper, we numerically analyzed a dynamic poro-thermo-viscoelastic problem within the type III thermoelasticity. The weak form led to a linear system composed of parabolic variational equations written in terms of the velocity, the volume fraction speed and the temperature. Then, using the finite element method to approximate the spatial variable and the implicit Euler scheme to discretize the time derivatives, we introduced a fully discrete scheme. A priori error estimates were proved by using a discrete version of Gronwall's inequality. Finally, we presented a one-dimensional numerical simulation to show the convergence of the algorithm and the exponential decay of the discrete energy (Example 1), the effect of the application of a surface force (Example 2) and the dependence on the type III thermal coefficient (Example 3).

Author Contributions: Conceptualization and Methodology N.B., J.A.L.-C., M.L., J.R.F. and A.S.; Software, Formal Analysis and Data Curation N.B. and J.R.F.; Validation N.B., M.L. and J.R.F.; Supervision J.R.F. and A.S.; Writing Original Draft Preparation N.B. and J.A.L.-C.; Writing Review and Editing M.L., J.R.F. and A.S.; Funding Acquisition J.R.F. and A.S.

Funding: This work has been partially funded by the research project PGC2018-096696-B-I00 (Ministerio de Ciencia, Innovación y Universidades, Spain) and by Xunta de Galicia, Spain, under the program Grupos de Referencia Competitiva with Ref. ED431C2019/21.

Conflicts of Interest: The authors declare no conflict of interest. The funders had no role in the design of the study; in the collection, analysis, or interpretation of data; in the writing of the manuscript, or in the decision to publish the results.

References

1. Cowin, S.C.; Nunziato, J.W. Linear elastic materials with voids. *J. Elast.* **1983**, *13*, 125–147. [CrossRef] [CrossRef]
2. Nunziato, J.W.; Cowin, S. A nonlinear theory of elastic materials with voids. *Arch. Ration. Mech. Anal.* **1979**, *415*, 175–201. [CrossRef] [CrossRef]
3. Aouadi, M. Uniqueness and existence theorems in thermoelasticity with voids without energy dissipation. *J. Frankl. Inst.* **2012**, *349*, 128–139.
4. Aouadi, M.; Ciarletta, M.; Iovane, G. A porous thermoelastic diffusion theory of types II and III. *Acta Mech.* **2017**, *228*, 931–949. [CrossRef] [CrossRef]
5. Apalara, T.A. Exponential decay in one-dimensional porous dissipation elasticity. *Quart. J. Mech. Appl. Math.* **2017**, *70*, 360–372. [CrossRef] [CrossRef]
6. Bazarra, N.; Fernández, J.R. Numerical analysis of a contact problem in poro-thermoelasticity with microtemperatures. *Z. Angew. Math. Mech.* **2018**, *98*, 1190–2009. [CrossRef] [CrossRef]
7. Birsan, M.; Altenbach, H. The Korn-type inequality in a Cosserat model for thin thermoelastic porous rods. *Meccanica* **2012**, *47*, 789–794. [CrossRef] [CrossRef]
8. Casas, P.S.; Quintanilla, R. Exponential decay in one-dimensional porous-thermoelasticity. *Mech. Res. Comm.* **2012**, *40*, 652–658.
9. Chirita, S.; Ciarletta, M.; Straughan, B. Structural stability in porous elasticity. *Proc. R. Soc. Lond. Ser. A Math. Phys. Eng. Sci.* **2006**, *462*, 2593–2605. [CrossRef] [CrossRef]
10. Ciarletta, M.; Svanadze, M.; Buonnano, L. Plane waves and vibrations in the theory of micropolar thermoelasticity for materials with voids. *Eur. J. Mech. A Solids* **2009**, *28*, 897–903. [CrossRef] [CrossRef]
11. Fernández, J.R.; Masid, M. A porous thermoelastic problem: An a priori error analysis and computational experiments. *Appl. Math. Comput.* **2017**, *305*, 117–135. [CrossRef] [CrossRef]
12. Iesan, D. On the nonlinear theory of thermoviscoelastic materials with voids. *J. Elast.* **2017**, *128*, 1–16. [CrossRef] [CrossRef]
13. Klinkel, S.; Reichel, R. A finite element formulation in boundary representation for the analysis of nonlinear problems in solid mechanics. *Comput. Methods Appl. Mech. Eng.* **2019**, *347*, 295–315. [CrossRef] [CrossRef]
14. Magaña, A.; Quintanilla, R. On the time decay of solutions in one-dimensional theories of porous materials. *Int. J. Solids Struct.* **2006**, *43*, 3414–3427. [CrossRef] [CrossRef]
15. Marin, M. Some basic theorems in elastostatics of micropolar materials with voids. *J. Comput. Appl. Math.* **1996**, *70*, 115–126. [CrossRef] [CrossRef]
16. Marin, M. Weak solutions in elasticity of dipolar porous materials. *Math. Probl. Eng.* **2008**, *2008*, 158908. [CrossRef] [CrossRef]
17. Marin, M. An approach of a heat-flux dependent theory for micropolar porous media. *Meccanica* **2016**, *51*, 1127–1133. [CrossRef] [CrossRef]
18. Marin, M.; Chirila, A.; Öchsner, A.; Vlase, S. About finite energy solutions in thermoelasticity of micropolar bodies with voids. *Bound. Value Probl.* **2019**, *2019*, 89. [CrossRef] [CrossRef]
19. Pamplona, P.X.; Muñoz Rivera, J.E.; Quintanilla, R. On the decay of solutions for porous-elastic systems with history. *J. Math. Anal. Appl.* **2011**, *379*, 682–705. [CrossRef] [CrossRef]
20. Pamplona, P.X.; Muñoz Rivera, J.E.; Quintanilla, R. Analyticity in porous-thermoelasticity with microtemperatures. *J. Math. Anal. Appl.* **2012**, *394*, 645–655. [CrossRef] [CrossRef]
21. Iesan, D. *Thermoelastic Models of Continua*; Kluwer: Alphen aan den Rijn, The Netherlands, 2004.
22. Straughan, B. *Mathematical Aspects of Multi-Porosity Continua*; Springer: Berlin, Germany, 2017.
23. Green, A.E.; Naghdi, P.M. A unified procedure for contruction of theories of deformable media. I Classical continuum physics, II Generalized continua, III Mixtures of interacting continua. *Proc. R. Soc. A* **1995**, *448*, 335–356, 357–377, 379–388. [CrossRef] [CrossRef]
24. Fareh, A.; Messaoudi, S.A. Stabilization of a type III thermoelastic Timoshenko system in the presence of a time-distributed delay. *Math. Nachr.* **2017**, *290*, 1017–1032. [CrossRef] [CrossRef]

25. Jorge, M.; Pinheiro, S.B. Improvement on the polynomial stability for a Timoshenko system with type III thermoelasticity. *Appl. Math. Lett.* **2019**, *96*, 95–100. [CrossRef] [CrossRef]
26. Magaña, A.; Quintanilla, R. Exponential stability in type III thermoelasticity with microtemperatures. *Z. Angew. Math. Phys.* **2018**, *69*, 129. [CrossRef] [CrossRef]
27. Miranville, A.; Quintanilla, R. Exponential decay in one-dimensional type III thermoelasticity with voids. *Appl. Math. Lett.* **2019**, *94*, 30–37. [CrossRef] [CrossRef]
28. Mustafa, M.I. A uniform stability result for thermoelasticity of type III with boundary distributed delay. *J. Math. Anal. Appl.* **2014**, *415*, 148–158. [CrossRef] [CrossRef]
29. Prasad, R.; Das, S.; Mukhopadhyay, S. A two-dimensional problem of a mode I crack in a type III thermoelastic medium. *Math. Mech. Solids* **2013**, *18*, 506–523. [CrossRef] [CrossRef]
30. Bazarra, N.; Fernández, J.R.; Quintanilla, R. Analysis of a thermoelastic problem of type III. **2019**, 277–285, unpublished work.
31. Iesan, D.; Quintanilla, R. On thermoelastic bodies with inner structure and microtemperatures. *J. Math. Anal. Appl.* **2009**, *354*, 12–23. [CrossRef] [CrossRef]
32. Bazarra, N.; Fernández, J.R.; Leseduarte, M.C.; Magaña, A.; Quintanilla, R. On the uniqueness and analyticity in viscoelasticity with double porosity. *Asymptot. Anal.* **2019**, *112*, 151–164. [CrossRef] [CrossRef]
33. Clement, P. Approximation by finite element functions using local regularization. *RAIRO Math. Model. Numer. Anal.* **1975**, *9*, 77–84. [CrossRef] [CrossRef]
34. Andrews, K.T.; Fernández, J.R.; Shillor, M. Numerical analysis of dynamic thermoviscoelastic contact with damage of a rod. *IMA J. Appl. Math.* **2005**, *70*, 768–795. [CrossRef]
35. Campo, M.; Fernández, J.R.; Kuttler, K.L.; Shillor, M.; Viaño, J.M. Numerical analysis and simulations of a dynamic frictionless contact problem with damage. *Comput. Methods Appl. Mech. Eng.* **2006**, *196*, 476–488. [CrossRef] [CrossRef]

© 2019 by the authors. Licensee MDPI, Basel, Switzerland. This article is an open access article distributed under the terms and conditions of the Creative Commons Attribution (CC BY) license (http://creativecommons.org/licenses/by/4.0/).

Article

Synthesis on the Acceleration Energies in the Advanced Mechanics of the Multibody Systems

Iuliu Negrean * and Adina-Veronica Crișan *

Faculty of Machine Building, Department of Mechanical Systems Engineering,
Technical University of Cluj-Napoca, 400641 Cluj-Napoca, Romania
* Correspondence: iuliu.negrean@mep.utcluj.ro (I.N.); adina.crisan@mep.utcluj.ro (A.-V.C.)

Received: 16 July 2019; Accepted: 22 August 2019; Published: 27 August 2019

Abstract: The present paper's objective is to highlight some new developments of the main author in the field of advanced dynamics of systems and higher order dynamic equations. These equations have been developed on the basis of the matrix exponentials which prove to have undeniable advantages in the matrix study of any complex mechanical system. The present paper proposes some new approaches, based on differential principles from analytical mechanics, by using some important dynamics notions, regarding the acceleration energies of the first, second and third order. This study extended the equations of the higher order, which provide the possibility of applying the initial motion conditions in the positions, velocities and accelerations of the first and second order. In order to determine the time variation laws for the generalized variables, the driving forces and acceleration energies of the higher order are applied by the time polynomial functions of the fifth order. According to inverse kinematics also named control kinematics of the robots, the applications of polynomial functions lead to the kinematic control functions of mechanical motions, especially the transitory motions. They influence the dynamic behavior of multibody systems, in which robot structures are included.

Keywords: advanced mechanics; analytical dynamics; acceleration energies; robotics

1. Introduction

Until the present, the dynamic behavior of rigid body and of the multibody systems can be studied by considering the fundamental theorems from dynamics or the differential principles of analytical mechanics [1–4]. Both are based on angular momentum, the mechanical work and kinetic energy, as well as on their fundamental theorems. In mechanics, the defining expressions for the advanced notions are based on fundamental input parameters that are represented by kinematic parameters along with their differential transformations, as well as by the mass properties highlighted by the inertia tensor and corresponding generalized variation laws.

The advanced concepts include higher-order accelerating energies known in the scientific literature [5–7] as kinetic energies of higher-order accelerations. The kinetic energy is used, in Newtonian mechanics, as a central function of Lagrange Euler equations. The existence of higher-order accelerations has been proven by the fact that they occur every time the bodies/multi-body systems are subjected to fast and sudden motions or, when particularly, this type of excitation produces vibration modes with more frequencies of resonance. The acceleration itself cannot produce vibration being just a static load. In fact, the sudden motions are generated by higher order accelerations. For example, during the operation of a cam and cam follower mechanism at high speeds and accelerations, the jumping off of the cam follower from the camshaft causing accidents may occur. Another example is the case of manufacturing processes that require rapid changes in the acceleration of a cutting tool. This leads to a quick tool wear and results in poor quality of the manufactured surfaces [8,9].

The two situations presented above are generated by the effect of higher order accelerations. These effects have to be also considered each time when a mechanical system is in a transient motion phase (take-off/landing, acceleration/deceleration, etc.). The novelty of the paper consists in presenting a new approach regarding the dynamic modeling of rigid bodies and multibody mechanical systems. Based on scientific literature [10–20], the input kinematic and dynamic parameters have been formulated. These parameters are compulsorily included in the dynamic equations of the higher order, corresponding to fast and sudden motions. This type of motion characterizes the multibody systems, a category where the robots are also included. The important formulations regarding the acceleration energies of the first, second, third and higher orders are also presented in explicit and matrix forms. The expressions for the acceleration energies of the higher order were determined, among others, based on matrix exponential functions [14,19]. Although at first sight they seem difficult to put into practice, they proved to have undisputable advantages in terms of their use in the kinematic and dynamic study of complex mechanical systems.

These expressions are implemented into the differential equations of the higher order, thus resulting the time variations of the generalized forces, which influences the dynamic behavior of rigid body or of multibody systems. To illustrate the validity of the proposed mathematical model, an experimental study, in which the higher order accelerations are highlighted, is performed.

2. Input Parameters in Advanced Dynamic Modeling

2.1. Position and Orientation Parameters

In Newtonian mechanics, the dynamics of the rigid bodies are studied based on the assumption that they do not deform under the action of the external forces.

To study the dynamical behavior of a rigid body, its linear and angular position in the Cartesian space must be known at each moment during the motion. To define the linear and angular position, [13,20], the last one being also known as orientation, the simplest mechanical model, which is the material point, is analyzed (Figure 1).

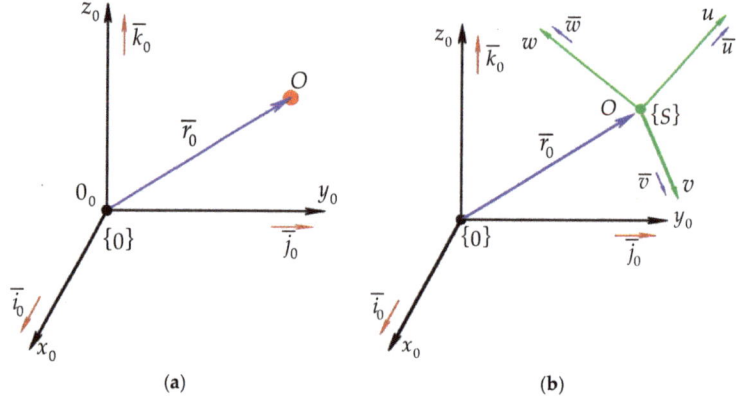

Figure 1. (**a**) The position and orientation for a point; (**b**) The position and orientation of a system.

The following notations are considered:

$$\chi = \{u; v; w\}; \chi_0 = \{u_0; v_0; w_0\} \text{ where } u = \{x; y; z\}; v = \{y; z; x\} \neq u; w = \{z; x; y\} \neq v \quad (1)$$

$$\overline{\chi} = \{\overline{u}; \overline{v}; \overline{w}\}; \overline{\chi}_0 = \{\overline{u}_0; \overline{v}_0; \overline{w}_0\} \text{ where } \overline{u} = \{\overline{i}; \overline{j}; \overline{k}\}; \overline{v} = \{\overline{j}; \overline{k}; \overline{i}\} \neq \overline{u}; \overline{w} = \{\overline{k}; \overline{i}; \overline{j}\} \neq \overline{v} \quad (2)$$

$$\delta_\chi = \{\alpha_\chi; \beta_\chi; \gamma_\chi\}; \cos \delta_\chi = c\delta_\chi; \sin \delta_\chi = s\delta_\chi \quad (3)$$

$$O_0x_0y_0z_0 \equiv \{0\}; \quad O'_0x'_0y'_0z'_0 \equiv \{0'\}; \quad Oxyz \equiv \{S\} \tag{4}$$

In the Expression (1), the coordinates or axes of the Cartesian frame, symbols from (2) representing the unit vectors, and (3) the angles and the direction cosines are defined.

The dynamic study of the rigid solid is carried out, according to Figure 1, by considering two reference systems: $O_0x_0y_0z_0 = \{0\}$, which is considered fixed and $Oxyz = \{S\}$ which is a mobile reference system, invariably linked to the body, having the origin in an arbitrary point O belonging to the rigid body. The system $Ox'_0y'_0z'_0 = \{0'\}$, represented in the same figure, is a system which originated in point O, and whose orientation is maintained constant all throughout the movement and is identical with the orientation of the fixed system, $\{0\}$, that is $\{0'\}_{OR} \equiv \{0\}_{OR}$.

The geometrical state of any material point (for example a certain point O) characterizes the position, according to Figure 1a. The position is usually defined by means of the position vector:

$$\bar{r}_0 = \begin{bmatrix} x_0 & y_0 & z_0 \end{bmatrix}^T, \text{ relative to } \{0\} \text{ frame} \tag{5}$$

The three linear coordinates from (5) are independent in the case of a free material point, and represents the degrees of freedom (d.o.f.). Further, the study is extended to a vector or an axis belonging to the Cartesian frame (see Figure 1b). This geometrical state is called orientation (angular position). Orientation (angular position) is defined through unit vectors. For any given unit vector $\bar{\chi} \in \{S\}$, in relation to one of the frames $\{0\}/\{0'\}$, the orientation is defined by the direction cosines:

$$\bar{\chi} = \bar{\chi}^T \cdot \begin{pmatrix} \bar{i}_0 \\ \bar{j}_0 \\ \bar{k}_0 \end{pmatrix} = \begin{pmatrix} c\alpha_\chi \\ c\beta_\chi \\ c\gamma_\chi \end{pmatrix} \equiv \begin{pmatrix} c\alpha \\ c\beta \\ c\gamma \end{pmatrix}_\chi, \text{ where } \bar{\chi}^T \cdot \bar{\chi} = c^2\alpha_\chi + c^2\beta_\chi + c^2\gamma_\chi = 1 \tag{6}$$

According to linear algebra, in expression (6), the symbol $\bar{\chi}^T$ defines the transposed matrix.

Further, considering the second expression from (6), the orientation of any vector or axis is characterized by two independent angles. The above geometrical aspects are extended on an orthogonal and right oriented frame (see Figure 1b, $Ouvw \equiv Oxyz \equiv \{S\}$) relative to $\{0\}$. In this case, the geometrical state is defined by position and orientation. The position is defined by means of position vector (5), while for orientation the rotation matrix is considered [10]:

$${}^0_S[R] = \begin{bmatrix} \bar{i} & \bar{j} & \bar{k} \end{bmatrix} = \begin{bmatrix} \begin{pmatrix} c\alpha \\ c\beta \\ c\gamma \end{pmatrix}_x & \begin{pmatrix} c\alpha \\ c\beta \\ c\gamma \end{pmatrix}_y & \begin{pmatrix} c\alpha \\ c\beta \\ c\gamma \end{pmatrix}_z \end{bmatrix}. \tag{7}$$

The resultant orientation matrix defined with (7), contains the unit vectors of $\{S\}$ frame relative to $\{0\}$. The rotation matrix or the matrix of direction cosines describes the orientation of each axis of the mobile frame attached to the rigid body, relative to a fixed frame. Between the nine direction cosines contained by the rotation matrix, six mathematical relations can be established. Thus, the result is that the orientation of a mobile frame with respect to a fixed frame is defined by up to three independent parameters, which are defining the orienting angles. Therefore, the resultant orientation of a frame $\{S\}$ with respect to another frame, $\{0\}/\{0'\}$ is characterized by three independent orientation angles (three d.o.f), [13]:

$$\bar{\bar{\psi}}(t) = \begin{bmatrix} \alpha_u(t) & \beta_v(t) & \gamma_w(t) \end{bmatrix}^T \tag{8}$$
$$(3x1)$$

The angles from (8) are the components of the column matrix of orientation $\bar{\bar{\psi}}(t)$, and they describe, from a geometrical point of view, dihedral angles between two geometrical planes:

$$\chi_0 = \{u_0; v_0; w_0\} = cst - \text{fixed plane} \in \{0'\}/\{0\} \text{ and } \chi = \{u; v; w\} = 0 - \text{mobile plane} \in \{S\}$$

Physically, the three angles defined with (8) expresses a simple rotation around one of the axes of the Cartesian reference system: $\chi = \{u; v; w\}$. Based on research from [10–14], when combining the three simple rotations, there is a result of twelve sets of orientation angles (8). Taking $\chi = \{x; y; z\}$, the expressions of definition are further developed for the three simple rotation matrices symbolized as:

$$R(\overline{\chi}; \delta_\chi) = \{R(\overline{x}; \alpha_x); R(\overline{y}; \beta_y); R(\overline{z}; \gamma_z)\}. \tag{9}$$

The paper proposes the following form of mathematical representation for the generalized matrix:

$$R(\overline{\chi}; \delta_\chi) = \{R(\overline{x}; \alpha_x); R(\overline{y}; \beta_y); R(\overline{z}; \gamma_z)\} = \begin{bmatrix} c(\delta_\chi \cdot \Delta_{yz}) & -s(\delta_\chi \cdot \Delta_z) & s(\delta_\chi \cdot \Delta_y) \\ s(\delta_\chi \cdot \Delta_z) & c(\delta_\chi \cdot \Delta_{zx}) & -s(\delta_\chi \cdot \Delta_x) \\ -s(\delta_\chi \cdot \Delta_y) & s(\delta_\chi \cdot \Delta_x) & c(\delta_\chi \cdot \Delta_{xy}) \end{bmatrix}, \tag{10}$$

where $\underset{\{\chi=\{u;v\}\}}{\Delta_{uv}} = \{\Delta_{yz}; \Delta_{zx}; \Delta_{xy}\} = \left\{ \begin{Bmatrix} 1 \\ 0 \end{Bmatrix} \text{ if } \delta_\chi = \left\{ \begin{Bmatrix} (\beta_y; \gamma_z) \\ \alpha_x \end{Bmatrix}; \begin{Bmatrix} (\gamma_z; \alpha_x) \\ \beta_y \end{Bmatrix}; \begin{Bmatrix} (\alpha_x; \beta_y) \\ \gamma_z \end{Bmatrix} \right\} \right\}, \tag{11}$

and $\underset{\{\chi=u\}}{\Delta_u} = \{\Delta_x; \Delta_y; \Delta_z\} = 1 - \underset{\{\chi=\{v;w\}\}}{\Delta_{vw}}. \tag{12}$

By substituting (11) and (12) in the generalized expression (10), the simple rotation matrices (9) are obtained. The generalized matrix can be written in a new formulation as follows:

$$R(\overline{\chi}; \delta_\chi) = Diag[\overline{\overline{\Delta_{uv}}}] + [\overline{\overline{\Delta_u}} \times], \tag{13}$$

where $\underset{(3\times 1)}{\overline{\overline{\Delta_{uv}}}} = [c(\delta_\chi \cdot \Delta_{yz}) \; c(\delta_\chi \cdot \Delta_{zx}) \; c(\delta_\chi \cdot \Delta_{xy})]^T, \tag{14}$

and $\underset{(3\times 1)}{\overline{\overline{\Delta_u}}} = [s(\delta_\chi \cdot \Delta_x) \; s(\delta_\chi \cdot \Delta_y) \; s(\delta_\chi \cdot \Delta_z)]^T, \tag{15}$

In the expression (13), the symbol $[\overline{\overline{\Delta_u}} \times]$ defines the skew-symmetric matrix associated to (15), while $Diag[\overline{\overline{\Delta_{uv}}}]$ represents the diagonal matrix, written below:

$$\underset{(3\times 3)}{Diag[\overline{\overline{\Delta_{uv}}}]} = \begin{bmatrix} c(\delta_\chi \cdot \Delta_{yz}) & 0 & 0 \\ 0 & c(\delta_\chi \cdot \Delta_{zx}) & 0 \\ 0 & 0 & c(\delta_\chi \cdot \Delta_{xy}) \end{bmatrix}. \tag{16}$$

By applying (14) and (15), an expression identical with the classical Rodriguez formula is obtained [17]:

$$R(\overline{\chi}; \delta_\chi) = \overline{\chi} \cdot \overline{\chi}^T \cdot (1 - c\delta_\chi) + I_3 \cdot c\delta_\chi + (\overline{\chi} \times s\delta_\chi). \tag{17}$$

According to research from [3–10], the three simple rotations from (8) are performed either around the moving axes or fixed axes belonging to $\{S\}$ or $\{0'\}/\{0\}$. The resultant rotation matrix, symbolized as $_S^0[R]$, defines the orientation of the system $\{S\}$(attached to the rigid body), with respect to a fixed system $\{0\}$, and $\chi = \{u; v; w\}$ is represented in the following matrix form:

$$_S^0[R] = R(\overline{u}; \alpha_u) \cdot R(\overline{v}; \beta_v) \cdot R(\overline{w}; \gamma_w), \tag{18}$$

$$_S^0[R] = \prod_{\{\overline{\chi}; \delta_\chi\}} R(\overline{\chi}; \delta_\chi) = \prod_{\{\chi=\{u;v;w\}\}} \exp[\overline{\chi} \times \delta_\chi] = \prod_{\{u;v\}} \{I_3 \cdot \overline{\overline{\Delta_{uv}}} + [\overline{\overline{\Delta_u}} \times]\}. \tag{19}$$

Using the research in the field of matrix exponentials [10,14], the following form for the resultant rotation matrix (18) is written as:

$$_S^0[R] = \prod_{\{\overline{\chi};\delta_\chi\}} R(\overline{\chi};\delta_\chi) = \prod_{\{\chi=\{u;v;w\}\}} \exp[\overline{\chi} \times \delta_\chi] = \exp[\overline{u} \times \alpha_u] \cdot \exp[\overline{v} \times \beta_v] \cdot \exp[\overline{w} \times \gamma_w] \quad (20)$$

The above mathematical Expressions (5) and (8), which define the position and orientation in the case of a rigid body, are written in a general form. Based on the definitions from the first paragraph of this section, a rigid body consists of an infinite number of material points and also, an infinite number of geometrical axes, parallel and perpendicular one to another, characterized by a continuous distribution in the entire volume of the rigid (S) [14,15]. The rigid body is also made up of an infinite number of assemblies consisting of three orthogonal geometrical plans, continuously distributed in the entire volume of the rigid body.

In terms of geometry, in order to define a right oriented reference frame having the origin in an arbitrary point O of the rigid body, a single ensemble composed of three orthogonal geometrical plans is chosen. According to (4) and Figure 2, this reference frame is symbolized as $Oxyz \equiv \{S\}$. The reference frame $\{S\}$ is linked to the body due to its rigid nature. Expressions (5) and (8) define the position and orientation of this frame.

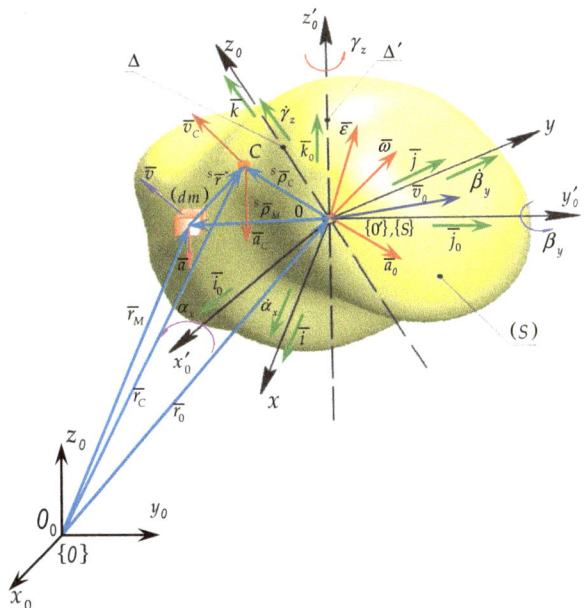

Figure 2. The representation of a free rigid body in the Cartesian frame.

Whereas the present paper proposes a study on advanced dynamics, it considers two material points belonging to a rigid body, so that the following conditions are met: $M \neq O$ and $C \neq \{O; M\}$. In this case, the expressions defining the position can be written as time functions, according to [12]:

$$\overline{r}_M(t) = \overline{r}_0(t) + \overline{\rho}_M(t) = \overline{r}_0(t) + {}_S^0[R](t) \cdot {}^S\overline{\rho}_{M'} \quad (21)$$

$$\overline{r}_C(t) = \overline{r}_0(t) + \overline{\rho}_C(t) = \overline{r}_0(t) + {}_S^0[R](t) \cdot {}^S\overline{\rho}_C, \text{ where } \overline{\rho}_M(t) \neq \overline{\rho}_C(t) \text{ and } \overline{r}_M(t) \neq \overline{r}_C(t) \quad (22)$$

Based on (18) and (19), the position equation is written, by considering the matrix exponentials:

$$\bar{r}_M(t) = \bar{r}_0(t) + \prod_{\{x=\{u;v;w\}\}} \exp[\bar{x} \times \delta_x] \cdot {}^S\bar{\rho}_M = \bar{r}_0(t) + \{\exp[\bar{u} \times \alpha_u] \cdot \exp[\bar{v} \times \beta_v] \cdot \exp[\bar{w} \times \gamma_w]\} \cdot {}^S\bar{\rho}_M. \quad (23)$$

The position equation for any material point from the rigid body can be determined only when the position $\bar{r}_0(t)$ and orientation ${}^0_S[R](t)$ of the moving frame $Oxyz \equiv \{S\}$ are well known. By analyzing the Expressions (21) and (22), the result is that the orientation is invariant for all the points of the rigid solid. Thus, by considering the geometrical aspects, the body is substituted by the moving frame $Oxyz \equiv \{S\}$. This is defined from a geometrically point of view by means of the six independent parameters (six d.o.f.), which are included in the following symbol:

$$\bar{\bar{X}}_{(6\times1)}(t) = \begin{bmatrix} \bar{r}_0(t) \\ \bar{\psi}(t) \end{bmatrix} = \begin{bmatrix} [x_0(t) \ y_0(t) \ z_0(t)]^T \\ [\alpha_u(t) \ \beta_v(t) \ \gamma_w(t)]^T \end{bmatrix} = \begin{bmatrix} q_i(t) \cdot \delta_i, \ i=1 \to 6 \\ \text{where } \delta_i = \begin{Bmatrix} (1-\Delta_i), \text{ for } q_i - \text{linear,} \\ \Delta_i, \text{ for } q_i - \text{angular} \end{Bmatrix} \end{bmatrix} \quad (24)$$

where $q_j(t)$ is the generalized coordinate and $\Delta_i = \{(1, q_i - \text{angular}), (0, q_i - \text{linear})\}$ is an operator which highlights the type of driving joint. In this paper, the symbol $\bar{\bar{X}}_{(6\times1)}(t)$, is substituted as:

$$\bar{\bar{X}}_{(6\times1)}(t) = \bar{\theta}(t) = \begin{bmatrix} q_i(t), \ i=1 \to 6 \end{bmatrix}^T, \text{ where } \bar{\theta} \neq \bar{\theta}^{(0)} \quad (25)$$

Considering Expression (24), the following notations are implemented:

$$\left\{\bar{\theta}(t); \dot{\bar{\theta}}(t); \ddot{\bar{\theta}}(t); \cdots ; \bar{\theta}^{(m)}(t)\right\} = \left\{\begin{array}{l} q_i(t); \dot{q}_i(t); \ddot{q}_i(t); \cdots ; q_i^{(m)}(t) \\ i = 1 \to 6, \ m \geq 1 \end{array}\right\}. \quad (26)$$

where $m \geq 1$ represents the order of the time derivative.

In advanced mechanics, instead of (8), in the case of the angular vector of orientation, the following expression of definition is used:

$$\bar{\psi}(t) = {}^0J_\psi[\alpha_u(t) - \beta_v(t) - \gamma_w(t)] \cdot \dot{\bar{\psi}}(t) = \bar{\psi}[q_j(t) \cdot \Delta_j; \ j = 1 \to k^* = 6, \ t], \quad (27)$$

$${}^0J_\psi[\alpha_u(t) - \beta_v(t) - \gamma_w(t)] = \begin{bmatrix} \bar{u} & R(\bar{u}; \alpha_u) \cdot \bar{v} & R(\bar{u}; \alpha_u) \cdot R(\bar{v}; \beta_v) \cdot \bar{w} \end{bmatrix}. \quad (28)$$

where ${}^0J_\psi$ represents the angular transfer matrix which is a function of the orienting angles.

In mechanics, the position and orientation of a moving frame $Oxyz \equiv \{S\}$, with respect to another frame, for example a fixed frame $\{0\}$, is represented in a matrix form, according to [5,8] by means of homogeneous transformations which are developed using matrix exponentials:

$${}^0_S[T]_{(4\times4)}(t) = \begin{bmatrix} {}^0_S[R](t) & \bar{p}(t) \\ 0 \ 0 \ 0 & 1 \end{bmatrix} = \begin{bmatrix} \prod_{\{x=\{u;v;w\}\}} \exp[\bar{x} \times \delta_x] & \bar{p}(t) \\ 0 \ 0 \ 0 & 1 \end{bmatrix}, \quad (29)$$

$$\bar{p}(t) = \sum_{\{x=\{u;v;w\}\}} \left\{\prod_{\{x=\{u;v;w\}\}} \exp[\bar{x} \times \delta_x]\right\} \cdot \bar{b}_x + \left\{\prod_{\{x=\{u;v;w\}\}} \exp[\bar{x} \times \delta_x]\right\} \cdot \bar{p}^{(0)} \cdot \Delta_p, \quad (30)$$

where $\Delta_p = \{\{0; \bar{p} = \bar{r}_0\}; \{1; \bar{p} = \bar{r}_M\}\}$,

$$\bar{b}_x = \left[\bar{x}^{(0)} \cdot \bar{x}^{(0)T} \cdot (\delta_x - s\delta_x) + I_3 \cdot s\delta_x + (\bar{x}^{(0)} \times) \cdot (1 - c\delta_x)\right] \cdot \left[\bar{p}^{(0)} \times \bar{x}^{(0)} \cdot \Delta_x\right]. \quad (31)$$

Expression (30) defines the position vector of the mobile system $\{S\}$ relative to fixed system $\{0\}$, while (31) represents a position vector, which according to [11,20], is defined by means of homogenous coordinates as a function of screw parameters. The conclusions and expressions of definition, synthetically disseminated in this introductory section, are compulsory applied in the advanced kinematics and dynamics of mechanical systems. For example, a kinetic element from this mechanical structure of a robot is considered (Figure 3).

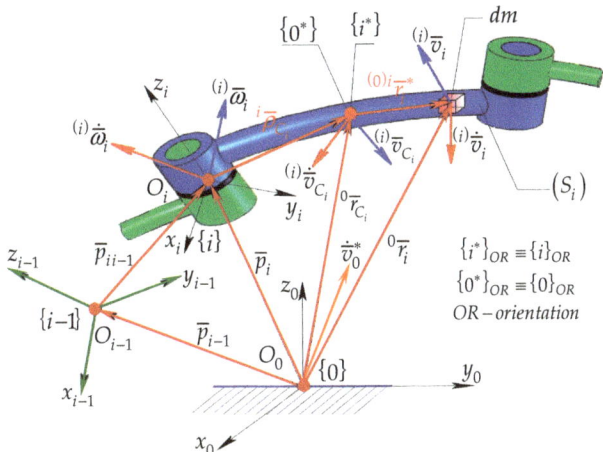

Figure 3. The representation of a kinetical element from robot mechanical structure.

In case of both rigid bodies and multibody systems, the dynamical study of current and sudden motions is performed. The study is based on the differential principles from analytical dynamics of systems, as well as on the advanced notions from the dynamics of solid bodies: Momentum and angular momentum, kinetic energy, acceleration energies of different orders and their absolute time derivatives of the high order.

2.2. The Kinematic Parameters of Higher Order

The parametric equations of motion are defined according to (24). The absolute position Equation (21) shows the variable distribution from a point to another material point of the body. By applying the first order time derivative on Expression (21), the result is:

$$\dot{\overline{r}}_M(t) = \dot{\overline{r}}_0(t) + \dot{\overline{p}}_M(t) = \dot{\overline{r}}_0(t) + {}^0_S[\dot{R}](t) \cdot {}^S\overline{p}_M = \dot{\overline{r}}_0(t) + {}^0_S[\dot{R}](t) \cdot {}^0_S[R]^T(t) \cdot {}^0_S[R](t) \cdot {}^S\overline{p}_M, \qquad (32)$$

The following property is mandatory for the motion equations defined with (32):

$$ {}^0_S[\dot{R}](t) \cdot {}^0_S[R]^T(t) = (\overline{\omega} \times) \qquad (33)$$

where considering [1,10], $(\overline{\omega} \times)$ is the skew symmetric matrix associated to the angular velocity vector.

The time derivatives of the higher order, applied on the property (32), are defined as follows:

$$ {}^0_S[R] \cdot {}^0_S[R]^T \overset{(k)}{} = \sum_{j=1}^{k-3} \frac{d^j}{dt^j} \left\{ {}^0_S[R] \overset{[k-(j+1)]}{} \cdot {}^0_S[R]^T \cdot [\overline{\omega} \times] \right\} \cdot \delta_{k-1} + \frac{d^{k-2}}{dt^{k-2}} \{[\overline{\omega} \times] \cdot [\overline{\omega} \times]\} \cdot \delta_k + [\overline{\omega} \times] \overset{(k-1)}{} + {}^0_S[R] \overset{(k-1)}{} \cdot {}^0_S[R]^T \cdot [\overline{\omega} \times], \qquad (34)$$

$$\text{where } k \geq 2, \text{ and } \delta_k = \{(0, \text{ when } k \leq 2) \text{ and } (1, \text{ when } k > 2)\}$$

In advanced kinematics and dynamics, the time derivatives of the higher order for rotation matrices and position vectors must be used according to [10–13] as follows:

$$ {}^0_S[R] \overset{(k)}{} [q_j(t) \cdot \Delta_j] = \sum_{j=1}^{6} \left\{ \frac{\partial}{\partial q_j} \{{}^0_S[R]\} \cdot \Delta_j \cdot \overset{(k)}{q_j} \right\} + \sum_{j=1}^{6} \sum_{r=1}^{k-1} \left\{ \frac{\prod_{p=1}^{r}(k-p)}{p!} \cdot \left\{ \frac{p! \cdot m!}{(m+p)!} \cdot \frac{\partial}{\partial q_j^{(m)}} \{{}^0_S[R]\} \cdot \Delta_j \cdot \overset{(k-p)}{q_j} \right\} \right\}, \qquad (35)$$

$$\overset{(k-1)}{\overline{v}_M} = \frac{d^k \overline{r}_M(t)}{dt^k} = \sum_{j=1}^{6} \left[\frac{\partial \overset{(m)}{\overline{r}_M}}{\partial \overset{(m)}{q_j}} \cdot \overset{(k)}{q_j} \right] + \sum_{j=1}^{6} \sum_{r=1}^{k-1} \left\{ \frac{\prod_{p=1}^{r}(k-p)}{p!} \cdot \left[\frac{p! \cdot m!}{(m+p)!} \cdot \frac{\partial \overset{(m+p)}{\overline{r}_M}}{\partial \overset{(m)}{q_j}} \cdot \overset{(k-p)}{q_j} \right] \right\} \quad (36)$$

where $k \geq 1$; $k = \{1; 2; 3; 4; 5; \ldots\}$ and $m \geq (k+1)$; $m = \{2; 3; 4; 5; \ldots\}$ (37)

The symbols: (k) and (m) represent the order of the time derivatives applied on expressions (35) and (36).

The expressions of definition for angular velocities, and then for the angular accelerations are:

$$\dot{\overline{\omega}}[\alpha_u(t) - \beta_v(t) - \gamma_w(t)] = \tfrac{d}{dt}\{\overline{\omega}[\alpha_u(t) - \beta_v(t) - \gamma_w(t)]\} = \tfrac{d}{dt}\{\dot{\alpha}_u(t) \cdot \{\exp[0]\} \cdot \overline{u}^{(0)} + \dot{\beta}_v(t) \cdot \{\exp[\overline{u}(t) \times \alpha_u(t)]\} \cdot \overline{v}^{(0)} + \dot{\gamma}_w(t) \cdot \{\exp[\overline{u}(t) \times \alpha_u(t)] \cdot \exp[\overline{v}(t) \times \beta_v(t)]\} \cdot \overline{w}^{(0)}\}. \quad (38)$$

By analyzing (38), it is noticed that the expressions of the kinematic parameters that define the resultant rotation motion are written based on the matrix exponential functions.

Both velocities and accelerations of the higher order are established based on the following vectors:

$$\overline{r}_0(t) + \exp\left[\prod_{\{\chi=\{u;v;w\}\}} \exp[\overline{\chi} \times \delta_\chi]\right] \cdot {}^S\overline{p}_C = \overline{r}_C\big[q_j(t); \; j = 1 \to k^* = 6, \; t\big] = \overline{r}_C(t), \quad (39)$$

$$\overline{\psi}(t) = \overline{\psi}\big[q_j(t) \cdot \Delta_j; \; j = 1 \to k^* = 6, \; t\big] = \alpha_u(t) \cdot \{\exp[0]\} \cdot \overline{u}^{(0)} + \beta_v(t) \cdot \{\exp[\overline{u}(t) \times \alpha_u(t)]\} \cdot \overline{v}^{(0)} + \gamma_w(t) \cdot \{\exp[\overline{u}(t) \times \alpha_u(t)] \cdot \exp[\overline{v}(t) \times \beta_v(t)]\} \cdot \overline{w}^{(0)} \quad (40)$$

where the orientation vector is defined according to the following expressions: (8), (19), (27) and (28).

An essential component (28) included in (40), known as the angular transfer matrix, can be defined as a function of a set of the orientation angles. Considering (39) and (40), it is observed that they are functions of generalized variables (24) and (25). Actually, the six generalized variables are the independent parameters of the position and orientation from (24). Using the research from [10–14], the on-time functions associated to position (39) and orientation (40) vectors and the differential properties which are compulsory and applied in advanced kinematics and dynamics, have been developed:

$$\frac{d^{k-1}}{dt^{k-1}}\left(\frac{\partial \overline{r}_C}{\partial q_j}\right) = \frac{(k-1)! \cdot m!}{(m+k-1)!} \cdot \frac{\partial \overset{(m+k-1)}{\overline{r}_C}}{\partial \overset{(m)}{q_j}}, \quad k \geq 1; \; m \geq (k+1); \; m = \{2; 3; 4; 5; \ldots\}, \quad (41)$$

$$\frac{d^{k-1}}{dt^{k-1}}\left(\frac{\partial \overline{\psi}_i}{\partial q_j} \cdot \Delta_j\right) = \frac{(k-1)! \cdot m!}{(m+k-1)!} \cdot \frac{\partial \overset{(m+k-3)}{\overline{\varepsilon}}}{\partial \overset{(m)}{q_j}} \cdot \Delta_j = \frac{(k-1)! \cdot m!}{(m+k-1)!} \cdot \frac{\partial \overset{(m+k-1)}{\overline{\psi}}}{\partial \overset{(m)}{q_j}} \cdot \Delta_j, \quad (42)$$

$$\overset{(k-1)}{\overline{\varepsilon}}(t) = \overset{(k)}{\overline{\omega}}(t) = \sum_{j=1}^{k^*=6} \frac{d^{k-1}}{dt^{k-1}}\left[\frac{\partial \overset{(m)}{\overline{\psi}(t)}}{\partial \overset{(m)}{q_j}} \cdot \Delta_j \cdot \ddot{q}_j(t)\right] + \frac{1}{m+1} \cdot \sum_{j=1}^{k^*=n} \frac{d^{k-1}}{dt^{k-1}}\left[\frac{\partial \overset{(m+1)}{\overline{\psi}}}{\partial \overset{(m)}{q_j}} \cdot \Delta_j \cdot \dot{q}_j(t)\right] = \overset{(k+1)}{\overline{\psi}}(t), \quad (43)$$

$$\overset{(k-1)}{\overline{a}_C}(t) = \overset{(k)}{\overline{v}_C}(t) = \sum_{j=1}^{k^*=6} \frac{d^{k-1}}{dt^{k-1}}\left[\frac{\partial \overset{(m)}{\overline{r}_C(t)}}{\partial \overset{(m)}{q_j}} \cdot \ddot{q}_j(t)\right] + \sum_{j=1}^{k^*=6} \frac{d^{k-1}}{dt^{k-1}}\left[\frac{1}{m+1} \cdot \frac{\partial \overset{(m+1)}{\overline{r}_C(t)}}{\partial \overset{(m)}{q_j}} \cdot \dot{q}_j(t)\right] = \overset{(k+1)}{\overline{r}_C}(t), \quad (44)$$

The symbols k and m from (41) and (42) define the deriving orders with respect to time.

The expression (44) defines the angular and linear accelerations of the higher order for a rigid solid in general motion. The input expressions and parameters of the higher order are compulsory and applied in the definition of the dynamic notions of the higher order, such as the momentum, angular momentum, kinetic energy and acceleration energy of the higher order.

They have been included in the dynamics theorems which applies to current and sudden motions.

The next step consists of establishing the time variation law for the generalized coordinates and generalized variables of the higher order. Considering [9], the result is that all parameters from advanced kinematics are functions of generalized variables as well as of their time derivatives and they can be developed by using polynomial interpolation functions [13,20]. The following general form for the polynomial interpolation functions of the higher order is used:

$$q_{ji}^{(m-p)}(\tau) = (-1)^p \cdot \frac{(\tau_i - \tau)^{p+1}}{t_i \cdot (p+1)!} \cdot q_{ji-1}^{(m)} + \frac{(\tau - \tau_{i-1})^{p+1}}{t_i \cdot (p+1)!} \cdot q_{ji}^{(m)} + \delta_p \cdot \sum_{k=1}^{p} \frac{\tau^{p-k}}{(p-k)!} \cdot a_{jik}, \quad (45)$$

$$\left\{ \begin{array}{l} \text{where } p = 0 \to m, \ m - \text{deriving order}, \ m \geq 2, \ m = 2,3,4,5,\ldots, \quad \delta_p = \{(0, \ p=0);(1; \ p \geq 1)\} \\ j = 1 \to n, \ (\text{degrees of freedom} - (d.o.f.)); \ i = 1 \to s, \ (\text{intervals of motion trajectories}) \\ \tau - \text{actual time variable}, \quad t_i = \tau_i - \tau_{i-1} \ (\text{time to each trajectory interval}) \end{array} \right\} \quad (46)$$

For every trajectory interval $(i = 1 \to s)$, the number of unknowns is $(m+1)$, and their significance is:

$$\left\{ \begin{array}{l} (a_{jik}) \text{ for } k = 1 \to m; \text{ and } \left(q_{ji-1}^{(m)}\right) \text{ for } i = 2 \to s \\ \text{where } (a_{jik}) - \text{integration constants}, \text{ and } \left(q_{ji-1}^{(m)}\right) - \text{generalized accelerations of } (m) \text{ order} \end{array} \right\}, \quad (47)$$

To determine the unknowns, (50) is required to apply geometrical and kinematical constraints:

$$\left\{ \begin{array}{l} (\tau_0) \Rightarrow q_{j0}^{(m-p)}, \ p = 0 \to m; \quad (\tau_s) \Rightarrow \left\{q_{js}^{(m)}, \ q_{js}\right\} \\ (\tau_i) \Rightarrow \left\{ \begin{array}{l} \ddot{q}_{ji}^{(2)} - \text{generalized accelerations} \\ q_{ji}^{(m-p)}(\tau^+) = q_{ji+1}^{(m-p)}(\tau^-), \quad p = 0 \to m, \ \text{continuity conditions} \\ \text{all conditions are applied to each } (\tau_i), \ \text{where } i = 1 \to s-1 \end{array} \right. \end{array} \right., \quad (48)$$

The kinematical constrains that can be applied in the case of polynomials of the higher order are:

$$(\tau_0) \Rightarrow \left\{h_{0k} = q_{j0k}; \ v_{0k} = \dot{q}_{j0k}; \ a_{0k} = \ddot{q}_{j0k}; \ \dot{a}_{0k} = \dddot{q}_{j0k}\right\}, \quad (\tau_n) \Rightarrow \left\{h_{nk} = q_{jnk}; \ \ddot{a}_{nk} = \dddot{q}_{jnk}\right\}, \quad (49)$$

$$(\tau_i) \Rightarrow \left\{ \begin{array}{l} a_{ik} = \ddot{q}_{jik}; \ i = 1 \to n-1; \ i \subset k; \ k = 1 \to m \\ \{h_{ik}(t^+) = h_{i+1k}(t^-); \ v_{ik}(t^+) = v_{i+1k}(t^-); \ \dot{a}_{ik}(t^+) = \dot{a}_{i+1k}(t^-)\} \\ \text{where } k = 1 \to m \text{ is the sequence of the working process} \end{array} \right. \quad (50)$$

Finally, Expression (45) is substituted in the advanced notions of kinematics and dynamics.

2.3. The Mass Properties

In order to find an exact geometrical solution, it is considered that the rigid body consists of an infinite number of material particles for which the distance between two points is maintained constant, regardless of the forces that act upon it. The elementary particles, characterized by elementary and infinitesimal mass, have a continuous distribution in the geometrical shape of the solid body.

When the density is constant inside the rigid structure, the solid rigid is homogeneous. When the integration limits applied on the geometrical outline are well defined, it results in a homogeneous body with a simple or regular shape. In this last case, the geometrical and mass integrals are applied.

An essential aspect in advanced dynamics is represented by mass properties [13] The mass and the position of the mass center is determined in relation with $O'_0 x'_0 y'_0 z'_0 \equiv \{0'\}$ frame, as presented:

$$M = \int dm; \; where \, dm = \{\rho_\sigma \cdot d\sigma\}; \; \sigma = \{V; A; L\}, \tag{51}$$

$$\overline{\rho}_C(t) = \frac{\int \overline{\rho}_M(t) \cdot dm}{\int dm} = \frac{\int \overline{\rho}_M(t) \cdot dm}{M}, \tag{52}$$

In the expression (51), ρ_σ defines the density of the material, where the symbol σ represents, by case, the volume (V), the area (A) or the length (L) of the rigid body subjected to study

The position of the mass center, with respect to the $O_0 x_0 y_0 z_0 \equiv \{0\}$ reference frame, is expressed at first in a classical form, and then by using matrix exponentials as:

$$\overline{r}_C(t) = \frac{\int \overline{r}_M(t) \cdot dm}{\int dm} = \frac{\int \overline{r}_M(t) \cdot dm}{M} = \overline{r}_0(t) + \overline{\rho}_C(t) = \overline{r}_0(t) + {}^0_S[R](t) \cdot {}^S\overline{\rho}_C =$$

$$= \overline{r}_0(t) + \exp\left[\prod_{\{\chi=\{u;v;w\}\}} \exp[\overline{\chi} \times \delta_\chi]\right] \cdot {}^S\overline{\rho}_C = \overline{r}_0(t) + \{\exp[\overline{u} \times \alpha_u] \cdot \exp[\overline{v} \times \beta_v] \cdot \exp[\overline{w} \times \gamma_w]\} \cdot {}^S\overline{\rho}_C. \tag{53}$$

By applying the time derivative on (53), the linear velocity and acceleration of the mass center are determined:

$$\overline{v}_C(t) = \dot{\overline{r}}_C(t) = \dot{\overline{r}}_0(t) + \dot{\overline{\rho}}_C(t) = \dot{\overline{r}}_0(t) + {}^0_S[\dot{R}](t) \cdot {}^0_S[R]^T(t) \cdot {}^0_S[R](t) \cdot {}^S\overline{\rho}_C =$$

$$= \overline{v}_0(t) + \overline{\omega}(t) \times \overline{\rho}_C(t) = \frac{d}{dt}\left\{\overline{r}_0(t) + \left\{\exp\left[\prod_{\{\chi=\{u;v;w\}\}} \exp[\overline{\chi} \times \delta_\chi]\right]\right\} \cdot {}^S\overline{\rho}_C\right\} = \tag{54}$$

$$= \frac{d}{dt}\{\overline{r}_0(t) + \{\exp[\overline{u} \times \alpha_u] \cdot \exp[\overline{v} \times \beta_v] \cdot \exp[\overline{w} \times \gamma_w]\} \cdot {}^S\overline{\rho}_C\}.$$

Using classical Formulations (52) and (53), the result is the expressions for the acceleration of mass center:

$$\overline{a}_C = \dot{\overline{v}}_C(t) = \dot{\overline{v}}_0(t) + \frac{d}{dt}\left[\overline{\omega}(t) \times \overline{\rho}_C(t)\right] = \overline{a}_0 + \overline{\varepsilon} \times \overline{\rho}_C + \overline{\omega} \times \overline{\omega} \times \overline{\rho}_C \tag{55}$$

In dynamic modeling, the following expression is necessary:

$$\frac{d}{dt}(\overline{\rho}_M(t)) = \frac{d}{dt}(\overline{\rho}_C(t) + \vec{r}^*(t)) = \frac{d}{dt}\left({}^0_S[R](t) \cdot ({}^S\overline{\rho}_C + {}^S\vec{r}^*)\right) =$$
$$= \overline{\omega}(t) \times \overline{\rho}_M(t) = \overline{\omega}(t) \times \overline{\rho}_C(t) + \overline{\omega}(t) \times \vec{r}^*(t). \tag{56}$$

In the case of rigid bodies involved in rotation motion, the inertia property is highlighted by the mechanical moments of inertia [12–14]. According to Figure 2, the position of the elementary mass (dm) relative to the mass center is defined by means of the vector: $\vec{r}^* = \vec{r}^*(t)$.

Considering (52), the following mass property is obtained:

$$\int \vec{r}^*(t) \cdot dm = {}^0_S[R](t) \cdot \int {}^S\vec{r}^* \cdot dm = 0. \tag{57}$$

Based on research [12,13], the inertial tensor and its variation law relative to the concurrent frames applied in the mass center: $\{S^*\}$ and $\{0^*\}_{OR} \equiv \{0'\}_{OR}$ is established, according to:

$$I_S^* = \int (\vec{r}^* \times) \cdot (\vec{r}^* \times)^T \cdot dm = {}^0_S[R] \cdot {}^SI_S^* \cdot {}^0_S[R]^T, \tag{58}$$

where $^S I_S^*$ is inertial tensor axial and centrifugal of the body (S) in with relation with $\{S^*\}$ frame, applied in the mass center (C), having the property: $\{S^*\}_{OR} \equiv \{S\}_{OR}$:

$$^S I_S^* = \int \left(^S \bar{r}^* \times\right) \cdot \left(^S \bar{r}^* \times\right)^T \cdot dm, \tag{59}$$

Since the dynamic study refers to the absolute motion, the use of the following expressions of the inertial tensor are required:

$$I_S' = {}_S^0[R] \cdot {}^S I_S \cdot {}_S^0[R]^T = I_{SC}' + I_{S'}^*, \int \left(^S \bar{\rho}_M \times\right) \cdot \left(^S \bar{\rho}_M \times\right)^T \cdot dm = {}^S I_S \text{ and } I_{SC}' = M \cdot \left(\bar{\rho}_C \times\right) \cdot \left(\bar{\rho}_C \times\right)^T. \tag{60}$$

The matrix expression (60) represents the generalized variation law of the inertial tensor axial and centrifugal with respect to the frame $\{0'\}$. Further, in the expression (60), I_{SC}' is the inertia matrix axial and centrifugal of the mass center relative to $\{0'\}$. The inertial tensor axial and centrifugal is defined with respect to the absolute frame $\{0\}_{OR} = \{0'\}_{OR}$, as presented below:

$$I_S = \int \left(\bar{r}_M \times\right) \cdot \left(\bar{r}_M \times\right)^T \cdot dm = M \cdot \left(\bar{r}_0 \times\right) \cdot \left(\bar{r}_0 \times\right)^T + I_S' = I_{SO} + I_S' = I_{SO} + I_{SC}' + I_{S'}^* \tag{61}$$

In the Expression (61), I_{SO} represents the inertia matrix containing axial and centrifugal inertia moments, which is determined with respect to the fixed system $\{0\}$.

3. Higher Order Acceleration Energies

The term, advanced notions, from analytical dynamics presented in this paper refers to the motion energies whose central functions are the higher order accelerations. They have been developing in any sudden and transitory motion of mechanical systems. By having Appell's function as the starting point [1,2], a function which was highlighted in 1899 and that is also known as the kinetic energy of accelerations, new mathematical formulations on the expressions for acceleration energies of the first, second, and third order have been developed [10,14]. In this section, they are presented only in an explicit form. From a physical point of view, the acceleration energy is expressed as $\langle J/s^2 \rangle$ and represents the mechanical power developed by a mechanical system per time unit. By applying certain constraints, the acceleration energy can also be defined as the second order time derivative applied on the kinetic energy corresponding to the whole mechanical system. In the following, based on [10,14], the acceleration energies of the higher order are defined. They characterize the complex mechanical systems and include among others, the robots. The author established the acceleration energy corresponding to a rigid body involved in general motion, in a generalized form [10,14], also called the first order acceleration energy and defined with:

$$E_A^{(1)} = \tfrac{1}{2} \cdot \int a_M^2 \cdot dm = \tfrac{1}{2} \cdot \int \dot{\bar{v}}_M^T \cdot \dot{\bar{v}}_M \cdot dm = \tfrac{1}{2} \cdot \int Trace\left[\dot{\bar{v}}_M \cdot \dot{\bar{v}}_M^T\right] \cdot dm, \quad \bar{a}_M = \bar{a}_0 + \bar{\varepsilon} \times \bar{\rho}_M + \bar{\omega} \times \bar{\omega} \times \bar{\rho}_M \tag{62}$$

By developing Expression (71), the first three terms are obtained:

$$\tfrac{1}{2} \cdot \int \bar{a}_0^T \cdot \bar{a}_0 \cdot dm = \tfrac{1}{2} \cdot M \cdot \bar{a}_0^T \cdot \bar{a}_0 = \tfrac{1}{2} \cdot M \cdot a_0^2, \tag{63}$$

$$\tfrac{1}{2} \cdot \int \bar{a}_0^T \cdot \left(\bar{\varepsilon} \times \bar{\rho}_M\right) \cdot dm = \tfrac{1}{2} \cdot \int \left(\bar{\varepsilon} \times \bar{\rho}_M\right)^T \cdot \bar{a}_0 \cdot dm = \tfrac{1}{2} \cdot M \cdot \bar{a}_0 \cdot \left(\bar{\varepsilon} \times \bar{\rho}_C\right), \tag{64}$$

$$\tfrac{1}{2} \cdot \int \bar{a}_0^T \cdot \left(\bar{\omega} \times \bar{\omega} \times \bar{\rho}_M\right) \cdot dm = \tfrac{1}{2} \cdot \int \left(\bar{\omega} \times \bar{\omega} \times \bar{\rho}_M\right)^T \cdot \bar{a}_0 \cdot dm = \tfrac{1}{2} \cdot M \cdot \bar{a}_0 \cdot \left(\bar{\omega} \times \bar{\omega} \times \bar{\rho}_C\right). \tag{65}$$

The following three components define the resultant rotation. Among these, the first two contain the angular velocity and acceleration. The defining expressions are shown below:

$$\frac{1}{2} \cdot \int \left(\bar{\varepsilon} \times \bar{\rho}_M\right)^T \cdot \left(\bar{\varepsilon} \times \bar{\rho}_M\right) \cdot dm = \frac{1}{2} \cdot \bar{\varepsilon}^T \cdot \left\{ \int \left(\bar{\rho}_M \times\right) \cdot \left(\bar{\rho}_M \times\right)^T \cdot dm \right\} \cdot \bar{\varepsilon} = \frac{1}{2} \cdot \bar{\varepsilon}^T \cdot I_S' \cdot \bar{\varepsilon} \quad (66)$$

Performing some transformations and developing the terms from (76), the result is:

$$E_A^{(1\omega\varepsilon)} = \frac{1}{2} \cdot \int \left(\bar{\varepsilon} \times \bar{\rho}_M\right)^T \cdot \left(\bar{\omega} \times \bar{\omega} \times \bar{\rho}_M\right) \cdot dm = \frac{1}{2} \cdot \bar{\varepsilon}^T \cdot (\bar{\omega} \times) \cdot \left[\int \left(\bar{\rho}_M \times\right) \cdot \left(\bar{\rho}_M \times\right)^T \cdot dm\right] \cdot \bar{\omega} = \frac{1}{2} \cdot \bar{\varepsilon}^T \cdot \left(\bar{\omega} \times I_S' \cdot \bar{\omega}\right). \quad (67)$$

The last component, corresponding to rotation motion, exclusively contains the angular velocity:

$$E_A^{(1\omega^4)} = \frac{1}{2} \cdot \bar{\omega}^T \cdot \left[\bar{\omega}^T \cdot I_S' \cdot \bar{\omega}\right] \cdot \bar{\omega} = \frac{1}{2} \cdot \bar{\omega}^T \cdot \left[\bar{\omega}^T \cdot Trace\left(I_{pS}'\right) \cdot \bar{\omega} - \bar{\omega}^T \cdot I_{pS}' \cdot \bar{\omega}\right] \cdot \bar{\omega}. \quad (68)$$

The acceleration energy of the first order can be written in the following final form:

$$E_A^{(1)} = \frac{1}{2} \cdot M \cdot \bar{a}_0^T \cdot \bar{a}_0 + M \cdot \bar{a}_0^T \cdot \left(\bar{\varepsilon} \times \bar{\rho}_C\right) + M \cdot \bar{a}_0^T \cdot \left(\bar{\omega} \times \bar{\omega} \times \bar{\rho}_C\right) + \frac{1}{2} \cdot \bar{\varepsilon}^T \cdot I_S' \cdot \bar{\varepsilon} + \\ + \bar{\varepsilon}^T \cdot \left(\bar{\omega} \times I_S' \cdot \bar{\omega}\right) + \frac{1}{2} \cdot \bar{\omega}^T \cdot \left[\bar{\omega}^T \cdot I_S' \cdot \bar{\omega}\right] \cdot \bar{\omega}. \quad (69)$$

If the following conditions are met: $O \equiv C$, $\bar{\rho}_C = 0$, and $I_S' \equiv I_S^*$, then the mobile reference frame is applied in the mass center of the rigid body and the position of the elementary particle (dm) with respect to this frame is known, as well as the inertial tensor axial and centrifugal $I_S' \equiv I_S^*$ with respect to the mass center. Thus, (69) defines the first order acceleration energy in a generalized form:

$$E_A^{(1)} = \frac{1}{2} \cdot M \cdot \bar{a}_C^T \cdot \bar{a}_C + \frac{1}{2} \cdot \bar{\varepsilon}^T \cdot I_S^* \cdot \bar{\varepsilon} + \bar{\varepsilon}^T \cdot \left(\bar{\omega} \times I_S^* \cdot \bar{\omega}\right) + \frac{1}{2} \cdot \bar{\omega}^T \cdot \left[\bar{\omega}^T \cdot I_S^* \cdot \bar{\omega}\right] \cdot \bar{\omega}, \quad (70)$$

The expression (70) represents a reinterpretation of König's theorem for the acceleration energy of the first order. In the case of multibody systems (which also includes robots), the expression of acceleration energy of the first order (69) is determined according to:

$$E_A^{(1)}\left[\bar{\theta}(t); \dot{\bar{\theta}}(t); \ddot{\bar{\theta}}(t)\right] = \\ = (-1)^{\Delta_M} \cdot \frac{1-\Delta_M}{1+3\cdot\Delta_M} \sum_{i=1}^n \left\{ \left\{ \frac{1}{2} \cdot M_i \cdot {}^{(i)}\dot{\bar{v}}_{C_i}^T \cdot {}^{(i)}\dot{\bar{v}}_{C_i} \right\} + \Delta_M^2 \cdot \left\{ \frac{1}{2} \cdot {}^{(i)}\dot{\bar{\omega}}_i^T \cdot \left[{}^{(i)}I_i^* \cdot {}^{(i)}\dot{\bar{\omega}}_i + \left({}^{(i)}\bar{\omega}_i \times {}^{(i)}I_i^* \cdot {}^{(i)}\bar{\omega}_i\right)\right] \right\} \right\} + \\ + \Delta_M^2 \cdot \sum_{i=1}^n \left\{ \frac{1}{2} \cdot {}^{(i)}\dot{\bar{\omega}}_i^T \cdot \left({}^{(i)}\bar{\omega}_i \times {}^{(i)}I_i^* \cdot {}^{(i)}\bar{\omega}_i\right) \right\} + \Delta_M^2 \cdot \sum_{i=1}^n \left\{ \frac{1}{2} \cdot {}^{(i)}\bar{\omega}_i^T \cdot \left[{}^{(i)}\bar{\omega}_i^T \cdot {}^{(i)}I_i^* \cdot {}^{(i)}\bar{\omega}_i\right] \cdot {}^{(i)}\bar{\omega}_i \right\} \quad (71)$$

where $\Delta_M = \{(-1, \text{General motion}); (0, \text{Translation motion}); (1, \text{Rotation motion})\}$, and $i = 1 \to n$ is the number of kinetic assemblies that compose the mechanical system.

Considering the advanced kinematics notions, the components of acceleration energy of the first order are:

$$E_A^{(1)TR} = \frac{1}{2} \cdot \sum_{i=1}^n M_i \cdot a_{C_i}^2 = \\ = \frac{1}{2} \cdot \sum_{i=1}^n M_i \cdot \left\{ \sum_{j=1}^n \sum_{p=1}^n \left[\frac{\partial^2 \bar{r}_{C_i}^{(m)}}{\partial q_j^{(m)} \cdot \partial q_p^{(m)}} \cdot \ddot{q}_j \cdot \ddot{q}_p + \frac{1}{(m+1)^2} \cdot \frac{\partial^2 \bar{r}_{C_i}^{(m+1)}}{\partial q_j^{(m)} \cdot \partial q_p^{(m)}} \cdot \dot{q}_j \cdot \dot{q}_p + \frac{1}{m+1} \cdot \frac{\partial \bar{r}_{C_i}^{(m)}}{\partial q_j^{(m)}} \cdot \frac{\partial \bar{r}_{C_i}^{(m+1)}}{\partial q_p^{(m)}} \cdot \ddot{q}_j \cdot \dot{q}_p \right] \right\} \quad (72)$$

$$E_A^{(1)ROT\varepsilon} = \frac{1}{2} \cdot \sum_{i=1}^n \bar{\varepsilon}_i^T \cdot I_i^* \cdot \bar{\varepsilon}_i = \left\{ \frac{1}{2} \cdot \sum_{i=1}^n \sum_{j=1}^n \left[\frac{\partial \bar{\psi}_i^{(m)}}{\partial q_j^{(m)}} \cdot \Delta_j \cdot \ddot{q}_j + \frac{1}{m+1} \cdot \frac{\partial \bar{\psi}_i^{(m+1)}}{\partial q_j^{(m)}} \cdot \Delta_j \cdot \dot{q}_j \right]^T \cdot I_i^* \cdot \bar{\varepsilon}_i \right\}, \quad (73)$$

$$I_i^* \cdot \overline{\varepsilon}_i = I_i^* \cdot \sum_{j=1}^{n} \left[\frac{\partial \overset{(m)}{\overline{\psi}_i}}{\partial \overset{(m)}{q_j}} \cdot \Delta_j \cdot \ddot{q}_j + \frac{1}{m+1} \cdot \frac{\partial \overset{(m+1)}{\overline{\psi}_i}}{\partial \overset{(m)}{q_j}} \cdot \Delta_j \cdot \dot{q}_j \right], \tag{74}$$

$$E_A^{(1)ROT\omega\varepsilon} = \sum_{i=1}^{n} \overline{\varepsilon}_i^T \cdot (\overline{\omega}_i \times I_i^* \cdot \overline{\omega}_i) = \sum_{i=1}^{n} \sum_{j=1}^{n} \left[\frac{\partial \overset{(m)}{\overline{\psi}_i}}{\partial \overset{(m)}{q_j}} \cdot \Delta_j \cdot \ddot{q}_j + \frac{1}{m+1} \cdot \frac{\partial \overset{(m+1)}{\overline{\psi}_i}}{\partial \overset{(m)}{q_j}} \cdot \Delta_j \cdot \dot{q}_j \right]^T \cdot E_A^{(1)ROT\omega\omega}, \tag{75}$$

$$E_A^{(1)ROT\omega\omega} = (\overline{\omega}_i \times I_i^* \cdot \overline{\omega}_i) = \sum_{i=1}^{n} \sum_{j=1}^{n} \sum_{p=1}^{n} \left[\frac{\partial \overset{(m)}{\overline{\psi}_i}}{\partial \overset{(m)}{q_j}} \right] \times \left[I_i^* \cdot \frac{\partial \overset{(m)}{\overline{\psi}_i}}{\partial \overset{(m)}{q_p}} \right] \cdot \Delta_j \cdot \Delta_p \cdot \dot{q}_j \cdot \dot{q}_p. \tag{76}$$

According research [10,14], the sudden motion of MBS (multibody system), the transient motion phases, as well as the mechanical systems subjected to the action of a system of external forces, with a time variation law [19,20], are characterized by linear and angular accelerations of the higher order. Thus, the acceleration energy of the second order was also developed. According to the same research, the explicit form for the acceleration energy of the second order is presented below:

$$E_A^{(2)} \left[\overline{\theta}(t); \dot{\overline{\theta}}(t); \ddot{\overline{\theta}}(t); \dddot{\overline{\theta}}(t) \right] =$$
$$= (-1)^{\Delta_M} \cdot \frac{1-\Delta_M}{1+3\cdot\Delta_M} \cdot \sum_{i=1}^{n} \left\{ \frac{1}{2} \cdot M_i \cdot {}^i\ddot{\overline{v}}_{C_i}^T \cdot {}^i\ddot{\overline{v}}_{C_i} \right\} + E_A^{(2)} \left[\overline{\theta}(t); \dot{\overline{\theta}}(t); \ddot{\overline{\theta}}^2(t) \right] + E_A^{(2)} \left[\overline{\theta}(t); \overset{.6}{\overline{\theta}}(t); \dddot{\overline{\theta}}(t) \right] + \tag{77}$$
$$+\Delta_M^2 \cdot \sum_{i=1}^{n} \left\{ \frac{1}{2} \cdot {}^i\dddot{\overline{\omega}}_i^T \cdot {}^iI_i^* \cdot {}^i\dddot{\overline{\omega}}_i + 2 \cdot {}^i\ddot{\overline{\omega}}_i^T \cdot ({}^i\ddot{\overline{\omega}}_i \times {}^iI_{pi}^* \cdot {}^i\overline{\omega}_i) + {}^i\dddot{\overline{\omega}}_i^T \cdot ({}^i\overline{\omega}_i \times {}^iI_{pi}^* \cdot {}^i\overline{\omega}_i) - {}^i\overline{\omega}_i^T \cdot \left[{}^i\dddot{\overline{\omega}}_i^T \cdot {}^iI_i^* \cdot {}^i\overline{\omega}_i \right] \cdot {}^i\overline{\omega}_i \right\},$$

where $E_A^{(2)} \left[\overline{\theta}(t); \dot{\overline{\theta}}(t); \ddot{\overline{\theta}}^2(t) \right] =$
$$= \Delta_M^2 \cdot \sum_{i=1}^{n} \left\{ 2 \cdot {}^i\ddot{\overline{\omega}}_i^T \cdot \left({}^i\overline{\omega}_i^T \cdot {}^iI_i^* \cdot {}^i\overline{\omega}_i \right) \cdot {}^i\overline{\omega}_i + 2 \cdot {}^i\ddot{\overline{\omega}}_i^T \cdot \left[{}^i\overline{\omega}_i^T \cdot {}^iI_{pi}^* \cdot {}^i\overline{\omega}_i \right] \cdot {}^i\overline{\omega}_i - 5 \cdot \left({}^i\ddot{\overline{\omega}}_i^T \cdot {}^iI_{pi}^* \right) \cdot \left({}^i\overline{\omega}_i^T \cdot {}^i\overline{\omega}_i \right) \cdot {}^i\overline{\omega}_i + \right. \tag{78}$$
$$\left. + \frac{5}{2} \cdot \left({}^i\overline{\omega}_i^T \cdot {}^i\overline{\omega}_i \right) \cdot Trace\left({}^iI_{pi}^* \right) \cdot \left({}^i\overline{\omega}_i^T \cdot {}^i\overline{\omega}_i \right) + \frac{1}{2} \cdot {}^i\overline{\omega}_i^T \cdot \left[{}^i\overline{\omega}_i^T \cdot {}^iI_{pi}^* \cdot {}^i\overline{\omega}_i \right] \cdot {}^i\overline{\omega}_i \right\},$$

and $E_A^{(2)} \left[\overline{\theta}(t); \overset{.6}{\overline{\theta}}(t); \dddot{\overline{\theta}}(t) \right] =$
$$= \Delta_M^2 \cdot \left\{ \sum_{i=1}^{n} {}^i\overline{\omega}_i^T \cdot \left[{}^i\overline{\omega}_i^T \cdot \left({}^i\overline{\omega}_i \times {}^iI_{pi}^* \cdot {}^i\overline{\omega}_i \right) \right] \cdot {}^i\overline{\omega}_i + \frac{1}{2} \cdot {}^i\overline{\omega}_i^T \cdot \left[{}^i\overline{\omega}_i^T \cdot \left({}^i\overline{\omega}_i^T \cdot {}^iI_i^* \cdot {}^i\overline{\omega}_i \right) \cdot {}^i\overline{\omega}_i \right] \cdot {}^i\overline{\omega}_i \right\} \tag{79}$$

The study of advanced dynamics is extended to the acceleration energy of the third order. According to [10,14], an explicit equation for the third order acceleration energy is developed:

$$E_A^{(3)} \left[\overline{\theta}(t); \dot{\overline{\theta}}(t); \ddot{\overline{\theta}}(t); \dddot{\overline{\theta}}(t); \overset{....}{\overline{\theta}}(t) \right] = (-1)^{\Delta_M} \cdot \frac{1-\Delta_M}{1+3\cdot\Delta_M} \cdot \sum_{i=1}^{n} \left\{ \frac{1}{2} \cdot M_i \cdot {}^i\dddot{\overline{v}}_{C_i}^T \cdot {}^i\dddot{\overline{v}}_{C_i} \right\} +$$
$$+\Delta_M^2 \cdot \sum_{i=1}^{n} \left\{ \frac{1}{2} \cdot {}^i\overset{....}{\overline{\omega}}_i^T \cdot {}^iI_i^* \cdot {}^i\overset{....}{\overline{\omega}}_i + 3 \cdot \overline{\omega}_i^T \cdot (\overline{\omega}_i \times I_{pi}^* \cdot \overline{\omega}_i) + 3 \cdot \dddot{\overline{\omega}}_i^T \cdot (\ddot{\overline{\omega}}_i \times I_{pi}^* \cdot \overline{\omega}_i) + 3 \cdot \ddot{\overline{\omega}}_i^T \cdot (\overline{\omega}_i \times I_{pi}^* \cdot \overline{\omega}_i) + \right. \tag{80}$$
$$\left. + 2 \cdot (\overline{\omega}_i \times \ddot{\overline{\omega}}_i)^T \cdot I_{pi}^* \cdot (\overline{\omega}_i \times \overline{\omega}_i) - 5 \cdot \dddot{\overline{\omega}}_i^T \cdot \left[\overline{\omega}_i^T \cdot I_i^* \cdot \ddot{\overline{\omega}}_i \right] \cdot \overline{\omega}_i - \overline{\omega}_i^T \cdot \left[\ddot{\overline{\omega}}_i^T \cdot I_i^* \cdot \ddot{\overline{\omega}}_i \right] \cdot \overline{\omega}_i + \right.$$
$$\left. + \overline{\omega}_i^T \cdot \left[\overline{\omega}_i^T \cdot I_{pi}^* \cdot (\ddot{\overline{\omega}}_i \times \overline{\omega}_i) \right] \cdot \overline{\omega}_i \right\}$$

An important aspect that has to be mentioned is that the component $E_A^{(3)} \left[\overline{\theta}(t); \dot{\overline{\theta}}(t); \ddot{\overline{\theta}}(t); \dddot{\overline{\theta}}^2(t) \right]$ of the acceleration energy of the third order is not included in the definition Equation (80). In the case of the multibody systems, the acceleration energies of the higher order are determined based on the following generalized expression:

$$E_A^{(p)}\left[\overline{\theta}(t); \dot{\overline{\theta}}(t); \cdots ; \overset{(p+1)}{\overline{\theta}}(t)\right] = \frac{1}{2} \sum_{i=1}^{n} \left\{ \int \overset{(p)}{\overline{v}_i}^T \cdot \overset{(p)}{\overline{v}_i} \cdot dm \right\} = \frac{1}{2} \sum_{i=1}^{n} \left\{ \int Trace\left[\overset{(p+1)}{\overline{r}_i} \cdot \overset{(p+1)}{\overline{r}_i}^T\right] \cdot dm \right\} =$$

$$= \frac{1}{2} \cdot \sum_{i=1}^{n} Trace\left\{ \overset{(p+1)}{_0^i[R]} \cdot \left[\int \overset{i}{\overline{r}_i^*} \cdot \overset{i}{\overline{r}_i^{*T}} \cdot dm + {}^i\overline{r}_{C_i} \cdot {}^i\overline{r}_{C_i}^T \cdot \int dm \right] \cdot \overset{(p+1)}{_0^i[R]}^T \right\} + \frac{1}{2} \cdot \sum_{i=1}^{n} Trace\left[\overset{(p+1)}{\overline{p}_i} \cdot \overset{(p+1)}{\overline{p}_i}^T\right] \cdot \int dm = \quad (81)$$

$$= \frac{1}{2} \cdot \sum_{i=1}^{n} Trace\left\{ \overset{(p+1)}{_0^i[R]} \cdot \left[{}^iI_{pi}^* + M_i \cdot {}^i\overline{r}_{C_i} \cdot {}^i\overline{r}_{C_i}^T \right] \cdot \overset{(p+1)}{_0^i[R]}^T \right\} + \frac{1}{2} \cdot \sum_{i=1}^{n} Trace\left[\overset{(p+1)}{\overline{p}_i} \cdot \overset{(p+1)}{\overline{p}_i}^T\right] \cdot M_i, \text{ where } p = 1, 2, 3, 4, \ldots$$

where $p \geq 1$, $k \geq 1$, $\{p; k\} = \{1; 2; 3; \ldots\}$, (p) — order of acceleration, (k) — derivation order

$$\text{and } E_A^{(p)}\left[\overset{(0)}{\overline{\theta}}(t); \dot{\overline{\theta}}(t); \cdots ; \overset{(p+k)}{\overline{\theta}}(t)\right] = E_A^{(p)}\left[\overline{\theta}(t); \dot{\overline{\theta}}(t); \cdots ; \overset{(p+1)}{\overline{\theta}}(t)\right] \quad (82)$$

Expression (81) includes the inertia tensor planar and centrifugal, ${}^iI_{pi}^* = \int {}^i\overline{r}_i^* \cdot {}^i\overline{r}_i^{*T} \cdot dm$.

The expressions for the acceleration energies of the higher order, according to [13] are further included in the equations of advanced dynamics corresponding to the mechanical systems which are subjected to fast and sudden movements.

4. Discussion on the Results

This section was introduced exclusively for highlighting the existence of the accelerations and acceleration energies of the higher order. For this purpose, according to Figure 4, the rotation motion of the arm of the serial robot Fanuc LR Mate 100 iB, on the angular interval $(0, \pi)$ was considered. The total time of the rotation motion was of 1.7 s.

Figure 4. Fanuc LR Mate 100 iB robot. The location of the transducer on the robot arm.

For highlighting the time variation law of the acceleration energies of the higher order, the polynomial interpolating functions of the fifth order have been applied in accordance with (48)–(51). This method requires, according to kinematical constraints (51), the establishment through measurements of the angular acceleration corresponding to robot arm, $\ddot{q}_3$. As a result, in the first step, the time variation law corresponding to the tangential component of the acceleration for the characteristic point belonging to the robot arm has been determined at every 0.033 s. The experimental data was collected by using a mono-axial accelerometer. The robot arm (more precisely the third kinetic ensemble of the robot) was engaged in rotation motion at high speeds and by using the data collected by the accelerometer. The time variation law for the tangential component of the acceleration of the characteristic point was graphically represented. Considering the rotation motion, in the next step the time variation law of the angular acceleration $\ddot{q}_3(\tau)$ was also represented. Both graphical representations are illustrated in Figure 5.

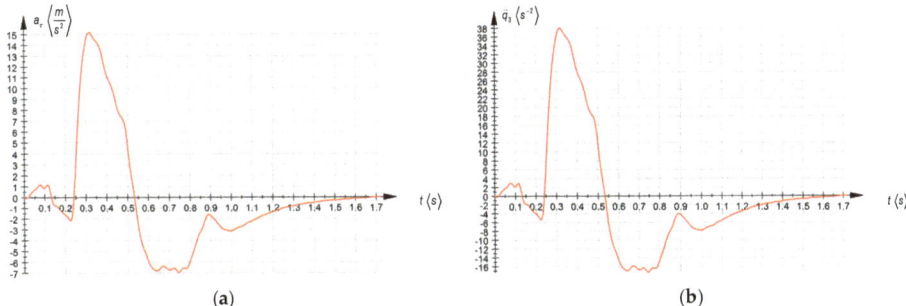

Figure 5. (a) Tangential accelerations variation law. (b) Generalized accelerations variation law. The above presented parameters were also determined on the same angular interval $(0, \pi)$, by applying the sudden stop. The graphical representation is illustrated in Figure 6.

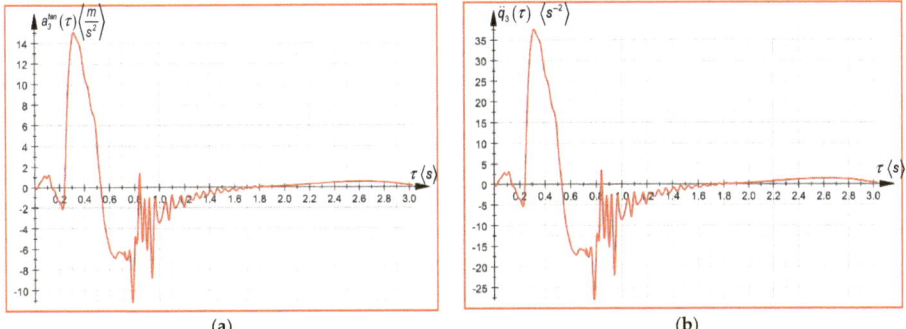

Figure 6. (a) The tangential first order accelerations variation law; (b) The generalized accelerations of first order variation law.

According to Figure 5, the maximum value for the linear acceleration exceeds $15 \langle m/s^2 \rangle$, which defines a sudden motion. The authors have considered, for this application, that the motion can be described by using the polynomial interpolating functions of the fifth order:

$$\dddot{q}_{ji}(\tau) = \frac{\tau_i - \tau}{t_i} \cdot \dddot{q}_{ji}(\tau_{i-1}) + \frac{\tau - \tau_{i-1}}{t_i} \cdot \dddot{q}_{ji}(\tau_i), \tag{83}$$

$$\ddot{q}_{ji}(\tau) = -\frac{(\tau_i - \tau)^2}{2 \cdot t_i} \cdot \dddot{q}_{ji-1} + \frac{(\tau - \tau_{i-1})^2}{2 \cdot t_i} \cdot \dddot{q}_{ji} + a_{ji1}, \tag{84}$$

$$\ddot{q}_{ji}(\tau) = \frac{(\tau_i - \tau)^3}{6 \cdot t_i} \cdot \dddot{q}_{ji-1} + \frac{(\tau - \tau_{i-1})^3}{6 \cdot t_i} \cdot \dddot{q}_{ji} + a_{ji1} \cdot \tau + a_{ji2}, \tag{85}$$

$$\dot{q}_{ji}(\tau) = -\frac{(\tau_i - \tau)^4}{24 \cdot t_i} \cdot \dddot{q}_{ji-1} + \frac{(\tau - \tau_{i-1})^4}{24 \cdot t_i} \cdot \dddot{q}_{ji} + a_{ji1} \cdot \frac{\tau^2}{2} + a_{ji2} \cdot \tau + a_{ji3}, \tag{86}$$

$$q_{ji}(\tau) = \frac{(\tau_i - \tau)^5}{120 \cdot t_i} \cdot \dddot{q}_{ji-1} + \frac{(\tau - \tau_{i-1})^5}{120 \cdot t_i} \cdot \dddot{q}_{ji} + a_{ji1} \cdot \frac{\tau^3}{6} + a_{ji2} \cdot \frac{\tau^2}{2} + a_{ji3} \cdot \tau + a_{ji4}. \tag{87}$$

where a_{jip}, $p = 1 \rightarrow 4$, are the integration constants which are determined from the geometrical and kinematical constraints (50), with an important role in ensuring the continuity of the rotation motion on the angular interval $(0, \pi)$, characterized by 51 interpolation segments. Every interpolation segment contains four unknowns, represented by the integration constants. It is obvious that the total number of the unknowns is 204. In order to determine the unknowns, according to (50), the numerical values for the angular acceleration (Figure 5b) were substituted in the functions (83)–(87) presented above. Considering the large number of results, more exactly 255 polynomial functions, only a few sequences are illustrated in Tables 1 and 2.

Table 1. The polynomial time functions for the generalized coordinates for the third rotation joint.

Time Interval τ $\langle s \rangle$	Polynomial Time Functions for Generalized Coordinates $\{k=1 \to 17, i=1 \to 51\}$	
	$q_{3ik} \langle rad \rangle$	$\dot{q}_{3ik} \langle s^{-1} \rangle$
[0, 0.0195]	$2478.916 \cdot \tau^5 - 2.736 \cdot \tau^3 + 0.047 \cdot \tau^2$	$12394.582 \cdot \tau^4 - 8.21 \cdot \tau^2 + 0.094 \cdot \tau$
[0.585, 0.605]	$3792.503 \cdot \tau^5 - 11239.525 \cdot \tau^4 +$ $+13301.344 \cdot \tau^3 - 7864.32 \cdot \tau^2 +$ $+2331.436 \cdot \tau - 277.457$	$18962.519 \cdot \tau^4 - 44958.1 \cdot \tau^3 +$ $+39904.034 \cdot \tau^2 - 15728.643 \cdot \tau +$ $+2331.436$
[1.191, 1.21]	$235.591 \cdot \tau^5 - 1429.192 \cdot \tau^4 +$ $+3470.091 \cdot \tau^3 - 4216.837 \cdot \tau^2 +$ $+2565.994 \cdot \tau - 622.767$	$1177.959 \cdot \tau^4 - 5716.771 \cdot \tau^3 +$ $+10410.273 \cdot \tau^2 - 8433.675 \cdot \tau +$ $+2565.994$
[1.698, 1.718]	$-62.678 \cdot \tau^5 + 538.206 \cdot \tau^4 -$ $-1849.467 \cdot \tau^3 + 3177.645 \cdot \tau^2 -$ $-2727.52 \cdot \tau + 939.274$	$-313.393 \cdot \tau^4 + 2152.825 \cdot \tau^3 -$ $-5548.403 \cdot \tau^2 + 6355.291 \cdot \tau - 2727.52$

Table 2. The Polynomial Time Functions for Generalized Variables in Case of the Third Rotation Joint.

Time Interval τ $\langle s \rangle$	Polynomial Time Functions for Generalized Variables $\{k=1 \to 17, i=1 \to 51\}$		
	$\ddot{q}_{3ik} \langle s^{-2} \rangle$	$\dddot{q}_{3ik} \langle s^{-3} \rangle$	$\ddddot{q}_{3ik} \langle s^{-4} \rangle$
[0, 0.0195]	$49578.331 \cdot \tau^3 -$ $-16.421 \cdot \tau + 0.094$	$148734.995 \cdot \tau^2 - 16.421$	$297469.99 \cdot \tau$
[0.585, 0.605]	$75850.076 \cdot \tau^3 -$ $-134874.302 \cdot \tau^2 +$ $+79808.069 \cdot \tau - 15728.643$	$227550.23 \cdot \tau^2 -$ $-269748.605 \cdot \tau + 79808.069$	$455100.461 \cdot \tau -$ -269748.605
[1.191, 1.21]	$4711.839 \cdot \tau^3 -$ $-17150.314 \cdot \tau^2 +$ $+20820.546 \cdot \tau - 8433.675$	$14135.518 \cdot \tau^2 -$ $-34300.628 \cdot \tau + 20820.546$	$28271.037 \cdot \tau -$ -34300.628
[1.698, 1.718]	$-1253.573 \cdot \tau^3 +$ $+6458.476 \cdot \tau^2 -$ $-11096.807 \cdot \tau + 6355.291$	$-3760.72 \cdot \tau^2 +$ $+12916.952 \cdot \tau - 11096.807$	$-7521.4407 \cdot \tau +$ $+12916.9529$

As a result, the numerical values for the angular acceleration (see Figure 5) were applied, finally obtaining all of the integration constants. In order to illustrate this step, some sequences containing the time variation laws of the functions described by Equations (83)–(87) are presented. These results are included in Tables 1 and 2. Similarly, the polynomial functions are obtained for all the 51 interpolation segments. As a result, the time variation law for the angular acceleration of the second and third order are also graphically represented according to Figure 7:

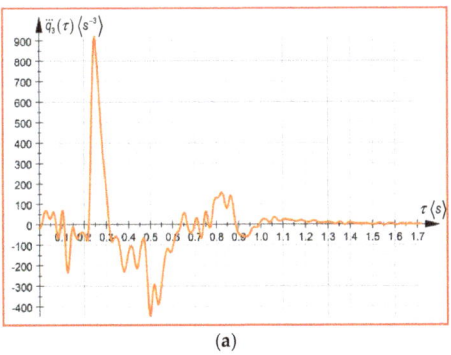 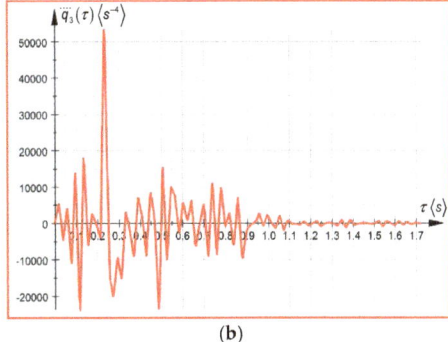

Figure 7. (a) The variation law of the angular accelerations of second order. (b) The variation law for the angular accelerations of third order.

Thus, by customizing Expressions (71), (77)–(80) for the considered robot structure, the following expressions characterizing the acceleration energies of the first, second and third order are:

$$E_{Aik}^{(1)}(\tau) = \frac{1}{2} \cdot \left(M_3 \cdot x_{C3}^2 + M_3 \cdot z_{C3}^2 + {}^3I_y\right) \cdot \left(\ddot{q}_{3ik}^2(\tau) + \dot{q}_{3ik}^4(\tau)\right), \qquad (88)$$

$$E_{Aik}^{(2)}(\tau) = \frac{1}{2} \cdot \left(M_3 \cdot x_{C3}^2 + M_3 \cdot z_{C3}^2 + {}^3I_y\right) \cdot \left[\dddot{q}_{3ik}^2(\tau) - 2 \cdot \dot{q}_{3ik}^3(\tau) \cdot \ddot{q}_{3ik}(\tau) + 9 \cdot \dot{q}_{3ik}^2(\tau) \cdot \ddot{q}_{3ik}^2(\tau) + \dot{q}_{3ik}^6(\tau)\right] \qquad (89)$$

$$\begin{aligned}E_{Aik}^{(3)}(\tau) = \frac{1}{2} \cdot \Big\{&\left(M_3 \cdot x_{C3}^2 + M_3 \cdot z_{C3}^2 + {}^3I_y\right) \cdot \Big[\dot{q}_{3ik}^8(\tau) - 8 \cdot \dot{q}_{3ik}^5(\tau) \cdot \ddot{q}_{3ik}(\tau) + 30 \cdot \dot{q}_{3ik}^4(\tau) \cdot \ddot{q}_{3ik}^2(\tau) - \\ &-12 \cdot \dot{q}_{3ik}^2(\tau) \cdot \ddot{q}_{3ik}(\tau) \cdot \dddot{q}_{3ik}(\tau) + 30 \cdot \dot{q}_{3ik}^4(\tau) \cdot \ddot{q}_{3ik}^2(\tau) - 12 \cdot \dot{q}_{3ik}^2(\tau) \cdot \ddot{q}_{3ik}(\tau) \cdot \dddot{q}_{3ik}(\tau) + \\ &+16 \cdot \dot{q}_{3ik}^2(\tau) \cdot \ddot{q}_{3ik}^2(\tau) + 24 \cdot \dot{q}_{3ik}(\tau) \cdot \ddot{q}_{3ik}^2(\tau) \cdot \dddot{q}_{3ik}(\tau) + 9 \cdot \dot{q}_{3ik}^4(\tau) + \dddot{q}_{3ik}^2(\tau)\Big]\Big\}\end{aligned} \qquad (90)$$

The obtained results (see example from Tables 1 and 2) regarding the higher order polynomial functions have been included in the above expressions of the acceleration energies that characterize the fast rotation motion of the robot arm. The results are represented in Figures 8 and 9:

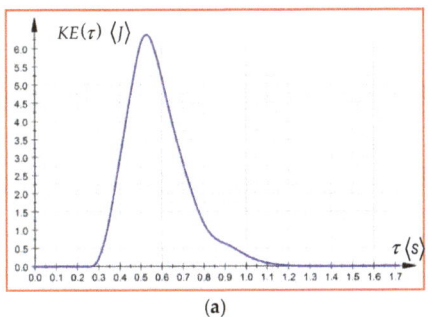 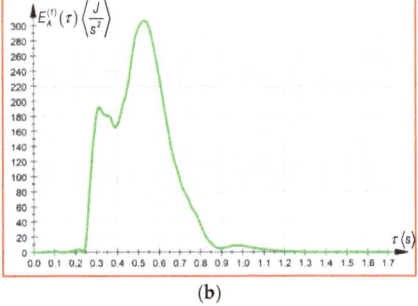

Figure 8. (a) The variation law for the kinetic energy. (b) The time variation law of the acceleration energy of the first order.

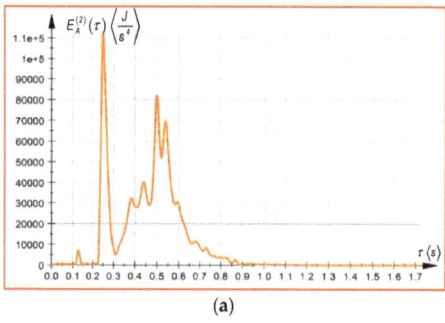 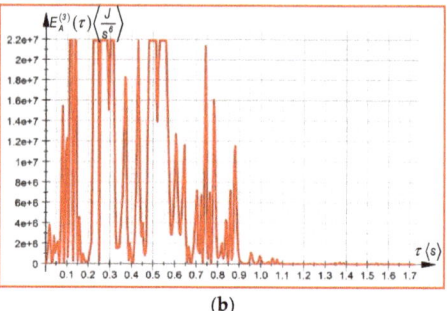

(a) (b)

Figure 9. (**a**) The variation law for the acceleration energy of second order; (**b**) The time variation law of the acceleration energy of third order.

In the above graphical representations (Figures 8 and 9), the time variation laws for the kinetic energy are illustrated, as well as the acceleration energies of the first, second and third order for an application in which the serial Fanuc robot, presented in the Figure 4, is implemented.

5. Conclusions

The main objective of this paper was to highlight some new developments in the field of advanced dynamics of mechanical systems and higher order dynamic equations. These equations were developed by using matrix exponentials which proved to have indisputable advantages in the matrix study of any complex mechanical system. In this paper, new approaches have been proposed based on differential principles from analytical mechanics by using some important dynamics notions, regarding the acceleration energies of the first, second and third order. The study was extended to the equations of the higher order, which provided the possibility of applying the initial motion conditions in the positions, velocities and accelerations of the first and second order. In order to determine the time variation laws for the generalized variables and acceleration energies of the higher order, the time polynomial functions of the fifth order were applied. In Newtonian dynamics, the establishment of dynamic equations as second order differential equations are performed by considering the kinetic energy as the main function. In the case of fast and sudden movements, the presence of higher order accelerations leads to higher order differential equations which define the variation laws in relation to time in the case of the forces and the driving moments that impose the motion of the mechanical systems. As a result, the application of acceleration energies highlights, in an explicit and direct form, the presence of higher-order accelerations, unlike the case where the study was performed considering kinetic energy.

In accordance to inverse kinematics, also known as the control kinematics of the robots, the use of polynomial functions leads to the control of the mechanical motion, especially of transitory motions. These aspects were exemplified in the previous section. Therefore, it was noticed that on the time interval of 1.7 s, a number of 51 measurements of the tangential acceleration were performed at every 0.033 s. Regarding the polynomial functions, this aspect led to a very good approximation of the mechanical motion. At the same time, the polynomial functions were substituted in the expressions of the acceleration energies of the higher order. Therefore, their time variation laws were obtained as shown in graphical representations from the previous section.

6. Contributions of the Authors

The purpose of this paper was to present some original defining expressions for acceleration energies of the higher order. These expressions also include the accelerations of the higher order which define the fast and sudden motions of mechanical systems. The novelty of this paper was in determining in an explicit form of the expressions for the acceleration energies based on advanced kinematic notions and using polynomial interpolation functions of the higher order.

Therefore, the contributions of the main author are highlighted by a great number of expressions included in the following sections of this paper:

- 2.1. Position and Orientation Parameters. The new expressions are: (10)–(16), (19) and (20), as well as (24), (27)–(30).
- 2.2. The Kinematic Parameters of Higher Order. The new equations are: (34)–(38) and (42)–(50);
- 3. Higher Order Acceleration Energies. This section contains the new expressions of definition for the acceleration energy of the first, second and third order in an explicit form: (69)–(82).

Author Contributions: Conceptualization, I.N.; methodology, I.N.; investigation and software A.-V.C; validation, I.N.; formal analysis, I.N; resources, I.N., A.-V.C.; data curation, I.N., A.-V.C.; writing—original draft preparation, A.-V.C; writing—review and editing, I.N., A.-V.C.; visualization, A.-V.C; supervision, I.N.

Funding: This research received no external funding.

Acknowledgments: The authors would like to thank Technical University of Cluj Napoca, Romania for providing the technical support in carrying out the experimental analysis.

Conflicts of Interest: The authors declare no conflicts of interest.

References

1. Appell, P. *Sur Une Forme Générale des Equations de la Dynamique*, 1st ed.; Gauthier-Villars: Paris, France, 1899.
2. Appell, P. *Traité de Mécanique Rationnelle*, 1st ed.; Garnier frères: Paris, France, 1903.
3. Craig, J.J. *Introduction to Robotics: Mechanics and Control*, 3rd ed.; Pearson Prentice Hall: Upper Sadelliffe, NJ, USA, 2005.
4. Yuan, B.S.; Book, W.J.; Huggins, J.D. Dynamics of Flexible Manipulator Arms: Alternative derivation, Verification, and Characteristics for Control. *ASME J. Dyn. Syst. Meas. Control* **1993**, *115*, 394–404. [CrossRef]
5. Bhatti, M.; Lu, D. Analytical Study of the Head-On Collision Process between Hydroelastic Solitary Waves in the Presence of a Uniform Current. *Symmetry* **2019**, *11*, 333. [CrossRef]
6. Hussain, F.; Ellahi, R.; Zeeshan, A. Mathematical Models of Electro-Magnetohydrodynamic Multiphase Flows Synthesis with Nano-Sized Hafnium Particles. *Appl. Sci.* **2018**, *8*, 275. [CrossRef]
7. Marin, M.; Vlase, S.; Ellahi, R.; Bhatti, M. On the Partition of Energies for the Backward in Time Problem of Thermoelastic Materials with a Dipolar Structure. *Symmetry* **2019**, *11*, 863. [CrossRef]
8. Thompson, P. *Snap, Crackle, and Pop, (AIAA Info.)*; Systems Technology: Hawthorne, CA, USA, 2011.
9. Eager, D.; Pendrill, A.M.; Reinstad, N. Beyond velocity and acceleration: jerk, snap and higher derivatives. *Eur. J. Phys.* **2016**, *37*, 065008. [CrossRef]
10. Negrean, I.; Negrean, D.C. The Acceleration Energy to Robot Dynamics. In Proceedings of the A&QT-R International Conference on Automation, Quality and Testing, Robotics, Cluj-Napoca, Romania, 23–25 May 2002; pp. 59–64.
11. Negrean, I.; Negrean, D.C. Matrix Exponentials to Robot Kinematics. In Proceedings of the 17th International Conference on CAD/CAM, Robotics and Factories of the Future, CARS&FOF, Durban, South Africa, 10–12 July 2001; pp. 1250–1257.
12. Negrean, I. New Formulations on Motion Equations in Analytical Dynamics. *Appl. Mech. Mater.* **2016**, *823*, 49–54. [CrossRef]
13. Negrean, I. Advanced Notions in Analytical Dynamics of Systems. *Acta Tech. Napoc. Ser. Appl. Math. Mech. Eng.* **2017**, *60*, 491–502.
14. Negrean, I. Energies of Acceleration in Advanced Robotics Dynamics. *Appl. Mech. Mater.* **2014**, *762*, 67–73. [CrossRef]
15. Vlase, S.; Teodorescu, P.P. ElastoDynamics of a Solid with A General "Rigid" Motion Using Fem Model Part I. Theoretical Approach. *Rom. J. Phys.* **2013**, *58*, 872–881.
16. Pars, L.A. *A Treatise on Analytical Dynamics*; Heinemann: London, UK, 2007; Volume 1, pp. 1–122.
17. Jazar, R.N. *Theory of Applied Robotics: Kinematics, Dynamics, and Control*, 2nd ed.; Springer Nature Switzerland AG: Basel, Switzerland, 2010.

18. Ardema, M.D. *Analytical Dynamics. Theory and Applications*; Springer Nature Switzerland AG: Basel, Switzerland, 2005.
19. Park, F.C. Computational Aspects of the Product-of-Exponentials Formula for Robot Kinematics. *IEEE Trans. Autom. Control* **1994**, *39*, 643–647. [CrossRef]
20. Brener, P. Technical Concepts. Orientation, Rotation, Velocity and Acceleration, and the SRM. User's manual, version 2.0; SEDRIS. 2008.

© 2019 by the authors. Licensee MDPI, Basel, Switzerland. This article is an open access article distributed under the terms and conditions of the Creative Commons Attribution (CC BY) license (http://creativecommons.org/licenses/by/4.0/).

Article

On the Partition of Energies for the Backward in Time Problem of Thermoelastic Materials with a Dipolar Structure

M. Marin [1,*], S. Vlase [2], R. Ellahi [3,4] and M.M. Bhatti [5,6]

1. Department of Mathematics and Computer Science, Transilvania University of Brasov, 500093 Brasov, Romania
2. Department of Mechanical Engineering, Transilvania University of Brasov, 500093 Brasov, Romania
3. Center for Modeling & Computer Simulation, Research Institute, King Fahd University of Petroleum & Minerals, Dhahran 31261, Saudi Arabia
4. Department of Mathematics & Statistics, Faculty of Basic and Applied Sciences (FBAS), International Islamic University (IIUI), Islamabad 44000, Pakistan
5. College of Mathematics and Systems Science, Shandong University of Science and Technology, Qingdao 266590, Shandong, China
6. Shanghai Institute of Applied Mathematics and Mechanics, Shanghai University, Shanghai 200072, China
* Correspondence: m.marin@unitbv.ro

Received: 12 June 2019; Accepted: 26 June 2019; Published: 2 July 2019

Abstract: We first formulate the mixed backward in time problem in the context of thermoelasticity for dipolar materials. To prove the consistency of this mixed problem, our first main result is regarding the uniqueness of the solution for this problem. This is obtained based on some auxiliary results, namely, four integral identities. The second main result is regarding the temporal behavior of our thermoelastic body with a dipolar structure. This behavior is studied by means of some relations on a partition of various parts of the energy associated to the solution of the problem.

Keywords: backward in time problem; dipolar thermoelastic body; uniqueness of solution; Cesaro means; partition of energies

---

1. Introduction

In our study, we approach a thermoelastic body having a dipolar structure. This kind of structure falls within a more general theory, namely, the theory of bodies with microstructure. The first studies in this context were published by Eringen (see, for instance, references [1,2]). One may deduce the importance of the dipolar structure due to the large number of published studies dedicated to this topic, of which we can mention [3–7]. As such, our present work can be considered a continuation in this respect.

A continuation of the theories of microstructure, is a theory that takes into account the voids in the materials. It is considered that the initiators of this theory were Nunziato and Cowin, in their known paper [8]. After that, the number of studies within this topic has grown impressively. We want to enumerate some of these [9–16]: The first result for the backward in time problem belongs to Serrin, who approached this problem in the context of Navier–Stokes equations (see [17]). In the paper [17], we find some uniqueness in the results with regards to the forward in time problem. After that, the number of studies dedicated to the backward in time problem has increased considerably. Of particular importance are the works [18–27]. We have to point out that the results obtained by Ciarletta in [23], and Ciarletta and Chirita in [24] were improved by Quintanilla in [25]. In addition, Quintanilla approached the question of location in time for solutions to the backward in time problem, in the context of thermoelasticity of Green and Naghdi [26,27] and the theory of porous thermoelastic bodies. The elastic porous bodies were also approached by Iovane and Passarella in [28]. The forward in time problem, in the context of theory for thermomicrostretch elastic solids, was approached by Passarella and Tibullo in [29]. It is worth noting that the idea of considering non-standard problems, in the context of the general theory of bodies having a dipolar structure, was inspired by Quintanilla and Straughan's work [30]. In [31,32] it is proved that the nonhomogenous temperature field has a profound influence on the nanobeam mechanics. Additionally, [33] is a recent contribution on stress-driven nonlocal modeling of thermoelastic nanostructures.

Here is the plane of our study. First of all, we summarize the main equations, the initial conditions, and the boundary data of the mixed problem. Then, we prove some estimates for the gradient of classical and dipolar displacements, and for the gradient of the function of voids. In the last part of our study we prove the main result, namely, the continuous dependence of solutions—with regards to the coefficients that couple the equations describing the dipolar deformation—with the equations that describe the behavior of voids. The description of the continuous dependence was possible due to the definition of an adequate measure.

2. Basic Equations and Conditions

In our paper, we approach a thermoelastic body having a dipolar structure. We will use an anisotropic body, which is situated in a regular domain D, included in the physical space E^3, that is, the three-dimensional Euclidean space. Consider that the boundary of the domain is a piecewise smooth surface ∂D. The closure of D is usual denoted by $\bar{D}$, $\bar{D} = D \cup \partial D$. An orthonormal system of references is introduced, and then tensors and vectors have components with Latin subscripts over 1, 2, 3. Typical conventions for summation over repeated indices and for derivation operations are implied. So, a subscript preceded by a comma is for a partial derivative with regards to corresponding spatial coordinate; while a superposed dot is for a derivative with regards to time variable. All the functions we use are assumed to be sufficiently regular as necessary. Additionally, if there is no possibility of confusion, then the dependence of function with regards to its spatial or time variables will be omitted. The evolution of the body with a dipolar structure will be described with the help of the following specific variables:

$$u_i(x,t), \; \phi_{ij}(x,t), \; \theta(x,t), \; (x,t) \in D \times [0, t_0). \tag{1}$$

Here, we denoted by u_i the components of the displacement vector field, by ϕ_{ij} the components of the dipolar displacement tensor field, and by θ the absolute temperature.

Using the above variables $u_i(x,t)$, and $\phi_{ij}(x,t)$ we will introduce the components of the tensors of strain, namely, ε_{ij}, κ_{ij}, and χ_{ijk}, as follows:

$$2\varepsilon_{ij} = u_{j,i} + u_{i,j}, \; \kappa_{ij} = u_{j,i} - \phi_{ij}, \; \chi_{ijk} = \phi_{ij,k}. \tag{2}$$

All our considerations are made within a linear theory, therefore it is natural to consider that the Helmholtz's free energy is a quadratic form with regards to its independent constitutive variables. The Helmholtz's free energy in the reference configuration will be denoted by W. So, in accordance with the principle of conservation of energy, we develop in series the function W and we keep the terms only until the second order. Because the reference state was assumed to be free of loadings, we deduce that the Helmholtz's free energy per mass can be considered of the following form (see [30]):

$$\begin{aligned}W = & \tfrac{1}{2}A_{ijmn}\varepsilon_{ij}\varepsilon_{mn} + D_{ijmn}\varepsilon_{ij}\kappa_{mn} + F_{ijmnr}\varepsilon_{ij}\chi_{mnr} + \tfrac{1}{2}B_{ijmn}\kappa_{ij}\kappa_{mn} \\ & +G_{ijmnr}\kappa_{ij}\chi_{mnr} + \tfrac{1}{2}C_{ijkmnr}\chi_{ijk}\chi_{mnr} - a_{ij}\varepsilon_{ij}\theta - b_{ij}\kappa_{ij}\theta - c_{ijk}\chi_{ijk}\theta - \tfrac{1}{2}c\theta^2.\end{aligned} \tag{3}$$

We will use this form of free energy used in the entropy production inequality and deduce the motion equations. In addition, from the same inequality, the constitutive equations are obtained. These equations express the tensors of stress with the help of the tensors of deformation. We will denote the components of the stress measures by τ_{ij}, η_{ij}, and μ_{ijk}. In this way, the constitutive equations establish a connection between the tensors τ_{ij}, η_{ij}, μ_{ijk} and the tensors ε_{ij}, κ_{ij}, χ_{ijk}.

We will use a procedure similar to that used by Green and Rivlin in [6], so that considering the Helmholtz' free energy (3) we deduce the next constitutive equations

$$\begin{aligned}
\tau_{ij} &= \frac{\partial W}{\partial \varepsilon_{ij}} = A_{ijmn}\varepsilon_{mn} + D_{mnij}\kappa_{mn} + F_{mnrij}\chi_{mnr} - a_{ij}\theta, \\
\eta_{ij} &= \frac{\partial W}{\partial \kappa_{ij}} = D_{ijmn}\varepsilon_{mn} + B_{ijmn}\kappa_{mn} + G_{ijmnr}\chi_{mnr} - b_{ij}\theta, \\
\mu_{ijk} &= \frac{\partial W}{\partial \chi_{ijk}} = F_{ijkmn}\varepsilon_{mn} + G_{mnijk}\kappa_{mn} + C_{ijkmnr}\chi_{mnr} - c_{ijk}\theta, \\
\eta &= -\frac{\partial W}{\partial \theta} = a_{ij}\varepsilon_{ij} + b_{ij}\kappa_{ij} + c_{ijk}\chi_{ijk} + c\theta,
\end{aligned} \quad (4)$$

which are satisfied in $D \times [0, t_0)$. Here, we denoted by η the entropy per unit mass.

For the vector of heat flux, having the components q_i we have a classical constitutive relation, namely,

$$q_i = K_{ij}\theta_{,j}, \quad (5)$$

where K_{ij} is the thermal conductivity symmetric tensor.

Also, we can deduce the main equations that govern the thermoelasticity of bodies with a dipolar structure, namely (see [5,6]):

- the motion equations:

$$\begin{aligned}
\left(\tau_{ij} + \eta_{ij}\right)_{,j} + \rho f_i &= \rho \ddot{u}_i, \\
\mu_{ijk,i} + \eta_{jk} + \rho g_{jk} &= I_{kr}\ddot{\phi}_{jr};
\end{aligned} \quad (6)$$

- the equation of energy:

$$\rho T_0 \dot{\eta} = q_{i,i} + \rho r. \quad (7)$$

The signification of the notations that we introduced in preceding equations is as follows: ρ, the density of mass, which is a constant; I_{ij}, the symmetric tensor of microinertia; k, the intrinsic inertia; ε_{ij}, κ_{ij}, χ_{ijk}, the strain tensors; τ_{ij}, η_{ij}, μ_{ijk}, the stress tensors; f_i, the body forces; g_{jk}, the dipolar charges; A_{ijmn}, B_{ijmn}, ..., a_{ij}, the functions that describe the properties of the material in terms of elasticity. Suppose the following symmetry relations take place:

$$\begin{aligned}
A_{ijmn} &= A_{jimn} = A_{mnij}, \quad B_{ijmn} = B_{mnij}, \quad a_{ij} = a_{ji}, \\
C_{ijkmnr} &= C_{mnrijk}, \quad F_{ijkmn} = F_{ijknm}, \quad D_{ijmn} = D_{ijnm}.
\end{aligned} \quad (8)$$

Assuming that there are no supply terms and taking into account the constitutive Equations (4) and (5) and the kinematic Equation (2), Equations (6) and (7) become

$$\rho \ddot{u}_i = \left[\left(C_{ijmn} + G_{ijmn}\right) u_{n,m} + \left(G_{mnij} + B_{ijmn}\right)\left(u_{n,m} - \phi_{mn}\right) + \right.$$
$$\left. + \left(F_{mnrij} + D_{ijmnr}\right)\phi_{nr,m} - \left(a_{ij} + b_{ij}\right)\theta\right]_{,j}, \quad (9a)$$

$$I_{kr}\ddot{\phi}_{jr} = \left[F_{ijkmn}u_{n,m} + D_{mnijk}\left(u_{n,m} - \phi_{mn}\right) + A_{ijkmnr}\phi_{nr,m} - c_{ijk}\theta\right]_{,i} +$$
$$+ G_{jkmn}u_{m,n} + B_{jkmn}\left(u_{n,m} - \phi_{mn}\right) + D_{jkmnr}\phi_{nr,m} - b_{jk}\theta, \quad (9b)$$

$$K_{ij}\left(\theta_{,j}\right)_{,i} = -T_0 \left[a_{ij}\dot{u}_{i,j} + b_{ij}\left(\dot{u}_{j,i} - \dot{\phi}_{ij}\right) + c_{ijk}\dot{\phi}_{ij,k} + c\dot{\theta}\right]. \quad (9c)$$

From now, we will assume that the Equations (2), (4), and (9) will be satisfied on the interval $(-\infty, 0]$.

The outward unit normal to the surface ∂D has the components n_i. With the help of this normal we can define the surface traction's of components t_i, the surface couple of components μ_{jk}, and the flux of heat, q. All of this makes sense in every point of regularity of the boundary ∂D and has the following expressions

$$t_i = (\tau_{ij} + \eta_{ij}) n_j, \ \mu_{jk} = \mu_{ijk} n_i, \ q = q_i n_i. \tag{10}$$

In close relation to these surface tractions, we consider the following homogeneous boundary conditions:

$$\begin{aligned}
&u_i(x,t) = 0, \ (x,t) \in \partial D_u \times (-\infty, 0], \ t_i = 0, \ (x,t) \in \partial D_u^c \times (-\infty, 0], \\
&\phi_{ij}(x,t) = 0, \ (x,t) \in \partial D_\phi \times (-\infty, 0], \ m_{jk} = 0, \ (x,t) \in \partial D_\phi^c \times (-\infty, 0], \\
&\theta(x,t) = 0, \ (x,t) \in \partial D_\theta \times (-\infty, 0], \ q = 0, \ (x,t) \in \partial D_\theta^c \times (-\infty, 0],
\end{aligned} \tag{11}$$

where the surfaces ∂D_u, ∂D_ϕ, ∂D_θ, and its complements ∂D_u^c, ∂D_ϕ^c, ∂D_θ^c are subsurfaces of the border ∂D, which are subject to the following restrictions:

$$\begin{aligned}
&\partial \bar{D}_u \cup \partial D_u^c = \partial \bar{D}_\phi \cup \partial D_\phi^c = \partial \bar{D}_\theta \cup \partial D_\theta^c = \partial D, \\
&\partial D_u \cap \partial D_u^c = \partial D_\phi \cap \partial D_\phi^c = \partial D_\theta \cap \partial D_\theta^c = \varnothing.
\end{aligned}$$

We still have to add the final restrictions. So, on the closed domain $\bar{D}$ we have:

$$\begin{aligned}
&u_i(x,0) = u_i^0(x), \ \dot{u}_i(x,0) = u_i^1(x), \ \theta(x,0) = \theta^0(x), \\
&\phi_{ij}(x,0) = \phi_{ij}^0(x), \ \dot{\phi}_{ij}(x,0) = \phi_{ij}^1(x),
\end{aligned} \tag{12}$$

where $u_i^0(x)$, $u_i^1(x)$, $\phi_{ij}^0(x)$, $\phi_{ij}^1(x)$, and $\theta^0(x)$ are continuous prescribed functions in all points where they are defined. Additionally, these functions are assumed be compatible with conditions (11) on the appropriate subsets of ∂D.

Let us consider the internal energy density Ψ (see [30]), which has the following expression:

$$\begin{aligned}
\Psi = &\tfrac{1}{2} A_{ijmn} \varepsilon_{ij} \varepsilon_{mn} + D_{ijmn} \varepsilon_{ij} \kappa_{mn} + F_{ijmnr} \varepsilon_{ij} \chi_{mnr} + \\
&+ \tfrac{1}{2} B_{ijmn} \kappa_{ij} \kappa_{mn} + G_{ijmnr} \kappa_{ij} \chi_{mnr} + \tfrac{1}{2} C_{ijkmnr} \chi_{ijk} \chi_{mnr}.
\end{aligned} \tag{13}$$

$\mathcal{P}$ is denoted the so-called boundary-final value problem, which consists of Equation (9), the boundary restrictions (11), and the final data (12).

To obtain the results we have proposed, we will have to impose some conditions on the functions we are dealing with.

So, if $J_m(x)$ is the minimum eigenvalue of the inertia tensor $I_{ij}(x)$, then we need to assume that J_m and ρ are continuous functions and the constitutive coefficients are of class $C^1(D)$. We also assume that:

(a) $\rho(x) \geq a_1$, $J_m(x) \geq a_2$, $c(x) \geq c_0$, where a_1, a_2, c_0 are real positive constants;
(b) the tensor K_{ij} is positive definite;
(c) the internal energy density Ψ is a positive definite quadratic form.

Based on hypothesis (b), we deduce that there exist two positive numbers, K_m and K_M, so that

$$K_m \theta_{,i} \theta_{,j} \leq K_{ij} \theta_{,i} \theta_{,j} \leq K_M \theta_{,i} \theta_{,j}, \tag{14}$$

and, as a consequence of the hypothesis (c), we can find the positive constants M_1 and M_2 so that the next inequality is satisfied:

$$\frac{M_1}{2} \left(\varepsilon_{ij} \varepsilon_{ij} + \kappa_{ij} \kappa_{ij} + \chi_{ijk} \chi_{ijk} \right) \leq \Psi \leq \frac{M_2}{2} \left(\varepsilon_{ij} \varepsilon_{ij} + \kappa_{ij} \kappa_{ij} + \chi_{ijk} \chi_{ijk} \right). \tag{15}$$

These hypotheses are not considered as very restrictive, as they are commonly imposed in mechanics of continuous media.

It is not difficult to equate our boundary-final value problem $\mathcal{P}$ with a boundary-initial problem, denoted by $\mathcal{P}'$, by a convenient change of variables. In this regard, we set $h'(t') = h(t)$, for $t' = -t$. But, to simplify writing, we will give up the sign "prime" so that the $\mathcal{P}'$ problem will be defined by the following conditions and equations:

- the motion Equations (9a) and (9b), satisfied in $D \times [0, \infty)$;
- the energy equation:

$$K_{ij}\left(\theta_{,j}\right)_{,i} = T_0\left[a_{ij}\dot{u}_{i,j} + b_{ij}\left(\dot{u}_{j,i} - \dot{\phi}_{ij}\right) + c_{ijk}\dot{\phi}_{ij,k} + c\dot{\theta}\right], \text{ in } D \times [0, \infty); \tag{16}$$

- the geometric Equation (2), satisfied in $D \times [0, \infty)$;
- the constitutive Equation (4), satisfied in $D \times [0, \infty)$;
- the initial conditions (11), satisfied in $\bar{D}$;
- the boundary conditions:

$$\begin{aligned}
u_i(x,t) &= 0, \ (x,t) \in \partial D_u \times [0,\infty), \ t_i = 0, \ (x,t) \in \partial D_u^c \times [0,\infty), \\
\phi_{ij}(x,t) &= 0, \ (x,t) \in \partial D_\phi \times [0,\infty), \ m_{jk} = 0, \ (x,t) \in \partial D_\phi^c \times [0,\infty), \\
\theta(x,t) &= 0, \ (x,t) \in \partial D_\theta \times [0,\infty), \ q = 0, \ (x,t) \in \partial D_\theta^c \times [0,\infty).
\end{aligned} \tag{17}$$

3. Main Result

We first establish some integral identities regarding a solution $\mathbf{u} = (u_i, \phi_{ij}, \theta)$ of the mixed problem $\mathcal{P}'$. These will be useful in obtaining the important results of our study.

Proposition 1. *If the ordered array $\mathbf{u} = (u_i, \phi_{ij}, \theta)$ satisfies the mixed problem $\mathcal{P}'$, then the following equality takes place:*

$$\begin{aligned}
\int_B &\left[\frac{1}{2}\left(\rho\dot{u}_i(t)\dot{u}_i(t) + I_{jk}\dot{\phi}_{jm}(t)\dot{\phi}_{km}(t)\right) + \Psi(t) + \frac{1}{2}c\theta^2(t)\right] dV = \\
&= \int_B \left[\rho\dot{u}_i(0)\dot{u}_i(0) + I_{jk}\dot{\phi}_{jm}(0)\dot{\phi}_{km}(0) + \Psi(0) + \frac{1}{2}c\theta^2(0)\right] dV + \\
&+ \int_0^t \int_D \frac{1}{T_0} K_{ij}\theta_{,i}(\tau)\theta_{,j}(\tau) dV d\tau, \ \forall t \in [0,\infty).
\end{aligned} \tag{18}$$

Proof. Taking into account the kinematics compatibility relations (2) and the differential conditions of equilibrium (9a) and (9b), we obtain the following equality:

$$\begin{aligned}
\frac{1}{2}\frac{\partial}{\partial t}\left(\rho\dot{u}_i(t)\dot{u}_i(t) + I_{jk}\dot{\phi}_{jm}(t)\dot{\phi}_{km}(t)\right) &= \\
= \left[\left(\tau_{ij} + \eta_{ij}\right)\dot{u}_j + \mu_{ijk}\dot{\phi}_{jk}\right]_{,i} &- \left(\tau_{ij}\dot{\varepsilon}_{ij} + \eta_{ij}\dot{\kappa}_{ij} + \mu_{ijk}\dot{\chi}_{ijk}\right).
\end{aligned} \tag{19}$$

Taking into account the constitutive Equation (4), the symmetry relations (8), and the expression of the internal energy density Ψ from (13), the last parentheses in the right-hand side of (19) becomes

$$\left(\tau_{ij}\dot{\varepsilon}_{ij} + \eta_{ij}\dot{\kappa}_{ij} + \mu_{ijk}\dot{\chi}_{ijk}\right) = \frac{\partial}{\partial t}\left(\Psi + \frac{1}{2}c\theta^2\right) + \left(\frac{1}{T_0}q_j\theta\right)_{,j} - \frac{1}{T_0}K_{ij}\theta_{,i}\theta_{,i}. \tag{20}$$

We substitute Equation (20) into Equation (19), then the resulting equality is integrated on cylinder $[0, t] \times D$. If we use the theorem of divergence and consider the conditions to the limit (17), we are led to equality (18), such that the proof of the proposition is finished. □

In a similar way, one can demonstrate the identity that follows, as a complement to identity (18):

$$\int_D \left[\frac{1}{2}\left(\rho \dot{u}_i(t)\dot{u}_i(t) + I_{jk}\phi_{jm}(t)\phi_{km}(t)\right) + \Psi(t) - \frac{1}{2}c\theta^2(t)\right]dV =$$
$$= \int_D \left[\rho \dot{u}_i(0)\dot{u}_i(0) + I_{jk}\phi_{jm}(0)\phi_{km}(0) + \Psi(0) - \frac{1}{2}c\theta^2(0)\right]dV - \quad (21)$$
$$- \int_0^t \int_D \left\{\dot{u}_i(\tau)\left[(a_{ij} + b_{ji})\,\theta(\tau)\right]_{,j} + \phi_{ij}(\tau)\left[c_{ijk}\theta(\tau)\right]_{,k} -\right.$$
$$\left. -b_{ij}\phi_{ij}(\tau)\theta(\tau) + \frac{1}{T_0}K_{ij}\theta_{,i}(\tau)\theta_{,j}(\tau)\right\}dVd\tau, \;\forall t \in [0, \infty).$$

To simplify writing, we enter the notation

$$2F(x,y) = A_{ijmn}\varepsilon_{ij}(x)\varepsilon_{mn}(y) + D_{ijmn}\left[\varepsilon_{ij}(x)\kappa_{mn}(y) + \varepsilon_{ij}(y)\kappa_{mn}(x)\right] +$$
$$+ F_{ijmnr}\left[\varepsilon_{ij}(x)\chi_{mnr}(y) + \varepsilon_{ij}(y)\chi_{mnr}(x)\right] + B_{ijmn}\kappa_{ij}(x)\kappa_{mn}(y) + \quad (22)$$
$$+ G_{ijmnr}\left[\kappa_{ij}(x)\chi_{mnr}(y) + \kappa_{ij}(y)\chi_{mnr}(x)\right] + C_{ijkmnr}\chi_{ijk}(x)\chi_{mnr}(y).$$

By using the symmetry relations (8), from (22) we deduce

$$F(x,y) = F(y,x). \quad (23)$$

By direct substitution in (22) and taking into account Equation (13), we also obtain

$$F(\tau,\tau) = \Psi(\tau). \quad (24)$$

Now, we can prove a result similar to (18), but in the case of homogeneous initial conditions.

Proposition 2. *Consider a solution (u_i, ϕ_{ij}, θ) of the problem backward in time $\mathcal{P}'$, which corresponds to null initial data. Then, the next identity takes place*

$$\int_B \left[\frac{1}{2}\left(\rho\dot{u}_i(t)\dot{u}_i(t) + I_{jk}\phi_{jm}(t)\phi_{km}(t)\right) - \frac{1}{2}c\theta^2(t)\right]dV =$$
$$= \int_B \left[\frac{1}{2}A_{ijmn}\varepsilon_{ij}(t)\varepsilon_{mn}(t) + D_{ijmn}\varepsilon_{ij}(t)\kappa_{mn}(t) +\right. \quad (25)$$
$$+ F_{ijmnr}\varepsilon_{ij}(t)\chi_{mnr}(t) + \frac{1}{2}B_{ijmn}\kappa_{ij}(t)\kappa_{mn}(t) +$$
$$\left. + G_{ijmnr}\kappa_{ij}(t)\chi_{mnr}(t) + \frac{1}{2}C_{ijkmnr}\chi_{ijk}(t)\chi_{mnr}(t)\right]dV,$$

for all $t \in [0, \infty)$.

Proof. By direct calculations, for a fixed $t \in (0, \infty)$, we get the identity:

$$\frac{\partial}{\partial t}\left(\rho\dot{u}_i(\tau)\dot{u}_i(2t-\tau) + I_{jk}\phi_{jm}(\tau)\phi_{km}(2t-\tau) - c\theta(\tau)\theta(2t-\tau)\right) =$$
$$= \rho\ddot{u}_i(\tau)\dot{u}_i(2t-\tau) + I_{jk}\ddot{\phi}_{jm}(\tau)\phi_{km}(2t-\tau) + c\theta(\tau)\dot{\theta}(2t-\tau) - \quad (26)$$
$$-\rho\dot{u}_i(\tau)\ddot{u}_i(2t-\tau) + I_{jk}\phi_{jm}(\tau)\ddot{\phi}_{km}(2t-\tau) - c\dot{\theta}(\tau)\theta(2t-\tau).$$

Taking into account the kinematic Equation (2), the constitutive Equation (4), the motion Equation (9), and the symmetry relations (8), we are led to the equality

$$\frac{\partial}{\partial t}\left(\rho\dot{u}_i(\tau)\dot{u}_i(2t-\tau) + I_{jk}\dot{\phi}_{jm}(\tau)\dot{\phi}_{km}(2t-\tau) - c\theta(\tau)\theta(2t-\tau)\right) =$$
$$= \left[(\tau_{ij} + \eta_{ij})(\tau)\dot{u}_i(2t-\tau) - (\tau_{ij} + \eta_{ij})(2t-\tau)\dot{u}_j(\tau) + \right.$$
$$+ \mu_{ijk}\dot{\phi}_{jk}(\tau)\dot{\phi}_{jk}(2t-\tau) - \mu_{ijk}\dot{\phi}_{jk}(2t-\tau)\dot{\phi}_{jk}(\tau) -$$
$$\left. - \frac{1}{T_0}\theta(\tau)q_i(2t-\tau) + \frac{1}{T_0}\theta(2t-\tau)q_i(\tau)\right]_{,i} + F(\tau, 2t-\tau), \quad (27)$$

the function $F(.,.)$ being defined in (22).

We just need to integrate this equality into $[0,t] \times D$, to keep in mind that the initial data are null and to use the definition (13), so we get the desired equality (25) and the proof of proposition is complete. □

The following two propositions are also useful in establishing the main outcomes of our study.

Proposition 3. *Consider that the ordered array* $\mathbf{u} = (u_i, \phi_{ij}, \theta)$ *satisfies the mixed problem* $\mathcal{P}'$. *Then, the following equality takes place*

$$\int_D \left[\rho u_i(t)\dot{u}_i(t) + I_{jk}\phi_{jm}(t)\dot{\phi}_{km}(t) - \frac{1}{2T_0}K_{ij}\left(\int_0^t \theta(\tau)d\tau\right)_{,i}\left(\int_0^t \theta(\tau)d\tau\right)_{,j}\right]dV =$$
$$= \int_D \left[\rho u_i(0)\dot{u}_i(0) + I_{jk}\phi_{jm}(0)\dot{\phi}_{km}(0)\right]dV + \int_0^t \int_D \rho\eta(0)\theta(\tau)dVd\tau + \quad (28)$$
$$+ \int_0^t \int_D \left[\rho\dot{u}_i(t)\dot{u}_i(t) + I_{jk}\dot{\phi}_{jm}(t)\dot{\phi}_{km}(t) - 2\Psi(\tau) - c\theta^2(\tau)\right]dVd\tau,$$

for all $t \in [0,\infty)$.

Proof. We will consider the constitutive relations (4), the geometric Equation (2), and the motion Equations (9a) and (9b), we deduce

$$\frac{\partial}{\partial t}\left(\rho u_i(t)\dot{u}_i(t) + I_{jk}\phi_{jm}(t)\dot{\phi}_{km}(t)\right) =$$
$$= \rho\dot{u}_i(t)\dot{u}_i(t) + I_{jk}\dot{\phi}_{jm}(t)\dot{\phi}_{km}(t) +$$
$$+ \left[(\tau_{ij}(t) + \eta_{ij}(t))u_j(t) + \mu_{ijk}(t)\phi_{jk}(t)\right]_{,i} - \quad (29)$$
$$- \left[(\tau_{ij}(t) + \eta_{ij}(t))\varepsilon_{ij}(t) + \eta_{ij}(t)\kappa_{ij}(t) + \mu_{ijk}(t)\chi_{ijk}(t)\right].$$

Taking into account definition (13), the last parentheses from the right-hand side of (29) can be restated in the following form:

$$\tau_{ij}(t)\varepsilon_{ij}(t) + \eta_{ij}(t)\kappa_{ij}(t) + \mu_{ijk}(t)\chi_{ijk}(t) =$$
$$= \left(\frac{1}{2}c\theta^2(t) + \Psi(t)\right) + \left(\frac{1}{T_0}\theta(t)\int_0^t q_i(\tau)d\tau\right)_{,i} - \quad (30)$$
$$- \frac{1}{T_0}K_{ij}\left(\int_0^t \theta(\tau)d\tau\right)_{,i}\left(\int_0^t \theta(\tau)d\tau\right)_{,i} - \rho\eta(0)\theta(t).$$

Here, the expression of the entropy η is obtained by integrating its equation of evolution (19), with regards to time variable.

We integrate the equality (30) on $[0, t] \times D$, and take into account the data on the border (17) and the theorem of divergence. As such, we arrive at the proposed equality (29), and the demonstration of proposition is over. □

Proposition 4. *Let us consider a solution* (u_i, ϕ_{ij}, θ) *of the mixed problem* $\mathcal{P}'$. *Then, the following equality takes place*

$$2\int_D \left[\rho u_i(t)\dot{u}_i(t) + I_{jk}\phi_{jm}(t)\dot{\phi}_{km}(t) - \frac{1}{2T_0}K_{ij}\left(\int_0^t \theta(\tau)d\tau\right)_{,i}\left(\int_0^t \theta(\tau)d\tau\right)_{,j}\right]dV$$
$$= \int_D \left[\rho \dot{u}_i(0)u_i(2t) + I_{jk}\dot{\phi}_{jm}(0)\phi_{km}(2t)\right]dV + \qquad (31)$$
$$+ \int_D \left[\rho u_i(0)\dot{u}_i(2t) + I_{jk}\phi_{jm}(0)\dot{\phi}_{km}(2t)\right]dV -$$
$$- \int_0^t \int_D \rho\eta(0)\left[\theta(t+\tau) - \theta(t-\tau)\right]dVd\tau,$$

for all $t \in [0, \infty)$.

Proof. We will take into account the geometric relations (2) and the motion Equations (9a) and (9b), we deduce the following identity

$$\frac{\partial}{\partial t}\left[\rho\left(\dot{u}_i(t+\tau)u_i(t-\tau) + u_i(t+\tau)\dot{u}_i(t-\tau)\right)\right] +$$
$$+ \frac{\partial}{\partial t}\left[I_{jk}\left(\dot{\phi}_{jm}(t+\tau)\phi_{km}(t-\tau) + \phi_{jm}(t+\tau)\dot{\phi}_{km}(t-\tau)\right)\right] =$$
$$= \left[(\tau_{ij} + \eta_{ij})(t+\tau)u_j(t-\tau) - (\tau_{ij} + \eta_{ij})(t-\tau)u_j(t+\tau) + \qquad (32) \right.$$
$$\left. + \mu_{ijk}(t+\tau)\phi_{jk}(t-\tau) - \mu_{ijk}(t-\tau)\phi_{jk}(t+\tau)\right]_{,i} -$$
$$- \left[(\tau_{ij} + \eta_{ij})(t+\tau)\varepsilon_{ij}(t-\tau) - (\tau_{ij} + \eta_{ij})(t-\tau)\varepsilon_{ij}(t+\tau)\right] -$$
$$- \left[\mu_{ijk}(t+\tau)\chi_{ijk}(t-\tau) - \mu_{ijk}(t-\tau)\chi_{jk}(t+\tau)\right].$$

On the other hand, by using the symmetry Equation (8) and the constitutive relations (4), the last two brackets receive the following form:

$$\left[(\tau_{ij} + \eta_{ij})(t+\tau)\varepsilon_{ij}(t-\tau) - (\tau_{ij} + \eta_{ij})(t-\tau)\varepsilon_{ij}(t+\tau)\right] +$$
$$+ \left[\mu_{ijk}(t+\tau)\chi_{ijk}(t-\tau) - \mu_{ijk}(t-\tau)\chi_{jk}(t+\tau)\right] =$$
$$= \frac{1}{T_0}\left[\theta(t+\tau)\int_0^{t-\tau}q_i(s)ds - \theta(t-\tau)\int_0^{t+\tau}q_i(s)ds\right]_{,i} - \qquad (33)$$
$$- \frac{1}{T_0}K_{ij}\left[\left(\int_0^{t+\tau}\theta(s)ds\right)_{,i}\left(\int_0^{t-\tau}\theta(s)ds\right)_{,j} - \left(\int_0^{t+\tau}\theta(s)ds\right)_{,i}\left(\int_0^{t-\tau}\theta(s)ds\right)_{,j}\right] -$$
$$- \rho\eta(0)\left[\theta(t+\tau) - \theta(t-\tau)\right].$$

Here, the expression of the entropy η is obtained by integrating its equation of evolution (19), with regards to time variable.

Let us integrate the equality (33) on $[0, t] \times D$, and take into account the border data (17) and the theorem of divergence. As such, we arrive at the proposed equality (31) and the demonstration of the proposition is over. □

If we combine the results from Equations (25) and (31), then we obtain a new useful equality

$$
\begin{aligned}
2\int_B \left[\left(\rho\dot{u}_i(t)\dot{u}_i(t) + I_{jk}\dot{\phi}_{jm}(t)\dot{\phi}_{km}(t)\right) - c\theta^2(t)\right]dV = \\
= -2\int_D \left[\rho u_i(0)\dot{u}_i(0) + I_{jk}\phi_{jm}(0)\dot{\phi}_{km}(0)\right]dV + \\
+ \int_D \left[\rho\dot{u}_i(0)u_i(2t) + I_{jk}\dot{\phi}_{jm}(0)\phi_{km}(2t)\right]dV + \\
+ \int_D \left[\rho u_i(0)\dot{u}_i(2t) + I_{jk}\phi_{jm}(0)\dot{\phi}_{km}(2t)\right]dV + \\
- \int_0^t \int_D \rho\eta(0)\left[2\theta(\tau) + \theta(t+\tau) - \theta(t-\tau)\right]dVd\tau,
\end{aligned}
\quad (34)
$$

for all $t \in [0, \infty)$.

Based on the previously demonstrated integral identities, we are able to address the main results of our study. First, we established a result of uniqueness for the solution of the backward in time problem. As a consequence, we approach the question of localization of the solutions of the backward in time problem.

Theorem 1. *At most, an ordered array* $\mathbf{u} = (u_i, \phi_{ij}, \theta)$ *can satisfy the equations and conditions of the backward problem* $\mathcal{P}'$.

Proof. As usual, we will assume, by absurdum, that the problem would admit two solutions. The difference of the two solutions is also a solution, because the problem $\mathcal{P}'$ is a linear one. Suffice it to show that this difference is null. For this we have to show that the problem $\mathcal{P}'$, for which the boundary and initial data are null, admits the null solution. It is clear that for the difference of two solutions, the boundary and initial conditions become homogeneous.

To simplify writing, we introduce the function M as a measure of the solution, defined by

$$
M(t) = \int_D \left[\frac{\varepsilon}{2}\left(\rho\dot{u}_i(t)\dot{u}_i(t) + I_{jk}\dot{\phi}_{jm}(t)\dot{\phi}_{km}(t)\right) + (\varepsilon+2)\Psi(t) + \frac{\varepsilon}{2}c\theta^2(t)\right]. \quad (35)
$$

Here, ε is a small positive number.

Based on assumptions (a), (b), and (c), it can be deduced that the function M is positive.

Because the initial data are zero, the identity (18) received the simpler form:

$$
\int_B \left[\frac{1}{2}\left(\rho\dot{u}_i(t)\dot{u}_i(t) + I_{jk}\dot{\phi}_{jm}(t)\dot{\phi}_{km}(t)\right) + \Psi(t) + \frac{1}{2}c\theta^2(t)\right]dV = \\
= \int_0^t \int_D \frac{1}{T_0}K_{ij}\theta_{,i}(\tau)\theta_{,j}(\tau)dVd\tau, \ \forall t \in [0,\infty).
\quad (36)
$$

Analogously, the identity (21) becomes

$$
\int_B \left[\frac{1}{2}\left(\rho\dot{u}_i(t)\dot{u}_i(t) + I_{jk}\dot{\phi}_{jm}(t)\dot{\phi}_{km}(t)\right) + \Psi(t) - \frac{1}{2}c\theta^2(t)\right]dV = \\
= -\int_0^t \int_D \left\{\dot{u}_i(\tau)\left[(a_{ij}+b_{ji})\theta(\tau)\right]_{,j} + \dot{\phi}_{ij}(\tau)\left[c_{ijk}\theta(\tau)\right]_{,k} - \right. \\
\left. -b_{ij}\dot{\phi}_{ij}(\tau)\theta(\tau) + \frac{1}{T_0}K_{ij}\theta_{,i}(\tau)\theta_{,j}(\tau)\right\}dVd\tau, \ \forall t \in [0,\infty).
\quad (37)
$$

If we use Equations (36) and (37), then the function M from (35) becomes

$$
M(t) = -\int_0^t \int_D \left\{2\dot{u}_i(\tau)\left[(a_{ij}+b_{ji})\theta(\tau)\right]_{,j} + \dot{\phi}_{ij}(\tau)\left[c_{ijk}\theta(\tau)\right]_{,k} - \right. \\
\left. -b_{ij}\dot{\phi}_{ij}(\tau)\theta(\tau) + \frac{1-\varepsilon}{T_0}K_{ij}\theta_{,i}(\tau)\theta_{,j}(\tau)\right\}dVd\tau, \ \forall t \in [0,\infty).
\quad (38)
$$

By direct derivation with regards to the variable t in (37), we are led to the following equality:

$$\frac{dM(t)}{dt} = -2 \int_D \left\{ \dot{u}_i(\tau) \left[(a_{ij} + b_{ji}) \theta(\tau) \right]_{,j} + \dot{\phi}_{ij}(\tau) \left[c_{ijk} \theta(\tau) \right]_{,k} - b_{ij} \dot{\phi}_{ij}(\tau) \theta(\tau) + \frac{1-\varepsilon}{2T_0} K_{ij} \theta_{,i}(\tau) \theta_{,j}(\tau) \right\} dV, \quad \forall t \in [0, \infty). \tag{39}$$

With the help of Schwarz' inequality and by using the arithmetic–geometric mean inequality, from (39) we can deduce the following inequality:

$$\frac{dM(t)}{dt} \leq C_1 \int_D \left[\rho \dot{u}_i(t) \dot{u}_i(t) + I_{jk} \dot{\phi}_{jm}(t) \dot{\phi}_{km}(t) + c\theta^2(t) \right] dV + \frac{\delta - 1 + \varepsilon}{T_0} \int_D K_{ij} \theta_{,i}(\tau) \theta_{,j}(\tau) dV, \quad \forall t \in [0, \infty). \tag{40}$$

Now, we take into account that the internal energy density Ψ is a positive definite quadratic form, according to the hypothesis (c), using the definition (35) of the function M and choose $\delta \leq 1 - \varepsilon$. Then, from (40) we obtain the inequality:

$$\frac{dM(t)}{dt} \leq \frac{C_1}{\varepsilon} \int_D \varepsilon \left[\rho \dot{u}_i(t) \dot{u}_i(t) + I_{jk} \dot{\phi}_{jm}(t) \dot{\phi}_{km}(t) + c\theta^2(t) \right] dV \leq \frac{C_1}{\varepsilon} M, \tag{41}$$

and it is then clear that a solution to this inequality meets the next inequality:

$$0 \leq M(t) \leq M(0) e^{C_1/\varepsilon}. \tag{42}$$

We recall that for the difference of the two supposed solutions, the initial conditions are homogeneous, then we have $M(0) = 0$, so from (42) we get

$$M(t) = 0, \quad \forall t \in [0, \infty)$$

and this together with the assumptions leads to the conclusion that our problem has only the solution

$$u_i(t) = 0, \quad \phi_{ij}(t) = 0, \quad \theta(t) = 0, \quad \forall t \in [0, \infty),$$

and the proof of Theorem 1 is concluded. □

Our final result is dedicated to the partition of various energies associated with the solution of the backward in time problem $\mathcal{P}^*$. We recall that this problem consists of the equations of motion (9), the constitutive relations (4), the kinematic Equation (2), the initial data (12), and the boundary restrictions in their homogeneous form (11).

First, using the procedure outlined at the end of Section 2, we transform the boundary-final value problem $\mathcal{P}^*$ into the boundary-initial value problem $\mathcal{P}'$. In this way, in what follows, we will be able to make the considerations only on the problem $\mathcal{P}'$. Let us denote by $\mathcal{T}$ the set of of those thermoelastodynamic processes defined in the cylinder $(-\infty, 0] \times D$, which satisfy the restriction

$$\int_D \frac{1}{T_0} K_{ij} \theta_{,i}(\tau) \theta_{,j}(\tau) dV \leq C_1, \tag{43}$$

for all $t \in [0, \infty)$. Here, C_1 is a given positive constant.

We will introduce the known Cesaro means necessary to evaluate the various types of energies that can be attached to a solution to the problem $\mathcal{P}'$. So, if (u_i, ϕ_{ij}, θ) is a solution of the mixed problem, then the Cesaro means are:

$$K(t) = \frac{1}{t}\int_0^t \int_D \left[\rho \dot{u}_i(\tau)\dot{u}_i(\tau) + I_{jk}\dot{\phi}_{jm}(\tau)\dot{\phi}_{jm}(\tau)\right] dV d\tau;$$

$$S(t) = \frac{1}{t}\int_0^t\int_D \left[\frac{1}{2}A_{ijmn}\varepsilon_{ij}(\tau)\varepsilon_{mn}(\tau) + D_{ijmn}\varepsilon_{ij}(\tau)\kappa_{mn}(\tau) + F_{ijmnr}\varepsilon_{ij}(\tau)\chi_{mnr}(\tau) + \right.$$
$$\left. +\frac{1}{2}B_{ijmn}\kappa_{ij}(\tau)\kappa_{mn}(\tau) + G_{ijmnr}\kappa_{ij}(\tau)\chi_{mnr}(\tau) + \frac{1}{2}C_{ijkmnr}\chi_{ijk}(\tau)\chi_{mnr}(\tau)\right]dVd\tau; \qquad (44)$$

$$R(t) = \frac{1}{t}\int_0^t \int_D \frac{1}{2}c\,\theta^2(\tau) dV d\tau;$$

$$D(t) = \frac{1}{t}\int_0^t \int_0^\tau \int_D \frac{1}{T_0}K_{ij}\theta_{,i}(s)\theta_{,j}(s)dVds\,d\tau.$$

In the particular case when meas(∂D_u)= 0 and meas(∂D_ϕ)= 0, it can be determined a rigid displacement, a rigid dipolar displacement, and a null temperature, which satisfy the equations of motion (9), the constitutive relations (4), the kinematic relations (2), and verify the homogeneous boundary conditions (17). In this case, the initial data can be decomposed as follows:

$$u_i^0 = u_i' + V_i^0, \; \dot{u}_i^0 = \dot{u}_i' + \dot{V}_i^0,$$
$$\phi_{ij}^0 = \phi_{ij}' + \Psi_{ij}^0, \; \dot{\phi}_{ij}^0 = \dot{\phi}_{ij}' + \dot{\Psi}_{ij}^0. \qquad (45)$$

The rigid displacements $u_i', \dot{u}_i', \phi_{ij}', \dot{\phi}_{ij}'$ can be computed using the functions

$$W_i(\omega) = \int_D \rho \omega_i Dv, \; W_{jk}(\psi) = \int_D I_{mn}\psi_{mj}\psi_{nk}dV,$$

such that we have

$$W_i(V_i^0) = 0, \; W_i(\dot{V}_i^0) = 0,$$
$$W_{jk}(\Psi_{ij}^0) = 0, \; W_{jk}(\dot{\Psi}_{ij}^0) = 0. \qquad (46)$$

Together with known notation $C^1(D)$, we will use the notation $W_n(D)$ for a Sobolev space defined on the domain D, and $\mathbf{W}_n(D) = [W_n(D)]^3$.

Other new notations:

$$\mathcal{C}^1(D) = \left\{(u_i, \phi_{ij}) \in C^1(D)^3 \times C^1(D)^9 : u_i = 0 \text{ on } \partial D_u, \; \phi_{ij} = 0 \text{ on } \partial D_\phi; \right.$$
$$\left. \text{if meas}(\partial D_u) = 0 \text{ and meas}(\partial D_\phi) = 0, \text{ then } W_i(u_i) = 0, \; W_{jk}(\phi_{ij}^0) = 0\right\};$$

$$\check{\mathcal{C}}^1(D) = \left\{\theta \in C^1(D) : \theta = 0 \text{ on } \partial D_\theta\right\};$$

$\mathcal{W}_1(D)$ the completion of $\mathcal{C}^1(D)$;

$\check{\mathcal{W}}_1(D)$ the completion of $\check{\mathcal{C}}^1(D)$,

the completion is by means of the original norm of the respective Sobolev space. Based on definition (13), and taking into account the hypothesis (15), we can obtain the following inequality (see [34]):

$$\int_D \Psi(\mathbf{u})dV \geq \frac{M_1}{2}\int_D \left[\rho u_i u_i + I_{jk}\phi_{jm}\phi_{km}\right]dV, \qquad (47)$$

for any $\mathbf{u} = (u_i, \phi_{ij}) \in \mathcal{W}_1(D)$. $M_1 > 0$ is a convenient chosen constant.

On the other hand, if we take into account the hypothesis (14), we can obtain the next inequality of Poincaré type:

$$\int_D K_{ij}\theta_{,i}\theta_{,j}dV \geq M_2 \int_D \theta^2 dV, \qquad (48)$$

for any $\theta \in \tilde{W}_1(D)$. $M_2 > 0$ is a convenient chosen constant.

Let us consider a solution (u_i, ϕ_{ij}, θ) to the problem $\mathcal{P}'$ in the particular case when meas$(\partial D_u) = 0$ and meas$(\partial D_\phi) = 0$. We can represent this solution in the following form:

$$u_i(t,x) = \bar{u}_i(t,x) + u'_i(t,x) + t\dot{u}'_i(t,x), \; \theta(t,x) = \bar{\theta}(t,x),$$
$$\phi_{ij}(t,x) = \bar{\psi}_{ij}(t,x) + \phi'_{ij}(t,x) + t\dot{\phi}'_{ij}(t,x), \; (t,x) \in [0,\infty) \times D, \qquad (49)$$

in which $(\bar{u}_i, \bar{\psi}_{ij}, \bar{\theta}) \in W_1(D) \times \tilde{W}_1(D)$. In addition, the ordered array $(\bar{u}_i, \bar{\psi}_{ij}, \bar{\theta})$ satisfies the problem $\mathcal{P}'$, which corresponds to the following initial data:

$$\bar{u}_i(0,x) = V_i(x), \; \dot{\bar{u}}_i(0,x) = \dot{V}_i(x), \; \bar{\theta}(t,x) = \vartheta(x),$$
$$\bar{\phi}_{ij}(0,x) = \Psi_{ij}(x), \; \dot{\bar{\phi}}_{ij}(0,x) = \dot{\Psi}_{ij}(x), \; \forall x \in D.$$

We now have everything prepared to address the problem of the equipartition of the various types of energies associated with the solution to the problem $\mathcal{P}'$.

Theorem 2. *Consider a solution (u_i, ϕ_{ij}, θ) of the backward in time problem $\mathcal{P}'$. If the initial data satisfy the following conditions:*

$$\mathbf{u} = (u_i) \in W_1(D), \; \dot{\mathbf{u}} = (\dot{u}_i) \in W_0(D),$$
$$\boldsymbol{\phi} = (\phi_{ij}) \in W_1(D), \; \dot{\boldsymbol{\phi}} = (\dot{\phi}_{ij}) \in W_0(D), \; \theta \in W_0(D),$$

then the following three statements are true:

i) The thermal component of energy, R, vanishes as $t \to \infty$:

$$\lim_{t \to \infty} R(t) = 0; \qquad (50)$$

ii) if meas$(\partial D_u) = 0$ and meas$(\partial D_\phi) = 0$, then we have

$$\lim_{t \to \infty} K(t) = \lim_{t \to \infty} S(t) + \frac{1}{2}\int_D \left[\rho \dot{u}'_i \dot{u}'_i + I_{jk}\dot{\phi}'_{jm}\dot{\phi}'_{km}\right] dV, \qquad (51)$$

$$\lim_{t \to \infty} D(t) = 2\lim_{t \to \infty} K(t) - \frac{1}{2}\int_D \left[\rho \dot{u}'_i \dot{u}'_i + I_{jk}\dot{\phi}'_{jm}\dot{\phi}'_{km}\right] dV - E(0) =$$
$$= 2\lim_{t \to \infty} S(t) + \frac{1}{2}\int_D \left[\rho \dot{u}'_i \dot{u}'_i + I_{jk}\dot{\phi}'_{jm}\dot{\phi}'_{km}\right] dV - E(0); \qquad (52)$$

iii) if meas$(\partial D_u) \neq 0$ or meas$(\partial D_\phi) \neq 0$, then we have

$$\lim_{t \to \infty} K(t) = \lim_{t \to \infty} S(t),$$
$$\lim_{t \to \infty} D(t) = 2\lim_{t \to \infty} K(t) - E(0) = \qquad (53)$$
$$= 2\lim_{t \to \infty} S(t) - E(0),$$

where, to simplify writing in (52) and (53), we used the notation:

$$E(t) = \int_D \left[\frac{1}{2}\left(\rho \dot{u}_i(t)\dot{u}_i(t) + I_{jk}\dot{\phi}_{jm}(t)\dot{\phi}_{km}(t)\right) + \Psi(t) + \frac{1}{2}c\theta^2(t)\right] dV, \qquad (54)$$

the internal energy density Ψ being defined in (13).

Proof. Taking into account the relations (18), (54), and (44) we immediately deduce the following equality:

$$K(t) + S(t) + R(t) = D(t) + E(0), \quad \forall t \in (0, \infty). \tag{55}$$

Now, we consider the restrictions (43) and (48), the definitions (44), and the notation (54) in order to deduce the following estimation:

$$R(t) \leq \frac{T_0 C_1}{2M_2 t} \max_{x \in D}\{c(x)\}, \quad \forall t \in (0, \infty). \tag{56}$$

Clearly, since $c(x)$ is a continuous function, we deduce that $\max_{x \in D}\{c(x)\}$ is bounded, such that after we pass to the limit in (56) for $t \to \infty$, we obtain the result (50).

If we consider the identity (34) and use the notations (44), we are led to the the following identity:

$$\begin{aligned}
K(t) - S(t) - R(t) = \\
= -\frac{1}{2t} \int_D \left[\rho u_i(0)\dot{u}_i(0) + I_{jk}\phi_{jm}(0)\dot{\phi}_{km}(0)\right] dV + \\
+ \frac{1}{4t} \int_D \left[\rho \dot{u}_i(0) u_i(2t) + I_{jk}\dot{\phi}_{jm}(0)\phi_{km}(2t)\right] dV + \\
+ \frac{1}{4t} \int_D \left[\rho u_i(0)\dot{u}_i(2t) + I_{jk}\phi_{jm}(0)\dot{\phi}_{km}(2t)\right] dV + \\
- \frac{1}{4t} \int_0^t \int_D \rho\eta(0) \left[2\theta(\tau) + \theta(t+\tau) - \theta(t-\tau)\right] dV d\tau,
\end{aligned} \tag{57}$$

for all $t \in [0, \infty)$.

On the other hand, taking into account the hypothesis (15), the inequalities (43) and (48), and the identity (18) we obtain the estimates

$$\int_D \rho \dot{u}_i(t)\dot{u}_i(t) dV \leq 2(C_1 + E(0)), \quad \int_D I_{jk}\dot{\phi}_{jm}(t)\dot{\phi}_{km}(t) dV \leq 2(C_1 + E(0)),$$

$$c_0 \int_D \rho\theta^2(t) dV \leq 2(C_1 + E(0)), \quad \int_D 2\Psi(t) dV \leq 2(C_1 + E(0)), \tag{58}$$

where c_0 is from hypothesis (a), C_1 is from (43), and E from (54).

In (57) we use the Schwarz's inequality and taking into account the estimates (58), we pass the limit—for $t \to \infty$—so we get the equality

$$\lim_{t \to \infty} K(t) = \lim_{t \to \infty} S(t) + \lim_{t \to \infty} \frac{1}{4t} \int_D \left[\rho \dot{u}_i(0) u_i(2t) + I_{jk}\dot{\phi}_{jm}(0)\phi_{km}(2t)\right] dV. \tag{59}$$

In this way, it will be easy to demonstrate the relation (53), if we show that the integral from the right-hand side of the identity (59) is bounded.

For this aim, we will use the fact that meas$(\partial D_u) \neq 0$ or meas$(\partial D_\phi) \neq 0$ and $(u_i, \psi_{ij}) \in \mathcal{W}_1(D)$. Furthermore, we consider the relations (47), (54), and (18) in order to get the following estimates:

$$M_1 \int_D \rho u_i(t) u_i(t) dV \leq \int_D 2\Psi(t) dV \leq 2(C_1 + E(0)),$$

$$M_1 \int_D I_{jk}\phi_{jm}(t)\phi_{km}(t) dV \leq \int_D 2\Psi(t) dV \leq 2(C_1 + E(0)),$$

such that it is easy, after we apply the Schwarz's inequality, to deduce

$$\lim_{t\to\infty} \frac{1}{4t} \int_D \left[\rho \dot{u}_i(0) u_i(2t) + I_{jk} \dot{\phi}_{jm}(0) \phi_{km}(2t) \right] dV = 0. \tag{60}$$

From (59) and (60) we obtain the first result from (53), and after we consider the equality (55), the second equality from (53) is proven.

Finally, we will prove the equalities (51) and (52). In this regard, we take into account that meas(∂D_u)= 0 and meas(∂D_ϕ)$\neq$ 0 such that with the help of the decompositions (45) and (49), and conditions (46), we are led to the equality

$$\begin{aligned}
\frac{1}{4t} \int_D \left[\rho \dot{u}_i(0) u_i(2t) + I_{jk} \dot{\phi}_{jm}(0) \phi_{km}(2t) \right] dV = \\
= \frac{1}{4t} \int_D \rho \dot{u}'_i u'_i(2t) dV + \frac{1}{4t} \int_D \rho \left(\dot{u}'_i + \dot{V}^0_i \right) u_i(2t) dV + \\
+ \frac{1}{2} \int_D \rho \dot{u}'_i \dot{u}'_i dV + \frac{1}{4t} \int_D I_{jk} \dot{\phi}'_{jm} \phi'_{km} dV + \\
+ \frac{1}{4t} \int_D I_{jk} \left(\dot{\phi}'_{jm} + \Psi^0_{jm} \right) \psi_{jk}(2t) dV + \frac{1}{2} \int_D I_{jk} \dot{\phi}'_{jm} \phi'_{km} dV.
\end{aligned} \tag{61}$$

Let us observe that relations (47), (54), and (18) involve the estimates

$$M_1 \int_D \rho u_i(t) u_i(t) dV \leq 2(C_1 + E(0)),$$

$$M_1 \int_D I_{jk} \psi_{jm}(t) \psi_{km}(t) dV \leq 2(C_1 + E(0)),$$

such that from (61) we are led to the equality

$$\begin{aligned}
\lim_{t\to\infty} \frac{1}{4t} \int_D \left[\rho \dot{u}_i(0) u_i(2t) + I_{jk} \dot{\phi}_{jm}(0) \phi_{km}(2t) \right] dV = \\
= \frac{1}{2} \int_D \rho \dot{u}'_i \dot{u}'_i dV + \frac{1}{2} \int_D I_{jk} \dot{\phi}'_{jm} \phi'_{km} dV.
\end{aligned} \tag{62}$$

Now, we substitute (62) into Equation (59) and then obtain the relation (51). Lastly, we consider the relations (50), (51), and (55) in order to obtain (52). With this, the proof of the theorem is completed. □

4. Conclusions

We first formulate the mixed backward in time problem in the context of thermoelasticity for dipolar materials. To prove the consistency of this mixed problem, our first main result is regarding the uniqueness of the solution for this problem. This is obtained based on some auxiliary results, namely, four integral identities. The second main result is regarding the temporal behavior of our thermoelastic body with dipolar structure. This behavior is studied by means of some relations on partition of various parts of the energy associated to the solution of the problem. After we introduce the Cesaro means for all parts of the total energy, we can evaluate the asymptotic partition of these parts. We have to say that the kinetic energy and potential energy become asymptotically equal when the variable time tends to infinity.

Author Contributions: All four authors conceived the framework and structured the whole manuscript, checked the results and completed the revision of the paper. The authors have equally contributed to the elaboration of this manuscript.

Funding: This research received no external funding.

Acknowledgments: The authors would like to express their gratitude to the anonymous reviewers for their constructive comments and suggestions, which led to an improved form of the manuscript.

Conflicts of Interest: The authors declare no conflict of interest.

References

1. Eringen, A.C. Theory of thermo-microstretch elastic solids. *Int. J. Eng. Sci.* **1990**, *28*, 1291–1301. [CrossRef]
2. Eringen, A.C. *Microcontinuum Field Theories*; Springer: New York, NY, USA, 1990
3. Hassan, M.; Marin, M.; Ellahi, R.; Alamri, S.Z. Exploration of convective heat transfer and flow characteristics synthesis by Cu–Ag/water hybrid-nanofluids. *Heat Transf. Res.* **2018**, *49*, 1837–1848. [CrossRef]
4. Marin, M.; Ellahi, R.; Chirila, A. On solutions of Saint-Venant's problem for elastic dipolar bodies with voids. *Carpathian J. Math.* **2017**, *33*, 219–232.
5. Mindlin, R.D. Micro-structure in linear elasticity. *Arch. Ration. Mech. Anal.* **1964**, 51–78. [CrossRef]
6. Green, A.E.; Rivlin, R.S. Multipolar continuum mechanics. *Arch. Ration. Mech. Anal.* **1964**, *17*, 113–147. [CrossRef]
7. Fried, E.; Gurtin, M.E. Thermomechanics of the interface between a body and its environment. *Contin. Mech. Therm.* **2007**, *19*, 253–271. [CrossRef]
8. Nunziato, J.W.; Cowin, S.C. A nonlinear theory of materials with voids. *Arch. Ration. Mech. Anal.* **1979**, *72*, 175–201. [CrossRef]
9. Cowin, S.C.; Nunziato, J.W. Linear elastic materials with voids. *J. Elast.* **1983**, *13*, 125–147. [CrossRef]
10. Goodman, M.A.; Cowin, S.C. A continuum theory of granular material. *Arch. Ration. Mech. Anal.* **1971**, *44*, 249–266. [CrossRef]
11. Othman, M.I.A.; Marin, M. Effect of thermal loading due to laser pulse on thermo-elastic porous medium under G-N theory. *Results Phys.* **2017**, *7*, 3863–3872. [CrossRef]
12. Abbas, I.A. A GN model based upon two-temperature generalized thermoelastic theory in an unbounded medium with a spherical cavity. *Appl. Math. Comput.* **2014**, *245*, 108–115. [CrossRef]
13. Abbas, I.A. Eigenvalue approach for an unbounded medium with a spherical cavity based upon two-temperature generalized thermoelastic theory. *J. Mech. Sci. Technol.* **2014**, *28*, 4193–4198. [CrossRef]
14. Othman, M.I.A.; Hasona, W.M.; Abd-Elaziz, E.M. Effect of Rotation on Micropolar Generalized Thermoelasticity with Two-Temperatures using a Dual-Phase-Lag Model. *Can. J. Phys.* **2014**, *92*, 149–158. [CrossRef]
15. Marin, M. Cesaro means in thermoelasticity of dipolar bodies. *Acta Mech.* **1997**, *122*, 155–168. [CrossRef]
16. Marin, M.; Craciun, E.M. Uniqueness results for a boundary value problem in dipolar thermoelasticity to model composite materials. *Compos. Part B Eng.* **2017**, *126*, 27–37. [CrossRef]
17. Serrin, J. The Initial Value Problem for the Navier-Stokes Equations. In *Nonlinear Problems Proc. Sympos.*; Univ. of Wisconsin Press: Madison, WI, USA, 1963; pp. 69–98.
18. Ciarletta, M.; Scalia, A. Some Results in Linear Theory of Thermomicrostretch Elastic Solids. *Meccanica* **2004**, *39*, 191–206. [CrossRef]
19. Knops R.J.; Payne, L.E. On the Stability of Solutions of the Navier-Stokes Equations Backward in Time. *Arch. Ration. Mech. Anal.* **1968**, *29*, 331–335. [CrossRef]
20. Galdi, G.P.; Straughan, B. Stability of Solutions of the Navier-Stokes Equations Backward in Time. *Arch. Ration. Mech. Anal.* **1988**, *101*, 107–114. [CrossRef]
21. Payne, L.; Straughan, B. Improperly Posed and Nonstandard Problems for Parabolic Partial Differential Equations. In *Elasticity: Mathematical Methods and Applications*; Ellis-Horwood Pub: Chichester, UK, 1990; pp. 273–300.
22. Ames, K.A.; Payne, L.E. Stabilizing Solutions of the Equations of Dynamical Linear Thermoelasticity Backward in Time. *Stab. Appl. Anal. Contin. Media* **1991**, *1*, 243–260.
23. Ciarletta, M. On the Uniqueness and Continuous Dependence of Solutions in Dynamical Thermoelasticity Backward in Time. *J. Therm. Stress.* **2002**, *25*, 969–984. [CrossRef]
24. Ciarletta, M.; Chirita, S. Spatial Behavior in Linear Thermoelasticity Backward in Time. In Proceedings of the Fourth International Congress on Thermal Stresses, Osaka, Japan, 8–11 June 2001; pp. 485–488.
25. Quintanilla, R. Impossibility of Localization in Linear Thermoelasticity with Voids. *Mech. Res. Commun.* **2007**, *34*, 522–527. [CrossRef]
26. Green A.E.; Naghdi, P.M. On Undamped Heat Waves in an Elastic Solid. *J. Therm. Stress.* **1992**, *15*, 253–264. [CrossRef]
27. Green A.E.; Naghdi, P.M. Thermoelasticity without Energy Dissipation. *J. Elast.* **1993**, *31*, 189–208. [CrossRef]

28. Iovane, G.; Passarella, F. Saint-Venant's Principle in Dynamic Porous Thermoelastic Media with Memory for Heat Flux. *J. Therm. Stress.* **2004**, *27*, 983–999. [CrossRef]
29. Passarella, F.; Tibullo, V. Some Results in Linear Theory of Thermoelasticity Backward in Time for Microstretch Materials. *J. Therm. Stress.* **2010**, *33*, 559–576. [CrossRef]
30. Quintanilla, R.; Straughan, B. Energy Bounds for Some Non-standard Problems in Thermoelasticity. *Proc. R. Soc. A Math. Phys. Eng. Sci.* **2005**, *461*, 1–15. [CrossRef]
31. De Sciarra, F.M.; Salerno, M. On thermodynamic functions in thermoelasticity without energy dissipation. *Eur. J. Mech. A Solids* **2014**, *46*, 84–95. [CrossRef]
32. Canadija, M.; Barretta, R.; de Sciarra, F.M. A gradient elasticity model of Bernoulli-Euler nanobeams in non-isothermal environments. *Eur. J. Mech. A Solids* **2016**, *55*, 243–255. [CrossRef]
33. Barretta, R.; Canadija, M.; Luciano, R.; de Sciarra, F.M. Stress-driven modeling of nonlocal thermoelastic behavior of nanobeams. *Int. J. Eng. Sci.* **2018**, *126*, 53–67. [CrossRef]
34. Hlavacek, I.; Hlavacek, M. On the Existence and Uniqueness of Solution and Some Variational Principles in Linear Theories of Elasticity with Couple-Stresses. I. Cosserat Continuum. *Appl. Math.* **1969**, *14*, 387–410.

© 2019 by the authors. Licensee MDPI, Basel, Switzerland. This article is an open access article distributed under the terms and conditions of the Creative Commons Attribution (CC BY) license (http://creativecommons.org/licenses/by/4.0/).

Article

Two-Dimensional Finite Element in General Plane Motion Used in the Analysis of Multi-Body Systems

Eliza Chircan [1], Maria-Luminița Scutaru [1,*] and Cătălin Iulian Pruncu [2,3]

1. Transilvania University of Brașov, B-dul Eroilor, 29, Brașov 500036, Romania
2. Mechanical Engineering, Imperial College, Exhibition Rd., London SW7 2AZ, UK
3. Mechanical Engineering, School of Engineering, University of Birmingham, Birmingham B15 2TT, UK
* Correspondence: lscutaru@unitbv.ro

Received: 5 June 2019; Accepted: 27 June 2019; Published: 1 July 2019

Abstract: A description of the motion equation of a single two-dimensional finite element used to model a multi-body system with elastic elements is made in the article. To establish them, the Lagrange's equations are used. Obtaining the dynamic response of a system with deformable components has become important for technical applications in recent decades. These engineering applications are characterized by high applied loads and high acceleration and velocities. A study of such mechanical systems leads to the identification of different mechanical phenomena (due to high deformations, resonance phenomena, and stability). Coriolis effects and relative motions significantly modify the motion equations and, implicitly, the dynamic response. These effects are highlighted in this paper for plane motion.

Keywords: multi-body system; finite element method (FEM); linear elastic elements; Lagrange's equations; two-dimensional finite element; plane motion

1. Introduction

Finite Element Analysis (FEA) of elastic bodies with overall rigid motion began in the 1970s. A number of papers analyzed different cases and types of finite elements used in this type of analysis [1–3]. One-dimensional and three-dimensional finite elements have been developed with plane motion or three-dimensional motion [4,5]. The field has remained interesting for researchers so far, with many papers on different aspects of the issue being published. The new field of multi-body systems (MBS), which has developed in recent decades, has reached the study of mechanical systems with (some) linear/nonlinear elastic elements. Why is this type of analysis necessary? If the velocities, accelerations, or loads that a constitutive element of a multi-body system supports are high, the elasticity of the elements may cause instability phenomena. The classical hypothesis of the rigid elements, commonly used in the dynamic analysis and synthesis of multi-body systems, no longer corresponds to the reality. The phenomena of resonance and loss of stability represent classic forms of manifestation of the elasticity properties. The classical approach involves the use of continuous mechanical models. Unfortunately, this type of analysis leads to a nonlinear system of differential equations that cannot be solved analytically in common engineering applications. The most convenient approach is FEA, a time-validated and frequently used method for analyzing mechanical systems with elastic elements. However, classical models of static or quasi-static systems do not correspond to the analysis of multi-body systems, which have rigid motion. The relatively large amount of motion between elements and Coriolis effects lead to the existence of additional terms in motion equations that cannot be neglected. These terms are usually strongly nonlinear. As a consequence, proper modeling of the problem is required [6,7]. The many aspects of a one-dimensional problem are described in [8]. In the following research, we develop a method for a two-dimensional finite element with plane motion. The motion equations for this type of element are established.

2. Two-Dimensional Finite Element

2.1. Two-Dimensional Model

Let us consider a two-dimensional finite element with a body with the general rigid plane motion of an MBS. To determine the governing equations for this element, the Lagrange's equations are used. The main phase in this type of approach is to obtain the Lagrangian for the one two-dimensional finite element. The terms necessary to build the Lagrangian are the kinetic energy of the two-dimensional finite element, the internal energy, and the work done by the concentrated and distributed loads. The shape function (determined in every case by the type of finite element chosen and the hypothesis made) will finally determine the form of the matrix coefficients and, as a consequence, the differential system of the governing motion equations. A basic hypothesis in this study is that the deformations of the elements of MBS are small enough not to influence the plane rigid motion of the system. Both the problem of the plane multi-body rigid motion of the system and velocities and the acceleration field distribution for each element can be solved using the classical method (see [9–11]).

The finite element chosen is related to the local reference frame Ox_1x_2 (participating with the finite element in the general plane rigid motion) and to the global reference frame $O'X_1X_2$. It is considered that the vector velocity and acceleration of the origin O related to the global reference system are known from a previous dynamical analysis. All the bodies are considered to be rigid as is the angular velocity and angular acceleration of the local reference system. In this way, the fields of velocity and acceleration are known for each point of each constitutive element of the MBS [12–18].

For one single two-dimensional finite element, the generalized independent coordinates can be the nodal displacements, depending on the shape function and hypothesis used to express the displacement of a point in the finite element. The final number of independent coordinates will depend on this hypothesis and shape function.

Consider the displacement $\{\delta(u, v)\}$ of an arbitrary point M of the domain of the finite element. If we use the shape function N_{ij} and the vector of the nodal displacements $\delta_{e,j}$, the displacements in the local coordinate system are

$$u = \delta_1 = N_{1j}\delta_{e,j}; \quad v = \delta_2 = N_{2j}\delta_{e,j}; \quad j = \overline{1,p}, \tag{1}$$

or

$$\delta_i = N_{ij}\delta_{e,j} \quad i = 1, 2; \quad j = \overline{1,p}, \tag{2}$$

where the nodal displacement vector of the finite element is labeled e, and $\delta_{e,j}$ depends on the independent coordinate chosen to define the element. The number of independent coordinates is p. For a triangular finite element we have

$$\delta_e^T = [u_1 \; v_1 \; u_2 \; v_2 \; u_3 \; v_3]. \tag{3}$$

The two lines of the shape function matrix **N** correspond to the displacements u and v and are named $N_{(u)} = N_{(1)}$ and $N_{(v)} = N_{(2)}$:

$$N = \begin{bmatrix} N_{(u)} \\ N_{(v)} \end{bmatrix} = \begin{bmatrix} N_{(1)} \\ N_{(2)} \end{bmatrix} = [N_{ij}], \quad i = 1, 2; \quad j = \overline{1,p}. \tag{4}$$

The displacement of the point $M(x_1, x_2)$, which is arbitrarily chosen, becomes, after deformation, $M'(x'_1, x'_2)$ and can be expressed through its components:

$$x'_1 = x_1 + u = x_1 + \delta_1; \quad x'_2 = x_2 + v = x_2 + \delta_2, \tag{5}$$

or with respect to the global reference system:

$$X'_1 = X_{10} + r_{1i}x'_i = X_{1o} + r_{1i}x_i + r_{1i}\delta_i = X_{1o} + r_{1i}x_i + r_{1i}N_{ij}\delta_{e,j},$$
$$X'_2 = X_{20} + r_{2i}x'_i = X_{2o} + r_{2i}x_i + r_{2i}\delta_i = X_{2o} + r_{2i}x_i + r_{2i}N_{ij}\delta_{e,j}, i = 1,2; \quad j = \overline{1,p}, \quad (6)$$

or

$$X'_k = X_{ko} + r_{ki}x_i + r_{ki}N_{ij}\delta_{e,j}, \quad k,i = 1,2, \quad j = \overline{1,p}. \quad (7)$$

2.2. Lagrangian of an Element

Using (6), the velocity of point M' can be obtained after differentiation with respect to the time:

$$\dot{X}'_k = \dot{X}_{ko} + \dot{r}_{ki}x_i + \dot{r}_{ki}N_{ij}\delta_{e,j} + r_{ki}N_{ij}\dot{\delta}_{e,j}, \quad k,i = 1,2, \quad j = \overline{1,p}. \quad (8)$$

The kinetic energy for one single element due to this velocity is

$$E_{ct} = \tfrac{1}{2}\int_0^L \rho A \dot{X}'_k \dot{X}'_k dA$$
$$= \tfrac{1}{2}\int_0^L \rho t \big(\dot{X}_{ko} + \dot{r}_{ki}x_i + \dot{r}_{ki}N_{ij}\delta_{e,j} + r_{ki}N_{ij}\dot{\delta}_{e,j}\big)\big(\dot{X}_{ko} + \dot{r}_{kl}x_l + \dot{r}_{kl}N_{lm}\delta_{e,m} + r_{kl}N_{lm}\dot{\delta}_{e,m}\big)dA \quad (9)$$

where t is the thickness of the element, and ρ is the mass density.

In plane motion, the rotation of the local system of coordinates is expressed by the plane rotation tensor r_{ij}. The columns of this tensor define the positions of the unit vectors of the local reference frame Oxyz. For plane rotation, these coefficients can be expressed as

$$r_{11} = r_{22} = \cos\theta; \quad r_{12} = r_{21} = -\sin\theta. \quad (10)$$

The ortho-normality condition of these unit vectors leads to

$$r_{ij}r_{kj} = r_{jk}r_{ji} = \delta_{ij}, \quad ijk = 1,2 \quad (11)$$

where δ_{ij} is the Kronecker delta (derived from the general three-dimensional transformation). By differentiation, the following equation will result:

$$\dot{r}_{ij}r_{kj} + r_{ij}\dot{r}_{kj} = 0, \quad i,k = 1,2. \quad (12)$$

Denote

$$\omega_{ik} = \dot{r}_{ij}r_{kj}, \quad (13)$$

the skew-symmetric tensor angular velocity. The relation (12) becomes

$$\omega_{ik} + \omega_{ki} = 0, \quad i,k = 1,2. \quad (14)$$

To this corresponds the angular velocity vector components defined by

$$\omega_3 = \omega_{21} = -\omega_{12} = \omega, \quad \omega_1 = \omega_{32} = -\omega_{23} = 0, \quad \omega_2 = \omega_{13} = -\omega_{31} = 0. \quad (15)$$

The angular velocity vector and the angular acceleration vector have, in plane motion, the same components in both the local reference system and the global reference system.

The angular acceleration skew symmetric operator it is defined as

$$\varepsilon_{ik} = \dot{\omega}_{ik} = \ddot{r}_{ij}r_{kj} + \dot{r}_{ij}\dot{r}_{kj}, \quad ijk = 1,2. \quad (16)$$

The angular acceleration vector is defined by

$$\varepsilon_3 = \varepsilon_{21} = -\varepsilon_{12} = \dot{\omega} = \varepsilon. \tag{17}$$

The other components are null. We shall have

$$\varepsilon_{ik} = \omega_{ik} = \ddot{r}_{ij}r_{kj} + \dot{r}_{ij}\dot{r}_{kj} = \ddot{r}_{ij}r_{kj} + \dot{r}_{ij}r_{jl}r_{ml}\dot{r}_{km} = \ddot{r}_{ij}r_{kj} - \omega_{il}\omega_{lk} \tag{18}$$

from where

$$\ddot{r}_{ij}r_{kj} = \varepsilon_{ik} + \omega_{il}\omega_{lk}, \; ijkl = 1,2. \tag{19}$$

The internal energy stored is obtained via the relation

$$E_p = \frac{1}{2}\int_0^L \sigma_{ij}\varepsilon_{ij}dV \quad i,j = 1,2. \tag{20}$$

From (20), it is possible to obtain the stiffness matrix depending on the hypothesis and type of finite element (and consequently, the shape functions) chosen. Here, we present a case as an example; any other case can be treated in the same manner. The relationship between stress and strain is

$$\sigma_{ij} = D_{ik}\varepsilon_{kj}. \tag{21}$$

If we choose, for example, a rectangular finite element with sides a and b, a displacement field of the form

$$u = \alpha_1 + \alpha_2 x + \alpha_3 xy + \alpha_4 yv = \beta_1 + \beta_2 x + \beta_3 xy + \beta_4 y \tag{22}$$

can be used to obtain the stiffness tensor.

The boundary conditions ($x = 0, y = 0 => u = u_1, v = v_1; x = a, y = 0 => u = u_2, v = v_2; x = a, y = b => u = u_3, v = v_3; x = 0, y = b => u = u_4, v = v_4;$) lead to the following fields for the displacements:

$$\begin{aligned} u &= (1-\xi)(1-\eta)u_1 + \xi(1-\eta)u_2 + \xi\eta u_3 + \eta(1-\xi)u_4 \\ v &= (1-\xi)(1-\eta)v_1 + \xi(1-\eta)v_2 + \xi\eta v_3 + \eta(1-\xi)v_4. \end{aligned} \tag{23}$$

Applying the classical laws of linear elasticity [2], the relation between the strain and nodal displacement are, in this case,

$$\begin{aligned} \varepsilon_{11} &= -\tfrac{1-\eta}{a}u_1 + \tfrac{1-\eta}{a}u_2 + \tfrac{\eta}{a}u_3 - \tfrac{\eta}{a}u_4, \\ \varepsilon_{22} &= -\tfrac{1-\xi}{b}v_1 + -\tfrac{\xi}{b}v_2 + \tfrac{\xi}{b}v_3 + \tfrac{1-\xi}{b}v_4, \\ \varepsilon_{12} = \varepsilon_{21} &= -\tfrac{1-\xi}{b}u_1 - \tfrac{1-\eta}{a}v_1 - \tfrac{\xi}{b}u_2 + \tfrac{1-\eta}{a}v_2 + \tfrac{\xi}{b}u_3 + \tfrac{\eta}{a}v_3 + \tfrac{1-\xi}{b}u_4 - \tfrac{\eta}{a}v_4. \end{aligned} \tag{24}$$

Using (20), (21), and (24), after some elementary calculus, it is possible to obtain the internal energy of the form (see [2])

$$E_p = \frac{1}{2}\delta_i k_{e,ij}\delta_j \quad i,j = \overline{1,p} \tag{25}$$

The external work of the concentrated load is

$$W = q_{e,j}\delta_{e,j}, \quad i = 1,2, \quad j = \overline{1,p}. \tag{26}$$

Here, q_{ej} is the vector of the concentrated loads acting in nodes.

The external work of the distributed loads is

$$W = \int_0^L (p_i \delta_i) dA = \left(\int_0^L p_i N_{ij} dA \right) \delta_{e,j} = q^*_{e,j} \delta_{e,j}, \quad i = 1, 2, \quad j = \overline{1, p}, \quad (27)$$

where the notation

$$q^*_{e,j} = \left(\int_0^L p_i N_{ij} dx \right) \delta_{e,j}, \quad i = 1, 2, \quad j = \overline{1, p} \quad (28)$$

is used. We have all the data to build the Lagrangian [8,19]

$$L = E_c - E_p + W^d + W^c, \quad (29)$$

or, taking into account the relations (9), (24), (25), and (26),

$$L = \tfrac{1}{2} \int_0^L \rho t \big(\dot{X}_{ko} + \dot{r}_{ki} x_i + \dot{r}_{ki} N_{ij} \delta_{e,j} + r_{ki} N_{ij} \dot{\delta}_{e,j} \big)\big(\dot{X}_{ko} + \dot{r}_{kl} x_l + \dot{r}_{kl} N_{lm} \delta_{e,m} + r_{kl} N_{lm} \dot{\delta}_{e,m} \big) dA \\ - \tfrac{1}{2} \delta_{e,i} k_{e,ij} \delta_{e,j} + q^*_{eL,j} \delta_{e,j} + q_{e,i} \delta_{e,i}. \quad (30)$$

3. Motion Equations

Theorem: *The motion equations written in the local coordinate system for two-dimensional finite element take the form*

$$m_{e,ij} \ddot{\delta}_{e,j} + 2 c^\omega_{e,ij} \dot{\delta}_{e,j} + \big(k_{e,ij} + k^\varepsilon_{e,ij} + k^{\omega^2}_{e,ij} \big) \delta_{e,j} = q_{e,i} + q^*_{e,i} - q^\varepsilon_{e,i} - q^{\omega^2}_{e,i} - m^o_{e,ij} \ddot{x}_{jo} \quad (31)$$

where

$$m_{e,ij} = \int_0^L \rho t N_{ki} N_{kj} dA, \quad i, j = \overline{1, p}, \quad k = 1, 2 c^\omega_{e,ij} = 2 \int_0^L N_{ki} \omega_{km} N_{mj} \rho t dA; \quad (32)$$

$$k^\varepsilon_{e,ij} = \int_0^L N_{ki} \varepsilon_{km} N_{mj} \rho t dA; \quad k^{\omega^2}_{e,ij} = \int_0^L N_{ki} \omega_{km} \omega_{ml} N_{lj} \rho t dA; \quad m^o_{e,ij} = \int_0^L \rho A N_{ji} dx, i, j = \overline{1, p}, \quad k, m, l = 1, 2. \quad (33)$$

Proof: We apply the Lagrange's equations in the form

$$\frac{d}{dt} \frac{\partial L}{\partial \dot{\delta}_{e,i}} - \frac{\partial L}{\partial \delta_{e,i}} = 0. \quad (34)$$

We obtain the motion equations (an extended presentation can be found in Appendix A):

$$\left[\int_0^L \rho t N_{ik} N_{jk} dA \right] \ddot{\delta}_{e,j} + 2 \left[\int_0^L N_{ki} \omega_{km} N_{mj} \rho t dA \right] \dot{\delta}_{e,j} + \left[\begin{array}{c} k_{e,ij} + \int_0^L N_{ki} \varepsilon_{km} N_{mj} \rho t dA + \\ + \int_0^L N_{ki} \omega_{km} \omega_{ml} N_{lj} \rho t dA \end{array} \right] \delta_{e,j} \\ = q_{e,i} + q^*_{e,i} - q^\varepsilon_{e,i} - q^{\omega^2}_{e,i} - \left(\int_0^L \rho A N_{ji} dx \right) \ddot{x}_{jo}. \quad (35)$$

Using the notation mentioned above, it is possible to obtain (30).

4. Conclusions

In engineering practice, there are frequent cases in which the elements of a device or a machine are required to operate with high forces or have high accelerations and speeds. In these cases, the elasticity of the elements can be manifested in such a way as to negatively influence the working process of the assembly. An analysis of the MBS motion in such conditions is required. In this work, we determined the motion equations for a two-dimensional finite element with plane motion. This case often occurs in practice. Most of the technical systems have elements operating in planar motion. In this case, a system of differential equations with matrix coefficients that can be strongly nonlinear is obtained. The mass tensor and the classical rigidity tensor are symmetrical matrices, while the Coriolis effects tensor is skew-symmetric. There are also terms that change the classical stiffness of the element and additional inertial terms. If we use FEA to obtain the motion equations for one single element in plane motion, we obtain a system of second-order differential equations, as presented in (30). Generally, this system is strongly nonlinear, with the matrix coefficients of the system depending on time and on the position of the system. For a usual engineering application, some aspects of these equations can cause difficulties. We highlight some properties of this system:

- The classical inertia tensor $m_{e,ij}$ is a symmetrical tensor;
- The damping tensor $c_{e,ij}^{\omega}$ is a skew symmetric tensor; this represents the effect of the accelerations due to the Coriolis effects (relative motions with respect to the mobile reference co-ordinate system;
- The stiffness tensor $k_{e,ij}$ is a symmetric tensor; this tensor is modify by additional terms depending on the general plane rotation of the element, becoming $k_{e,ij} + k_{e,ij}^{\varepsilon} + k_{e,ij}^{\omega^2}$;
- The vector of the generalized loads contains some supplementary terms due to inertia of finite elements being in rigid motion; these are $-q_{e,i}^{\varepsilon} - q_{e,i}^{\omega^2} - m_{e,ij}^{o}\ddot{x}_{jo}$.

The common method used to solve this system is to linearize this system, considering the tensor coefficients as being constant for very short time intervals (rigid motion freezing). In this way, it is possible to obtain a system of differential equations with constant coefficients. To solve this, usual and well-known methods can be used. There can be singularities due to the inertia term affecting the stiffness matrix [20].

Author Contributions: Conceptualization, M.L.S.; methodology, E.C., S.M.L. and C.P.; investigation and software M.L.S.; validation, E.C., M.L.S.; formal analysis, M.L.S.; resources, M.L.S., C.P.; data curation, M.L.S., C.P.; writing—original draft preparation, M.L.S.; writing—review and editing, M.L.S.; visualization, M.L.S.; supervision, C.P.

Funding: This research received no external funding.

Acknowledgments: The authors would like to thank to the Transylvania University of Brasov and the Imperial College of London for providing support to help us to successfully complete the work.

Conflicts of Interest: The authors declare no conflict of interest.

Appendix A

The derivatives of the Lagrangian:

$$\frac{\partial L}{\partial \dot{\delta}_{e,i}} = \int_0^L \rho \dot{X}_{ko} r_{kl} N_{lm} t dA + \int_0^L \rho A \dot{r}_{ki} x_i r_{kl} N_{lm} t dA + \int_0^L \rho \dot{r}_{ki} N_{ij} \delta_{e,j} r_{kl} N_{lm} t dA$$
$$+ \int_0^L \rho r_{ki} N_{ij} \dot{\delta}_{e,j} r_{kl} N_{lm} t dA$$
$$-\frac{\partial L}{\partial \delta_{e,i}} = -\dot{X}_{ko} \dot{r}_{kl} \left(\int_0^L N_{lm} \rho t dA \right) - \dot{r}_{ki} \dot{r}_{kl} \left(\int_0^L x_i N_{lm} \rho t dA \right) - \dot{r}_{ki} \dot{r}_{kl} \left(\int_0^L N_{ij} N_{lm} \rho t dA \right) \delta_{e,j} \quad \text{(A1)}$$
$$- \dot{r}_{ki} r_{kl} \left(\int_0^L N_{ij} N_{lm} \rho t dA \right) \dot{\delta}_{e,m}.$$

Applying $\frac{d}{dt}\left(\frac{\partial L}{\partial \dot{\delta}_{e,i}}\right) - \frac{\partial L}{\partial \delta_{e,i}} = 0$ obtains

$$\left(\int_0^L N_{lm}\rho t dA\right)r_{kl}\ddot{X}_{ko} + \ddot{r}_{ki}r_{kl}\left(\int_0^L x_i N_{lm}\rho t dA\right) + \ddot{r}_{ki}r_{kl}\left(\int_0^L N_{ij}N_{lm}\rho t dA\right)\delta_{e,j} + 2\dot{r}_{ki}r_{kl}\left(\int_0^L N_{ij}N_{ij}\rho t dA\right)\dot{\delta}_{e,j}$$
$$+ r_{ki}r_{kl}\left(\int_0^L N_{ij}N_{lm}\rho t dA\right)\ddot{\delta}_{e,j} - (q_{e,i} + q^*_{e,i}) + k_{e,ij}\delta_{e,j} = 0;$$
$$\left(\int_0^L \rho t N_{ik}N_{jk}dA\right)\ddot{\delta}_{e,j} + 2\omega\left(\int_0^L \rho t N_{2i}N_{1j}dA - \int_0^L \rho t N_{1i}N_{2j}dA\right)\dot{\delta}_{e,j} +$$
$$\left[k_{e,ij} + \varepsilon\left(\int_0^L \rho t N_{2i}N_{1j}dA - \int_0^L \rho t N_{1i}N_{2j}dA\right) - \omega^2\left(\int_0^L \rho t N_{1i}N_{1j}dA + \int_0^L \rho t N_{2i}N_{2j}dA\right)\right]\delta_{e,j}$$
$$= q_{e,i} + q^*_{e,i} - q^\varepsilon_{e,i} - q^{\omega^2}_{e,i} - m^o_{e,ij}\ddot{x}_{jo},$$

(A2)

or

$$m_{e,ij}\ddot{\delta}_{e,j} + 2\omega\left(m^{(21)}_{ij} - m^{(12)}_{ij}\right)\dot{\delta}_{e,j} + \left[k_{e,ij} + \varepsilon\left(m^{(21)}_{ij} - m^{(12)}_{ij}\right) - \omega^2\left(m^{(11)}_{ij} + m^{(22)}_{ij}\right)\right]\delta_{e,j} = \quad \text{(A3)}$$

References

1. Erdman, A.G.; Sandor, G.N.; Oakberg, A. A General Method for Kineto-Elastodynamic Analysis and Synthesis of Mechanisms. *J. Eng. Ind. ASME Trans.* **1972**, *94*, 1193–1205. [CrossRef]
2. Mayo, J.; Domínguez, J. Geometrically non-linear formulation of flexible multibody systems in terms of beam elements: Geometric stiffness. *Comput. Struct.* **1996**, *59*, 1039. [CrossRef]
3. Vlase, S.; Danasel, C.; Scutaru, M.L.; Mihalcica, M. Finite element analysis of two-dimensional linear elastic systems with a plane "Rigid motion". *Rom. J. Phys.* **2014**, *59*, 476–487.
4. Piras, G.; Cleghorn, W.L.; Mills, J.K. Dynamic finite-element analysis of a planar high speed, high-precision parallel manipulator with flexible links. *Mech. Mach. Theory* **2005**, *40*, 849–862. [CrossRef]
5. Thompson, B.S.; Sung, C.K. A survey of Finite Element Techniques for Mechanism Design. *Mech. Mach. Theory* **1986**, *21*, 351–359. [CrossRef]
6. Vlase, S.; Teodorescu, P.P.; Itu, C.; Scutaru, M.L. Elasto-dynamics of a solid with a general "rigid" motion using FEM model. Part II. Analysis of a double cardan joint. *Rom. J. Phys.* **2013**, *58*, 882–892.
7. Vlase, S. Dynamical Response of a Multibody System with Flexible Element with a general Three-Dimensional Motion. *Rom. J. Phys.* **2012**, *57*, 676–693.
8. Vlase, S.; Marin, M.; Oechsner, A.; Scutaru, M.L. Motion equation for a flexible one-dimensional element used in the dynamical analysis of a multibody system. *Contin. Mech.* **2019**, *31*, 715–724. [CrossRef]
9. Negrean, I. Advanced notions in Analytical Dynamics of Systems. *Acta Technica Napocensis—Applied Mathematics. Mech. Eng.* **2017**, *60*, 491–501.
10. Negrean, I. Generalized forces in Analytical Dynamics of Systems. *Acta Technica Napocensis—Applied Mathematics. Mech. Eng.* **2017**, *60*, 357–368.
11. Pennestri, E.; de Falco, D.; Vita, L. An Investigation of the Inuence of Pseudoinverse Matrix Calculations on Multibody Dynamics by Means of the Udwadia-Kalaba Formulation. *J. Aerosp. Eng.* **2009**, *22*, 365–372.
12. Deü, J.-F.; Galucio, A.C.; Ohayon, R. Dynamic responses of flexible-link mechanisms with passive/active damping treatment. *Comput. Struct.* **2008**, *86*, 258–265. [CrossRef]
13. Gerstmayr, J.; Schöberl, J.A. 3D Finite Element Method for Flexible Multibody Systems. *Multibody Syst. Dyn.* **2006**, *15*, 305–320. [CrossRef]
14. Ibrahimbegović, A.; Mamouri, S.; Taylor, R.L.; Chen, A.J. Finite Element Method in Dynamics of Flexible Multibody Systems: Modeling of Holonomic Constraints and Energy Conserving Integration Schemes. *Multibody Syst. Dyn.* **2000**, *4*, 195–223. [CrossRef]
15. Khang, N.V. Kronecker product and a new matrix form of Lagrangian equations with multipliers for constrained multibody systems. *Mech. Res. Commun.* **2011**, *38*, 294–299. [CrossRef]
16. Marin, M. Cesaro means in thermoelasticity of dipolar bodies. *Acta Mech.* **1997**, *122*, 155–168. [CrossRef]
17. Hassan, M.; Marin, M.; Ellahi, R.; Alamri, S.Z. Exploration of convective heat transfer and flow characteristics synthesis by Cu-Ag/water hybrid-nanofluids. *Heat Transf. Res.* **2018**, *49*, 1837–1848. [CrossRef]

18. Zhang, X.; Erdman, A.G. Dynamic responses of flexible linkage mechanisms with viscoelastic constrained layer damping treatment. *Comput. Struct.* **2001**, *79*, 1265–1274. [CrossRef]
19. Simeon, B. On Lagrange multipliers in flexible multibody dynamic. *Comput. Methods Appl. Mech. Eng.* **2006**, *195*, 6993–7005. [CrossRef]
20. Chircan, E.; Scutaru, M.L.; Toderita, A. Dynamical response of a beam in a centrifugal field using the finite element method. In Proceedings of the XVth International Conference Acoustics & Vibration of Mechanical Structures, Timisoara, Romania, 30–32 May 2019.

© 2019 by the authors. Licensee MDPI, Basel, Switzerland. This article is an open access article distributed under the terms and conditions of the Creative Commons Attribution (CC BY) license (http://creativecommons.org/licenses/by/4.0/).

Article

Inference about the Ratio of the Coefficients of Variation of Two Independent Symmetric or Asymmetric Populations

Zhang Yue [1] and Dumitru Baleanu [2,3,*]

[1] Teaching and Research Section of Public Education, Hainan Radio and TV University, No.20 Haidianerxi Road, Meilan District, Haikou 570208, Hainan, China; rezastat@yahoo.com
[2] Department of Mathematics, Faculty of Art and Sciences, Cankaya University, 0630 Ankara, Turkey
[3] Institute of Space Sciences, 077125 Magurele-Bucharest, Romania
* Correspondence: dumitru@cankaya.edu.tr

Received: 23 May 2019; Accepted: 20 June 2019; Published: 21 June 2019

Abstract: Coefficient of variation (CV) is a simple but useful statistical tool to make comparisons about the independent populations in many research areas. In this study, firstly, we proposed the asymptotic distribution for the ratio of the CVs of two separate symmetric or asymmetric populations. Then, we derived the asymptotic confidence interval and test statistic for hypothesis testing about the ratio of the CVs of these populations. Finally, the performance of the introduced approach was studied through simulation study.

Keywords: coefficient of variation; ratio; symmetric and asymmetric distributions; test of hypothesis

1. Introduction

Based on the literature, to describe a dataset (random variable), three main characteristics containing central tendencies, dispersion tendencies and shape tendencies, are used. A central tendency (or measure of central tendency) is a central or typical value for a random variable that describes the way in which the random variable is clustered around a central value. It may also be called a center or location of the distribution of the random variable. The most common measures of central tendency are the mean, the median and the mode. Measures of dispersion like the range, variance and standard deviation tell us about the spread of the values of a random variable. It may also be called a scale of the distribution of the random variable. The shape tendencies such as skewness and kurtosis describe the distribution (or pattern) of the random variable.

The division of the standard deviation to the mean of population, $CV = \frac{\sigma}{\mu}$, is called as coefficient of variation (CV) which is an applicable statistic to evaluate the relative variability. This free dimension parameter can be widely used as an index of reliability or variability in many applied sciences such as agriculture, biology, engineering, finance, medicine, and many others [1–3]. Since it is often necessary to relate the standard deviation to the level of the measurements, the CV is a widely used measure of dispersion. The CVs are often calculated on samples from several independent populations, and questions about how to compare them naturally arise, especially when the distributions of the populations are skewed. In real world applications, the researchers may intend to compare the CVs of two separate populations to understand the structure of the data. ANOVA and Levene tests can be used to investigate the equality of CVs of populations in case the means or variances of the populations are equal. It is obvious that in many situations two populations with different means and variances may have an equal CV. For the normal case, the problems of interval estimating the CV or comparison of two or several CVs have been well addressed in the literature. Due to possible small differences of two small CVs and no strong interpretation, the ratio of CVs is more accurate than the difference of CVs. Bennett [4] proposed a likelihood ratio test. Doornbos and Dijkstra [5] and Hedges and Olkin [6] presented two tests based on the non-central t test. A modification of Bennett's method was provided by Shafer and Sullivan [7]. Wald tests have been introduced by [8–10]. Based on Renyi's divergence, Pardo and Pardo [11] proposed a new method. Nairy and Rao [12] applied the likelihood ratio, score test and Wald test to check that the inverse CVs are equal. Verrill and Johnson [13] applied one-step Newton estimators to establish a likelihood ratio test. Jafari and Kazemi [14] developed a parametric bootstrap (PB) approach. Some statisticians improved these tests for symmetric distributions [15–23]. The problem of comparing two or more CVs arises in many practical situations [24–26]. Nam and Kwon [25] developed approximate interval estimation of the ratio of two CVs for lognormal distributions by using the Wald-type, Fieller-type, log methods, and the method of variance estimates recovery (MOVER). Wong and Jiang [26] proposed a simulated Bartlett corrected likelihood ratio approach to obtain inference concerning the ratio of two CVs for lognormal distribution.

In applications, it is usually assumed that the data follows symmetric distributions. For this reason, most previous works have focused on the comparison of CVs in symmetric distributions, especially normal distributions. In this paper, we propose a method to compare the CVs of two separate symmetric or asymmetric populations. Firstly, we propose the asymptotic distribution for the ratio of the CVs. Then, we derive the asymptotic confidence interval and test statistic for hypothesis testing about the ratio of the CVs. Finally, the performance of the introduced approach is studied through simulation study. The introduced approach seems to have many advantages. First, it is powerful. Second, it is not too computational. Third, it can be applied to compare the CVs of two separate symmetric or asymmetric populations. We apply a methodology similar to that which has been used in [27–33]. The comparison between the parameters of two datasets or models has been considered in several works [34–40].

2. Asymptotic Results

Assume that X and Y are uncorrelated variables with non-zero means μ_X and μ_Y, and the finite i^{th} central moments:

$$\mu_{iX} = E(X - \mu_X)^i, \quad \mu_{iY} = E(Y - \mu_Y)^i, \quad i \in \{2,3,4\},$$

respectively. Also assume two samples $X_1, \ldots, X_m$, and $Y_1, \ldots, Y_n$, distributed from X and Y, respectively. From the motivation given in the introduction, the parameter:

$$\gamma = \frac{CV_Y}{CV_X},$$

is interesting to inference, where CV_Y and CV_X are the CVs corresponding to Y and X, respectively.

Assume $m_{iX} = \frac{1}{m}\sum_{k=1}^{m}(x_k - \bar{x})^i$, $m_{iY} = \frac{1}{n}\sum_{k=1}^{n}(y_k - \bar{y})^i$, $i \in \{2,3,4\}$. CV_X and CV_Y, are consistently estimated [41] by $\hat{C}V_X = \frac{\sqrt{m_{2X}}}{\bar{X}}$ and $\hat{C}V_Y = \frac{\sqrt{m_{2Y}}}{\bar{Y}}$, respectively. So, it is obvious that

$$\hat{\gamma} = \frac{\hat{C}V_Y}{\hat{C}V_X}$$

can reasonably estimate the parameter γ. For simplicity, let $m = n$. When $n = m$, let $n^* = min(m,n)$ instead of m and n in the following discussions.

Lemma 1. *If the above assumptions are satisfied, then:*

$$\sqrt{n}(\hat{C}V_X - CV_X) \xrightarrow{L} N(0, \delta_X^2), \quad as \quad n \to \infty,$$

where:

$$\delta_X^2 = \left[\frac{\mu_{4X} - \mu_{2X}^2}{4\mu_X^2 \mu_{2X}} - \frac{\mu_{3X}}{\mu_X^3} + \frac{\mu_{2X}^2}{\mu_{4X}}\right],$$

is the asymptotic variance.

Proof. The outline of proof can be found in [41]. □

The next theorem corresponds to the asymptotic distribution of $\hat{\gamma}$. This theorem will be applied to construct the confidence interval and perform hypothesis testing for the parameter γ.

Theorem 1. *If the previous assumptions are satisfied, then:*

$$\sqrt{n}(\hat{\gamma} - \gamma) \xrightarrow{L} N(0, \lambda^2), \quad \text{as } n \to \infty,$$

where:

$$\lambda^2 = \frac{1}{CV_X^2}(\gamma^2 \delta_X^2 + \delta_Y^2),$$

and:

$$\delta_Y^2 = \left[\frac{\mu_{4Y} - \mu_{2Y}^2}{4\mu_Y^2 \mu_{2Y}} - \frac{\mu_{3Y}}{\mu_Y^3} + \frac{\mu_{2Y}^2}{\mu_{4Y}}\right].$$

Proof. By using Lemma 1, we have:

$$\sqrt{n}(\hat{C}V_X - CV_X) \xrightarrow{L} N(0, \delta_X^2), \quad \text{as } n \to \infty,$$

and:

$$\sqrt{n}(\hat{C}V_Y - CV_Y) \xrightarrow{L} N(0, \delta_Y^2), \quad \text{as } n \to \infty.$$

Slutsky's Theorem gives:

$$\sqrt{n}\left[\begin{pmatrix}\hat{C}V_X \\ \hat{C}V_Y\end{pmatrix} - \begin{pmatrix}CV_X \\ CV_Y\end{pmatrix}\right] \xrightarrow{L} N\left(0, \begin{pmatrix}\delta_X^2 & 0 \\ 0 & \delta_Y^2\end{pmatrix}\right),$$

for independent samples [41].

Now define $f: R^2 \to R$ as $f(x, y) = \frac{y}{x}$. Then we have:

$$\nabla f(x, y) = \left(-\frac{y}{x^2}, \frac{1}{x}\right),$$

where $\nabla f(x, y)$ is the gradient function. Consequently, we have $\nabla f(CV_X, CV_Y) \Sigma (\nabla f(CV_X, CV_Y))^T = \lambda^2$. Because of continuity of ∇f in the neighbourhood of (CV_X, CV_Y), by using Cramer's Rule:

$$\sqrt{n}(f(\hat{C}V_X, \hat{C}V_Y) - f(CV_X, CV_Y)) = \sqrt{n}(\hat{\gamma} - \gamma) \xrightarrow{L} N(0, \lambda^2), \quad n \to \infty,$$

the proof ends. □

Thus, the asymptotic distribution can be constructed as:

$$T_n = \sqrt{n}\left(\frac{\hat{\gamma} - \gamma}{\lambda}\right) \xrightarrow{L} N(0, 1), \quad \text{as } n \to \infty. \tag{1}$$

2.1. Constructing the Confidence Interval

As can be seen, the parameter λ depends on CV_X, δ_X^2, δ_Y^2 and γ which are unknown parameters in practice. The result of the next theorem can be applied to construct the confidence interval and to perform the hypothesis testing for the parameter γ.

Theorem 2. *If the previous assumptions are satisfied, then:*

$$T_n^* = \sqrt{n}\left(\frac{\hat{\gamma}-\gamma}{\hat{\lambda}}\right) \xrightarrow{L} N(0,1), \quad \text{as } \to \infty, \tag{2}$$

where:

$$\hat{\lambda}^2 = \frac{1}{\widehat{CV}_X^2}\left(\hat{\gamma}^2\hat{\delta}_X^2 + \hat{\delta}_Y^2\right),$$

$$\hat{\delta}_X^2 = \left[\frac{m_{4X} - m_{2X}^2}{4\overline{X}^2 m_{2X}} - \frac{m_{3X}}{\overline{X}^3} + \frac{m_{2X}^2}{m_{4X}}\right],$$

and:

$$\hat{\delta}_Y^2 = \left[\frac{m_{4Y} - m_{2Y}^2}{4\overline{Y}^2 m_{2Y}} - \frac{m_{3Y}}{\overline{Y}^3} + \frac{m_{2Y}^2}{m_{4Y}}\right].$$

Proof. From the *Weak Law of Large Numbers*, it is known that:

$$\overline{X} \xrightarrow{P} \mu_X, \quad \overline{Y} \xrightarrow{P} \mu_Y, \quad m_{iX} \xrightarrow{P} \mu_{iX}, \quad m_{iY} \xrightarrow{P} \mu_{iY}, \quad i \in \{2,3,4\},$$

as $n \to \infty$.

Consequently, by applying Slutsky's Theorem, we have $\hat{\lambda} \xrightarrow{P} \lambda$, as $n \to \infty$. Appliying Theorem 1 the proof is completed. □

Now, T_n^* is a pivotal quantity for γ. In the following, this pivotal quantity is used to construct asymptotic confidence interval for γ.

$$\left(\hat{\gamma} - \frac{\hat{\lambda}}{\sqrt{n}}Z_{\frac{\alpha}{2}}, \hat{\gamma} + \frac{\hat{\lambda}}{\sqrt{n}}Z_{\frac{\alpha}{2}}\right). \tag{3}$$

2.2. Hypothesis Testing

In real word applications, researchers are interested in testing about the parameter γ. For example, the null hypothesis $H_0 : \gamma = 1$ means that the CVs of two populations are equal. To perform the hypothesis test $H_0 : \gamma = \gamma_0$, the test statistic:

$$T_0 = \sqrt{n}\left(\frac{\hat{\gamma} - \gamma_0}{\lambda^*}\right), \tag{4}$$

is generally applied, such that:

$$\lambda^{*2} = \frac{1}{\widehat{CV}_X^2}\left(\gamma_0^2 \hat{\delta}_X^2 + \hat{\delta}_Y^2\right).$$

If the null hypothesis $H_0 : \gamma = \gamma_0$ is satisfied, then the asymptotic distribution of T_0 is standard normal.

2.3. Normal Populations

Naturally, many phenomena follow normal distribution. This distribution is very important in natural and social sciences. Many researchers focused on the comparison between the CVs of two independent normal distributions. Nairy and Rao [12] reviewed and studied several methods such as likelihood ratio test, score test and Wald test that could be used to compare the CVs of two independent normal distributions. If the parent distributions X and Y are normal, then:

$$\mu_{3X} = \mu_{3Y} = 0, \quad \mu_{4X} = 3\mu_{2X}^2, \quad \mu_{4Y} = 3\mu_{2Y}^2.$$

Consequently, for normal distributions, δ_X^2 and $\hat{\delta}_X^2$ can be rewritten as:

$$\delta_X^2 = \frac{\mu_{2X}\mu_X^2 + 2\mu_{2X}^2}{2\mu_X^4},$$

and:

$$\hat{\delta}_X^2 = \frac{m_{2X}\overline{X}^2 + 2m_{2X}^2}{2\overline{X}^4},$$

respectively.

3. Simulation Study

In this section, the accuracy of the given theoretical results is studied and analyzed by different simulated datasets. For the populations X and Y, we respectively simulated different samples from symmetric distribution (normal) and asymmetric distributions (gamma and beta) with different CV values, $(CV_X, CV_Y) \in \{(1,1),(1,2),(2,3),(2,5)\}$, which are equivalent to $\gamma \in \{1,2,1.5,2.5\}$. Figures 1–3 show the plots of probability density function (PDF) for the considered distributions.

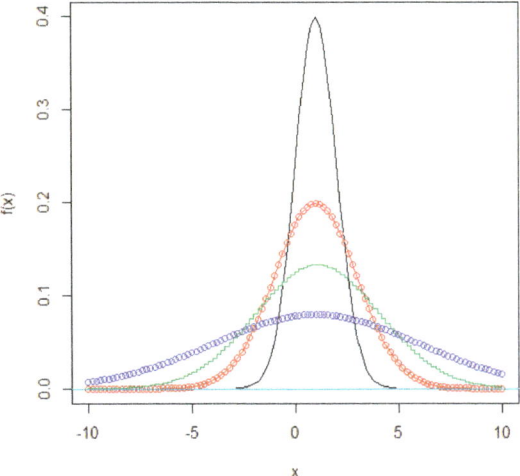

Figure 1. Probability density function (PDF) of normal (μ, σ^2) distribution with different coefficient of variation (CV) values. Black: $\mu = 1$, $\sigma = 1$, CV = 1; red: $\mu = 1$, $\sigma = 2$, CV = 2; green: $\mu = 1$, $\sigma = 3$, CV = 3; blue: $\mu = 1$, $\sigma = 5$, CV = 5.

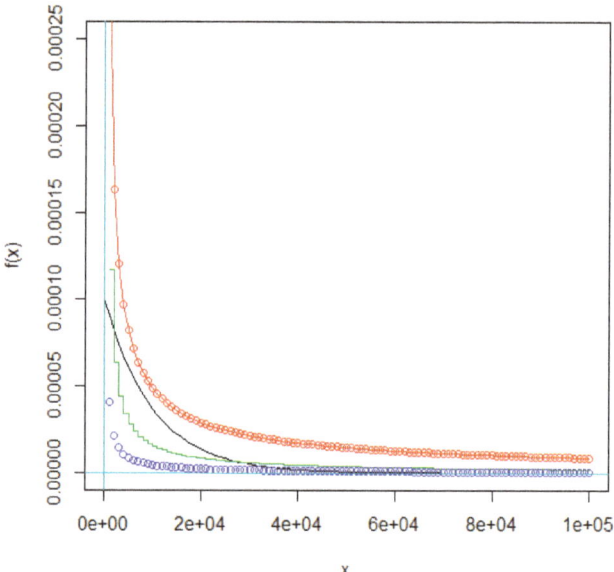

Figure 2. PDF of gamma (α, λ) distribution with different CV values. Black: $\alpha = 1, \lambda = 0.001$, CV = 1; red: $\alpha = 0.25, \lambda = 0.001$, CV = 2; green: $\alpha = 0.11, \lambda = 0.001$, CV = 3; blue: $\alpha = 0.04, \lambda = 0.001$, CV = 5.

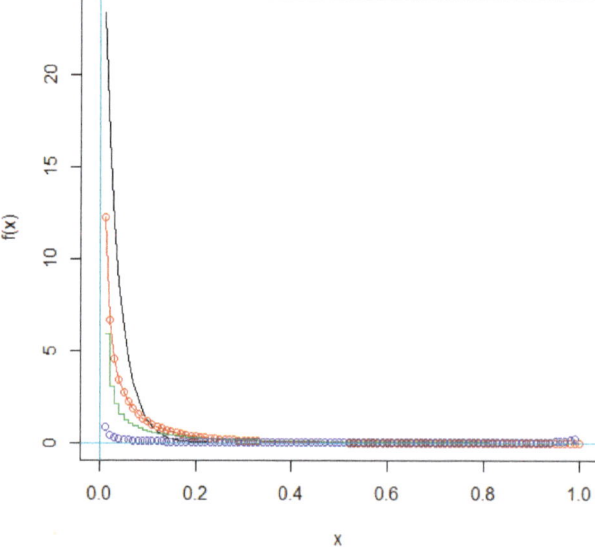

Figure 3. PDF of beta (α, β) distribution with different CV values. Black: $\alpha = 0.94, \beta = 30.39$, CV = 1; red: $\alpha = 0.21, \beta = 6.87$, CV = 2; green: $\alpha = 0.08, \beta = 2.51$, CV = 3; blue: $\alpha = 0.009, \beta = 0.285$, CV = 5.

The simulations are accomplished after 1000 runs using the *R 3.3.2* software (*R Development Core Team*, 2017) on a PC (Processor: Intel(R) CoreTM(2) Duo CPU T7100 @ 1.80GHz 1.80GHz, RAM: 2.00GB, System Type: 32-bit).

To check the accuracy of Equations (3) and (4), we estimated the coverage probability,

$$CP = \frac{\text{number of runs that Equation (3) contained true } \gamma}{1000},$$

for each parameter setting. We also computed the value of the test statistic in Equation (4), for each run. Then we considered the Shapiro–Wilk's normality test and the Q-Q plots to verify normality assumption for the proposed test statistic. Table 1 summarizes the CP values for different parameter settings.

Table 1. The CP values for different parameter settings.

Distribution	(CV_X, CV_Y)	(50,100)	(75,100)	(100,200)	(200,300)	(500,700)	(700,1000)
	(1,1)	0.945	0.947	0.951	0.953	0.959	0.960
Normal	(1,2)	0.945	0.948	0.952	0.953	0.958	0.959
	(2,3)	0.944	0.948	0.953	0.953	0.959	0.961
	(2,5)	0.946	0.950	0.950	0.955	0.956	0.960
	(1,1)	0.946	0.948	0.952	0.956	0.958	0.961
Gamma	(1,2)	0.947	0.949	0.951	0.954	0.958	0.961
	(2,3)	0.947	0.950	0.952	0.953	0.959	0.961
	(2,5)	0.945	0.949	0.952	0.956	0.958	0.962
	(1,1)	0.944	0.950	0.950	0.954	0.958	0.961
Beta	(1,2)	0.946	0.948	0.952	0.954	0.957	0.960
	(2,3)	0.945	0.947	0.952	0.954	0.958	0.959
	(2,5)	0.945	0.948	0.951	0.954	0.956	0.960

As Table 1 indicates, the CP are very close to the considered level $(1 - \alpha = 0.95)$, especially when sample size was increased, and consequently the proposed method controlled the type I error. In other words, about 95% of simulated confidence intervals contained true γ and consequently it can be accepted that Equation (3) is asymptotically confidence interval for γ. The values of CPU times (in seconds) for different parameter settings given in Table 2, verify that this approach is not too time consuming. Furthermore, Figure 4 and Table 3 illustrate the Q-Q plots and the p-values of Shapiro–Wilk's test, respectively, to study the normality of the introduced test statistic.

First column:

Up: $(CV_X, CV_Y) = (1,1)$ and $(m,n) = (50,100)$; down: $(CV_X, CV_Y) = (1,2)$ and $(m,n) = (75,100)$.

Second column:

Up: $(CV_X, CV_Y) = (1,2)$ and $(m,n) = (100,200)$; down: $(CV_X, CV_Y) = (2,3)$ and $(m,n) = (200,300)$.

Third column:

Up: $(CV_X, CV_Y) = (2,3)$ and $(m,n) = (500,700)$; down: $(CV_X, CV_Y) = (3,5)$ and $(m,n) = (700,1000)$.

Table 3 indicates that all p-values are more than 0.05 and consequently the Shapiro–Wilk's test verified the normality of the proposed test statistic. This result could also be derived from Q-Q plots. Since the points form almost a straight line, the observed quantiles are very similar to the quantiles of theoretical distribution (normal). Therefore, the simulation results verify that the asymptotic theoretical results seem to be quite satisfying for all parameter settings. Consequently our proposed approach is a good choice to perform hypothesis testing and to establish a confidence interval for the ratio of the CVs in two separate populations.

Table 2. The CPU times for running the introduced approach.

Distribution	(CV_X, CV_Y)	(50,100)	(75,100)	(100,200)	(200,300)	(500,700)	(700,1000)
Normal	(1,1)	8.64	10.08	14.08	23.09	51.92	68.67
	(1,2)	8.72	10.29	16.41	21.85	52.19	74.17
	(2,3)	9.52	9.50	15.42	21.10	51.05	65.87
	(2,5)	9.35	10.90	15.25	24.31	49.97	74.90
Gamma	(1,1)	9.45	9.02	15.16	22.13	47.05	74.92
	(1,2)	8.00	9.58	14.65	24.87	49.96	66.20
	(2,3)	9.63	9.29	14.47	21.91	52.52	66.84
	(2,5)	8.69	9.83	16.29	24.27	50.68	66.11
Beta	(1,1)	9.53	10.57	14.19	21.47	53.26	66.58
	(1,2)	9.20	9.50	14.15	24.85	48.67	75.00
	(2,3)	9.02	9.63	14.25	23.52	50.29	69.67
	(2,5)	8.75	9.17	15.73	22.89	50.79	73.95

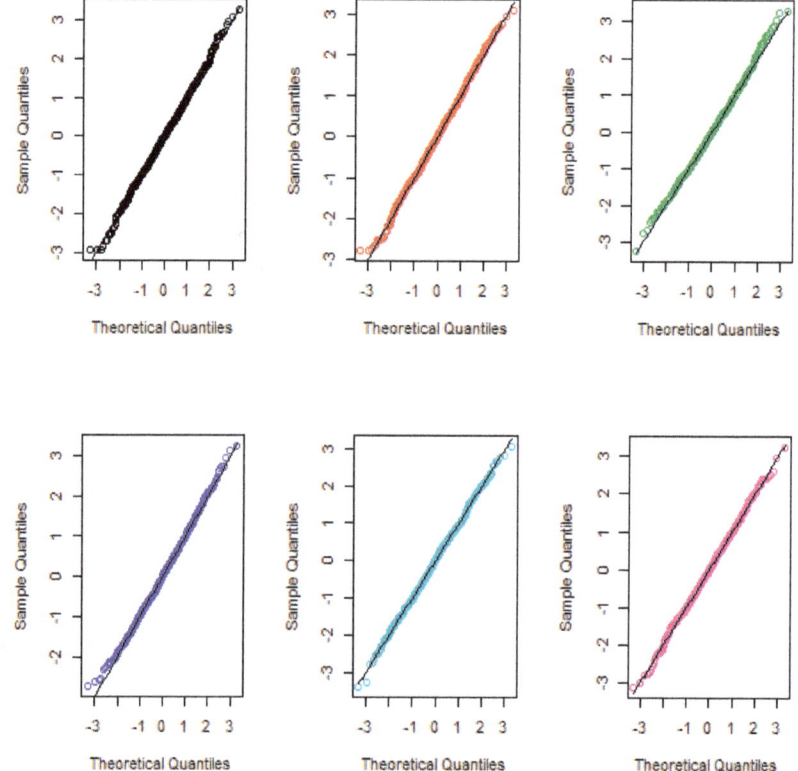

Figure 4. Q-Q plots to study the normality of the introduced test statistic.

Table 3. P-values for studying the normality of the introduced test statistic.

		(m,n)					
Distribution	(CV_X, CV_Y)	(50,100)	(75,100)	(100,200)	(200,300)	(500,700)	(700,1000)
Normal	(1,1)	0.444	0.551	0.662	0.701	0.899	0.977
	(1,2)	0.432	0.580	0.656	0.795	0.860	0.982
	(2,3)	0.408	0.600	0.602	0.718	0.859	0.943
	(2,5)	0.481	0.569	0.681	0.740	0.848	0.955
Gamma	(1,1)	0.428	0.545	0.677	0.760	0.851	0.905
	(1,2)	0.407	0.544	0.612	0.775	0.880	0,909
	(2,3)	0.484	0.508	0.611	0.708	0.855	0.940
	(2,5)	0.494	0.556	0.647	0.754	0.800	0.978
Beta	(1,1)	0.411	0.599	0.657	0.709	0.870	0.946
	(1,2)	0.489	0.585	0.652	0.763	0.841	0.978
	(2,3)	0.411	0.505	0.606	0.724	0.874	0.908
	(2,5)	0.461	0.527	0.671	0.757	0.847	0.933

4. Conclusions

Coefficient of variation is a simple but useful statistical tool to make comparisons about independent populations. In many situations two populations with different means and variances may have equal CVs. In real world applications, researchers may intend to study the similarity of the CVs in two separate populations to understand the structure of the data. Due to possible small differences of two small CVs and no strong interpretation, the ratio of CVs is more accurate than the difference of the CVs. In this study, we proposed the asymptotic distribution, derived the asymptotic confidence interval and established hypothesis testing for the ratio of the CVs in two separate populations. The results indicated that the coverage probabilities are very close to the considered level, especially when sample sizes were increased, and consequently the proposed method controlled the type I error. The values of CPU times also verified that this approach is not too time consuming. Shapiro–Wilk's normality test and Q-Q plots also verified the normality of the proposed test statistic. The results verified that the asymptotic approximations were satisfied for all simulated datasets and the introduced technique acted well in constructing CI and performing tests of hypothesis.

Author Contributions: Conceptualization, Z.Y. and D.B.; formal analysis, Z.Y. and D.B. investigation, Z.Y. and D.B.; methodology, Z.Y. and D.B.; resources, Z.Y. and D.B.; software, Z.Y. and D.B.; supervision, Z.Y. and D.B.; visualization, Z.Y. and D.B.; writing—original draft, Z.Y. and D.B.; writing—review & editing, D.B.

Funding: This research received no external funding.

Conflicts of Interest: The authors declare no conflict of interest.

References

1. Meng, Q.; Yan, L.; Chen, Y.; Zhang, Q. Generation of Numerical Models of Anisotropic Columnar Jointed Rock Mass Using Modified Centroidal Voronoi Diagrams. *Symmetry* **2018**, *10*, 618. [CrossRef]
2. Aslam, M.; Aldosari, M.S. Inspection Strategy under Indeterminacy Based on Neutrosophic Coefficient of Variation. *Symmetry* **2019**, *11*, 193. [CrossRef]
3. Iglesias-Caamaño, M.; Carballo-López, J.; Álvarez-Yates, T.; Cuba-Dorado, A.; García-García, O. Intrasession Reliability of the Tests to Determine Lateral Asymmetry and Performance in Volleyball Players. *Symmetry* **2018**, *10*, 16. [CrossRef]
4. Bennett, B.M. On an approximate test for homogeneity of coefficients of variation. In *Contribution to Applied Statistics*; Ziegler, W.J., Ed.; Birkhauser Verlag: Basel, Switzerland; Stuttgart, Germany, 1976; pp. 169–171.

5. Doornbos, R.; Dijkstra, J.B. A multi sample test for the equality of coefficients of variation in normal populations. *Commun. Stat. Simul. Comput.* **1983**, *12*, 147–158. [CrossRef]
6. Hedges, L.; Olkin, I. *Statistical Methods for Meta-Analysis*; Academic Press: Orlando, FL, USA, 1985.
7. Shafer, N.J.; Sullivan, J.A. A simulation study of a test for the equality of the coefficients of variation. *Commun. Stat. Simul. Comput.* **1986**, *15*, 681–695. [CrossRef]
8. Rao, K.A.; Vidya, R. On the performance of test for coefficient of variation. *Calcutta Stat. Assoc. Bull.* **1992**, *42*, 87–95. [CrossRef]
9. Gupta, R.C.; Ma, S. Testing the equality of coefficients of variation in k normal populations. *Commun. Stat. Theory Methods* **1996**, *25*, 115–132. [CrossRef]
10. Rao, K.A.; Jose, C.T. Test for equality of coefficient of variation of k populations. In Proceedings of the 53rd Session of International Statistical Institute, Seoul, Korea, 22–29 August 2001.
11. Pardo, M.C.; Pardo, J.A. Use of Rényi's divergence to test for the equality of the coefficient of variation. *J. Comput. Appl. Math.* **2000**, *116*, 93–104. [CrossRef]
12. Nairy, K.S.; Rao, K.A. Tests of coefficient of variation of normal population. *Commun. Stat. Simul. Comput.* **2003**, *32*, 641–661. [CrossRef]
13. Verrill, S.; Johnson, R.A. Confidence bounds and hypothesis tests for normal distribution coefficients of variation. *Commun. Stat. Theory Methods* **2007**, *36*, 2187–2206. [CrossRef]
14. Jafari, A.A.; Kazemi, M.R. A parametric bootstrap approach for the equality of coefficients of variation. *Comput. Stat.* **2013**, *28*, 2621–2639. [CrossRef]
15. Feltz, G.J.; Miller, G.E. An asymptotic test for the equality of coefficients of variation from k normal populations. *Stat. Med.* **1996**, *15*, 647–658. [CrossRef]
16. Fung, W.K.; Tsang, T.S. A simulation study comparing tests for the equality of coefficientsof variation. *Stat. Med.* **1998**, *17*, 2003–2014. [CrossRef]
17. Tian, L. Inferences on the common coefficient of variation. *Stat. Med.* **2005**, *24*, 2213–2220. [CrossRef] [PubMed]
18. Forkman, J. Estimator and Tests for Common Coefficients of Variation in Normal Distributions. *Commun. Stat. Theory Methods* **2009**, *38*, 233–251. [CrossRef]
19. Liu, X.; Xu, X.; Zhao, J. A new generalized p-value approach for testing equality of coefficients of variation in k normal populations. *J. Stat. Comput. Simul.* **2011**, *81*, 1121–1130. [CrossRef]
20. Krishnamoorthy, K.; Lee, M. Improved tests for the equality of normal coefficients of variation. *Comput. Stat.* **2013**, *29*, 215–232. [CrossRef]
21. Jafari, A.A. Inferences on the coefficients of variation in a multivariate normal population. *Commun. Stat. Theory Methods* **2015**, *44*, 2630–2643. [CrossRef]
22. Hasan, M.S.; Krishnamoorthy, K. Improved confidence intervals for the ratio of coefficients of variation of two lognormal distributions. *J. Stat. Theory Appl.* **2017**, *16*, 345–353. [CrossRef]
23. Shi, X.; Wong, A. Accurate tests for the equality of coefficients of variation. *J. Stat. Comput. Simul.* **2018**, *88*, 3529–3543. [CrossRef]
24. Miller, G.E. Use of the squared ranks test to test for the equality of the coefficients of variation. *Commun. Stat. Simul. Comput.* **1991**, *20*, 743–750. [CrossRef]
25. Nam, J.; Kwon, D. Inference on the ratio of two coefficients of variation of two lognormal distributions. *Commun. Stat. Theory Methods* **2016**, *46*, 8575–8587. [CrossRef]
26. Wong, A.; Jiang, L. Improved Small Sample Inference on the Ratio of Two Coefficients of Variation of Two Independent Lognormal Distributions. *J. Probab. Stat.* **2019**. [CrossRef]
27. Haghbin, H.; Mahmoudi, M.R.; Shishebor, Z. Large Sample Inference on the Ratio of Two Independent Binomial Proportions. *J. Math. Ext.* **2011**, *5*, 87–95.
28. Mahmoudi, M.R.; Mahmoodi, M. Inferrence on the Ratio of Variances of Two Independent Populations. *J. Math. Ext.* **2014**, *7*, 83–91.
29. Mahmoudi, M.R.; Mahmoodi, M. Inferrence on the Ratio of Correlations of Two Independent Populations. *J. Math. Ext.* **2014**, *7*, 71–82.
30. Mahmouudi, M.R.; Maleki, M.; Pak, A. Testing the Difference between Two Independent Time Series Models. *Iran. J. Sci. Technol. Trans. A Sci.* **2017**, *41*, 665–669. [CrossRef]
31. Mahmoudi, M.R.; Mahmoudi, M.; Nahavandi, E. Testing the Difference between Two Independent Regression Models. *Commun. Stat. Theory Methods* **2016**, *45*, 6284–6289. [CrossRef]

32. Mahmoudi, M.R.; Nasirzadeh, R.; Mohammadi, M. On the Ratio of Two Independent Skewnesses. *Commun. Stat. Theory Methods* **2018**, in press. [CrossRef]
33. Mahmoudi, M.R.; Behboodian, J.; Maleki, M. Large Sample Inference about the Ratio of Means in Two Independent Populations. *J. Stat. Theory Appl.* **2017**, *16*, 366–374. [CrossRef]
34. Mahmoudi, M.R. On Comparing Two Dependent Linear and Nonlinear Regression Models. *J. Test. Eval.* **2018**, in press. [CrossRef]
35. Mahmoudi, M.R.; Heydari, M.H.; Avazzadeh, Z. Testing the difference between spectral densities of two independent periodically correlated (cyclostationary) time series models. *Commun. Stat. Theory Methods* **2018**, in press. [CrossRef]
36. Mahmoudi, M.R.; Heydari, M.H.; Avazzadeh, Z. On the asymptotic distribution for the periodograms of almost periodically correlated (cyclostationary) processes. *Digit. Signal Process.* **2018**, *81*, 186–197. [CrossRef]
37. Mahmoudi, M.R.; Heydari, M.H.; Roohi, R. A new method to compare the spectral densities of two independent periodically correlated time series. *Math. Comput. Simul.* **2018**, *160*, 103–110. [CrossRef]
38. Mahmoudi, M.R.; Mahmoodi, M.; Pak, A. On comparing, classifying and clustering several dependent regression models. *J. Stat. Comput. Sim.* **2019**, in press. [CrossRef]
39. Mahmoudi, M.R.; Maleki, M. A New Method to Detect Periodically Correlated Structure. *Comput. Stat.* **2017**, *32*, 1569–1581. [CrossRef]
40. Mahmoudi, M.R.; Maleki, M.; Pak, A. Testing the Equality of Two Independent Regression Models. *Commun. Stat. Theory Methods* **2018**, *47*, 2919–2926. [CrossRef]
41. Ferguson, T.S. *A Course in Large Sample Theory*; Chapman & Hall: London, UK, 1996.

© 2019 by the authors. Licensee MDPI, Basel, Switzerland. This article is an open access article distributed under the terms and conditions of the Creative Commons Attribution (CC BY) license (http://creativecommons.org/licenses/by/4.0/).

Article

On Comparing and Classifying Several Independent Linear and Non-Linear Regression Models with Symmetric Errors

Ji-Jun Pan [1], Mohammad Reza Mahmoudi [2,*], Dumitru Baleanu [3] and Mohsen Maleki [4]

[1] College of Mathematics, Dianxi Science and Technology, Normal University, Lincang 677000, China; elinhd2013@gmail.com
[2] Department of Statistics, Faculty of Science, Fasa University, Fasa 74616 86131, Iran
[3] Department of Mathematics, Faculty of Art and Sciences, Cankaya University Balgat, Ankara 06530, Turkey; dumitru@cankaya.edu.tr
[4] Department of Statistics, Faculty of Science, Shiraz University, Shiraz 71946 85115, Iran; m.maleki.stat@gmail.com
* Correspondence: mahmoudi.m.r@fasau.ac.ir

Received: 30 May 2019; Accepted: 19 June 2019; Published: 20 June 2019

Abstract: In many real world problems, science fields such as biology, computer science, data mining, electrical and mechanical engineering, and signal processing, researchers aim to compare and classify several regression models. In this paper, a computational approach, based on the non-parametric methods, is used to investigate the similarities, and to classify several linear and non-linear regression models with symmetric errors. The ability of each given approach is then evaluated using simulated and real world practical datasets.

Keywords: comparison; Friedman test; linear regression; nonlinear regression; sign test; symmetric errors; Wilcoxon test

1. Introduction

In many situations, we aim to study the effects of variables $X_1, \ldots, X_k$ on variable Y. Simple and multiple regressions are data analysis techniques to model these effects. The authors of the references [1,2] applied simple and multiple linear regression models in different science fields, such as agriculture, biology, material, mechanical engineering, and signal processing. In many real world problems, scientists want to compare the relationship between the dependent variable and independent variables in several separate datasets.

The comparison of the correlation between the variables X and Y in two separate datasets, different techniques was provided by [3–5]. The comparison of the correlation between the variables X and Y in a dataset, and the correlation between the two variables X and W in another dataset, resulted in different methods developed by [6–10]. The correlation between the variables X and Y in a dataset, and the correlation between two variables W and Z in another dataset, were compared by different methods in [9,11,12]. The comparison and classification of two, and more simple linear regression models, have been considered in [13–16]. The comparison of two regression models has been reported in [14–22].

In the present research, we aim to compare and classify several linear and non-linear regression models that fitted on several independent datasets. The non-parametric methods are used to construct an approach to investigate the similarity and to classify the linear and non-linear regression models. A given approach is then evaluated using simulation and real world studies. The introduced approach is powerful and applicable in its ability to compare any linear or non-linear regression models.

2. Models Comparing and Classification

Assume $(X_{1j}, \ldots, X_{kj}, Y_j)$, $j = 1, \ldots, n_i$, is a sample dataset of size n_i, from $(X_1, \ldots, X_k, Y)$. The equations of m linear or non-linear regression models can be written by:

$$Y_{ij} = f_i(X_{1j}, \ldots, X_{kj}) + \varepsilon_{ij}, \; j = 1, \ldots, n_i, i = 1, \ldots, m, \qquad (1)$$

such that for $i = 1, \ldots, m$, ε_{ij}, $j = 1, \ldots, n_i$, are zero-mean symmetric random variables with unknown and equal variance σ_i^2.

By considering Equation (1), consequently, the conditional expectation of Y based on $f_i(X_1, \ldots, X_k)$, that we show it by $\theta_i(X_1, \ldots, X_k)$, is given by:

$$\theta_i(X_1, \ldots, X_k) = E\big(Y \big| f_i(X_1, \ldots, X_k)\big) = f_i(X_1, \ldots, X_k). \qquad (2)$$

In real-word problems the aim is to test the hypothesis $H_0 : \theta_1(X_1, \ldots, X_k) = \theta_2(X_1, \ldots, X_k) = \ldots = \theta_m(X_1, \ldots, X_k)$. Under the rejection of H_0, we conclude that at least two models of the m regression models are not statistically similar, and if H_0 is accepted then it can be concluded that the m regression models are statistically equal.

The regression equations can be represented by:

$$Y_i = f_i(X_1, \ldots, X_k) + \varepsilon_i, \; i = 1, \ldots, m, \qquad (3)$$

such that $Y_i = (y_1, \ldots, y_{n_i})^T$, $i = 1, \ldots, m$, are the values for the dependent variable Y, $X_1 = (x_{11}, \ldots, x_{1n_i})^T$, $\ldots$, $X_k = (x_{k1}, \ldots, x_{kn_i})^T$, $i = 1, \ldots, m$ are the values for the independent variables $(X_1, \ldots, X_k)$, $f_i(X_1, \ldots, X_k) = \big(f_i(x_{11}, \ldots, x_{k1}), \ldots, f_i(x_{1n_i}, \ldots, x_{kn_i})\big)^T$, and $\varepsilon_i = (\varepsilon_{i1}, \ldots, \varepsilon_{in_i})^T$, $i = 1, \ldots, m$, are zero-mean random variables with unknown and equal variance σ_i^2.

First, all m regression models are estimated by

$$\hat{Y}_i = \hat{f}_i(X_1, \ldots, X_k), \; i = 1, \ldots, m, \qquad (4)$$

for all $n = $ distinct points $(n_1 \cup n_2 \cup \ldots \cup n_m)$ values of $(X_1, \ldots, X_k)$, where $\hat{Y}_i = (\hat{y}_{i1}, \ldots, \hat{y}_{in})^T$, $i = 1, \ldots, m$, are the estimated values for dependent variable Y, based on ith regression model. Since ε_i, $i = 1, \ldots, m$, are zero-mean symmetric random variables, consequently, $\hat{y}_{i1}, \ldots, \hat{y}_{in}$, $i = 1, \ldots, m$, are unbiased estimators for $\theta_i(X_1, \ldots, X_k)$, $i = 1, \ldots, m$, respectively. In other words, $\hat{y}_{i1}, \ldots, \hat{y}_{in}$, $i = 1, \ldots, m$, are random variables with mean $\theta_i(X_1, \ldots, X_k)$, $i = 1, \ldots, m$.

Remark 1. *$n = $ distinct points $(n_1 \cup n_2 \cup \ldots \cup n_m)$ means that the repeated points are assumed once.*

Now, to compare the fitted regression models, the Friedman test [23–26] will be applied on n couples $(\hat{y}_{11}, \ldots, \hat{y}_{m1}), \ldots, (\hat{y}_{1n}, \ldots, \hat{y}_{mn})$.

The Friedman test that is a non-parametric alternative to the repeated measures is used to compare related datasets (datasets that are repeated on the same subjects). This test is commonly applied when dataset do not follow the parametric conditions, such as normality assumption.

Classification

In previous discussion, if H_0 is false, then we conclude that the mechanism of one model or mechanisms of some models are significantly different from the other models. However, to determine which models are significantly different from each other, the sign test or Wilcoxon test are applied in order to compare each of the regression model pairs.

3. Simulation Study

This section assesses the ability of the introduced approach simulation datasets. First, the different datasets from different regression models are produced. Then, we compute the values of the Estimated Type I error probability ($\hat{\alpha}$) and the Estimated Power ($\hat{\pi}$) of the introduced approach. For comparison, the Wilcoxon and Friedman tests are applied. The simulations are accomplished after 1000 runs and using the R 3.5.3 software (R Development Core Team, 2018) on a PC (Processor: Intel(R) CoreTM(2) Duo CPU T7100 @ 1.80GHz 1.80GHz, RAM: 2.00GB, System Type: 32-bit).

Example 1. *Assume the simple linear regression model:*

$$Y = \beta X + \varepsilon, \tag{5}$$

such that ε and X are independent.

Example 2. *Let*

$$Y = \beta_0 + \beta_1 X + \beta_2 X \varepsilon, \tag{6}$$

such that ε and X are independent.

Example 3. *Assume:*

$$Y = 1 + \beta X + \varepsilon, \tag{7}$$

such that ε and X are independent.

Example 4. *Assume the multiple linear regression model:*

$$Y = \beta_0 + \beta_1 X_1 + 2\beta_2 X_2 + \varepsilon, \tag{8}$$

such that ε, X_1 and X_2 are independent.

Example 5. *For the first dataset, assume the simple nonlinear regression model:*

$$Y = e^X + \varepsilon, \tag{9}$$

such that ε and X are independent.

For the second and the third datasets let $Y = \{e^X + \varepsilon, 1 + \beta X + \varepsilon\}$, and $Y = \{e^X + \varepsilon, 1 + \beta X + \varepsilon, 2X + \varepsilon\}$, respectively.

Figures 1 and 2 shows the density plots of the some parts of the response variable Y. As it can be seen in these figures, the density plots are symmetric, but not necessarily normal (Figure 2).

Symmetry 2019, 11, 820

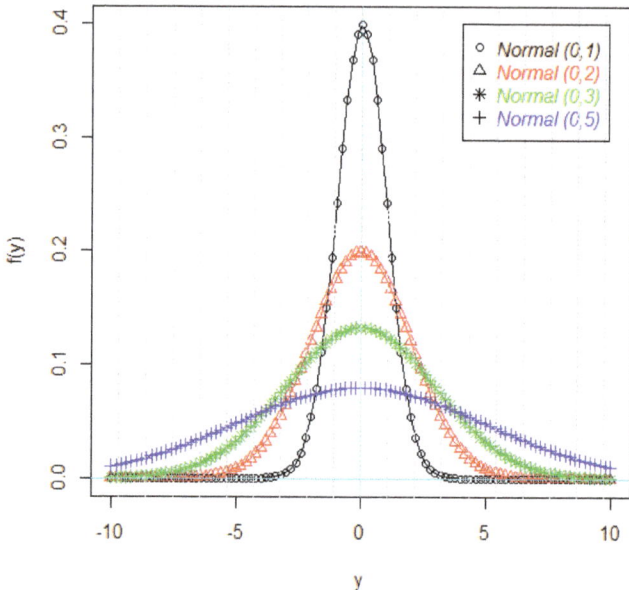

Figure 1. The density plots of the some parts of the response variable Y (Black or Pie: Normal (0,1); Red or Triangle: Normal (0,2); Green or Star: Normal (0,3); Blue or Plus: Normal (0,5)).

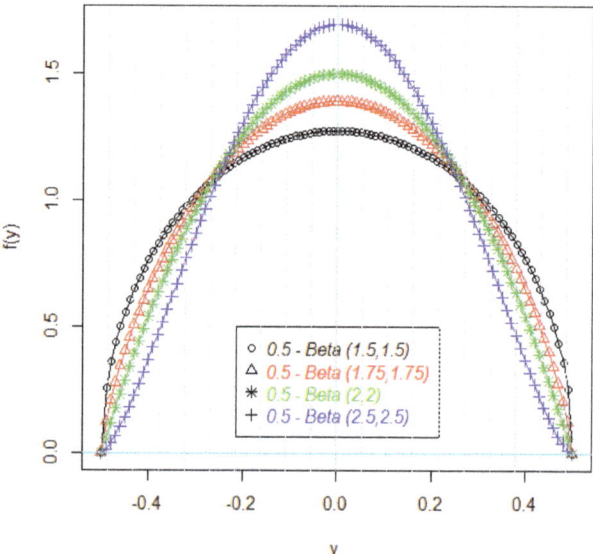

Figure 2. The density plots of the some parts of the response variable Y (Black or Pie: 0.5 − Beta (1.5,1.5); Red or Triangle: 0.5 − Beta (1.75,1.75); Green or Star: 0.5 − Beta (2,2); Blue or Plus: 0.5 − Beta (2.5,2.5)).

The values of $\hat{\alpha}$ (first four rows) and $\hat{\pi}$ (other rows) for Examples 1 to 5 are summarized in Tables 1–5, respectively. As Tables 1–5 indicate the values of $\hat{\alpha}$ are very close to size test ($\alpha = 0.05$), and consequently the introduced approach can be controlled the type I error. Also the values of $\hat{\pi}$ show that the given technique can distinguished between the null and alternative hypotheses.

Table 1. The values of $\hat{\alpha}$ and $\hat{\pi}$ for Example 1.

		β		(n_1, n_2, n_3)			
ε	X	Second	Third	(10, 10, 10)	(20, 40, 60)	(50, 75, 100)	(75, 100, 150)
Uniform $(-2,2)$	Normal $(0, 0.25)$	1	1	0.053	0.051	0.051	0.049
Uniform $(-2,2)$	Exponential (5)	1	1	0.052	0.052	0.051	0.048
Normal $(0, 0.5)$	Normal $(0, 0.25)$	1	1	0.053	0.052	0.050	0.049
Normal $(0, 0.5)$	Exponential (5)	1	1	0.053	0.052	0.050	0.049
Uniform $(-2,2)$	Normal $(0, 0.25)$	1	2	0.738	0.882	0.934	0.958
Uniform $(-2,2)$	Exponential (5)	1	2	0.753	0.801	0.950	0.981
Normal $(0, 0.5)$	Normal $(0, 0.25)$	1	2	0.754	0.854	0.945	0.972
Normal $(0, 0.5)$	Exponential (5)	1	2	0.749	0.889	0.913	0.970
Uniform $(-2,2)$	Normal $(0, 0.25)$	1	3	0.703	0.825	0.941	0.993
Uniform $(-2,2)$	Exponential (5)	1	3	0.710	0.859	0.917	0.975
Normal $(0, 0.5)$	Normal $(0, 0.25)$	1	3	0.728	0.824	0.910	0.984
Normal $(0, 0.5)$	Exponential (5)	1	3	0.707	0.864	0.934	0.953
Uniform $(-2,2)$	Normal $(0, 0.25)$	2	1	0.768	0.828	0.913	0.978
Uniform $(-2,2)$	Exponential (5)	2	1	0.703	0.824	0.928	0.951
Normal $(0, 0.5)$	Normal $(0, 0.25)$	2	1	0.794	0.846	0.930	0.968
Normal $(0, 0.5)$	Exponential (5)	2	1	0.794	0.800	0.903	0.955
Uniform $(-2,2)$	Normal $(0, 0.25)$	2	2	0.745	0.813	0.946	0.971
Uniform $(-2,2)$	Exponential (5)	2	2	0.718	0.858	0.937	0.981
Normal $(0, 0.5)$	Normal $(0, 0.25)$	2	2	0.784	0.866	0.901	0.953
Normal $(0, 0.5)$	Exponential (5)	2	2	0.726	0.821	0.944	0.999
Uniform $(-2,2)$	Normal $(0, 0.25)$	2	3	0.795	0.849	0.924	0.982
Uniform $(-2,2)$	Exponential (5)	2	3	0.755	0.856	0.928	0.961
Normal $(0, 0.5)$	Normal $(0, 0.25)$	2	3	0.763	0.845	0.936	0.988
Normal $(0, 0.5)$	Exponential (5)	2	3	0.710	0.865	0.914	0.975

Table 2. The values of $\hat{\alpha}$ and $\hat{\pi}$ for Example 2.

		$(\beta_0, \beta_1, \beta_2)$		(n_1, n_2, n_3)			
ε	X	Second	Third	(10, 10, 10)	(20, 40, 60)	(50, 75, 100)	(75, 100, 150)
Uniform $(-1,1)$	Normal $(0,1)$	$(2,1,2)$	$(2,1,2)$	0.052	0.052	0.051	0.049
Uniform $(-1,1)$	Exponential (1)	$(2,1,2)$	$(2,1,2)$	0.053	0.051	0.050	0.049
Normal $(0,2)$	Normal $(0,1)$	$(2,1,2)$	$(2,1,2)$	0.053	0.052	0.051	0.049
Normal $(0,2)$	Exponential (1)	$(2,1,2)$	$(2,1,2)$	0.052	0.052	0.051	0.049
Uniform $(-1,1)$	Normal $(0,1)$	$(2,1,2)$	$(0,2,1)$	0.770	0.843	0.903	0.981
Uniform $(-1,1)$	Exponential (1)	$(2,1,2)$	$(0,2,1)$	0.743	0.817	0.909	0.979
Normal $(0,2)$	Normal $(0,1)$	$(2,1,2)$	$(0,2,1)$	0.771	0.842	0.918	0.992
Normal $(0,2)$	Exponential (1)	$(2,1,2)$	$(0,2,1)$	0.791	0.855	0.934	0.967
Uniform $(-1,1)$	Normal $(0,1)$	$(2,1,2)$	$(3,2,1)$	0.737	0.891	0.941	0.997
Uniform $(-1,1)$	Exponential (1)	$(2,1,2)$	$(3,2,1)$	0.798	0.860	0.932	0.988
Normal $(0,2)$	Normal $(0,1)$	$(2,1,2)$	$(3,2,1)$	0.740	0.849	0.947	0.993
Normal $(0,2)$	Exponential (1)	$(2,1,2)$	$(3,2,1)$	0.712	0.827	0.916	0.997
Uniform $(-1,1)$	Normal $(0,1)$	$(0,2,1)$	$(2,1,2)$	0.782	0.837	0.932	0.966
Uniform $(-1,1)$	Exponential (1)	$(0,2,1)$	$(2,1,2)$	0.780	0.830	0.936	0.960
Normal $(0,2)$	Normal $(0,1)$	$(0,2,1)$	$(2,1,2)$	0.720	0.857	0.945	0.998
Normal $(0,2)$	Exponential (1)	$(0,2,1)$	$(2,1,2)$	0.767	0.897	0.902	0.958
Uniform $(-1,1)$	Normal $(0,1)$	$(0,2,1)$	$(0,2,1)$	0.790	0.809	0.921	0.992
Uniform $(-1,1)$	Exponential (1)	$(0,2,1)$	$(0,2,1)$	0.741	0.814	0.935	0.992
Normal $(0,2)$	Normal $(0,1)$	$(0,2,1)$	$(0,2,1)$	0.710	0.844	0.945	0.981
Normal $(0,2)$	Exponential (1)	$(0,2,1)$	$(0,2,1)$	0.760	0.871	0.906	0.972
Uniform $(-1,1)$	Normal $(0,1)$	$(0,2,1)$	$(3,2,1)$	0.776	0.807	0.919	0.969
Uniform $(-1,1)$	Exponential (1)	$(0,2,1)$	$(3,2,1)$	0.701	0.875	0.928	0.963
Normal $(0,2)$	Normal $(0,1)$	$(0,2,1)$	$(3,2,1)$	0.780	0.803	0.936	0.987
Normal $(0,2)$	Exponential (1)	$(0,2,1)$	$(3,2,1)$	0.720	0.886	0.923	0.960

Table 3. The values of $\hat{\alpha}$ and $\hat{\pi}$ for Example 3.

ε	X	β Second	β Third	(n_1, n_2, n_3) (10, 10, 10)	(20, 40, 60)	(50, 75, 100)	(75, 100, 150)
Uniform (−1, 1)	Geometric (0.4)	1	1	0.053	0.051	0.050	0.049
Uniform (−1, 1)	Binomial (2, 0.7)	1	1	0.053	0.051	0.051	0.050
Normal (0, 0.5)	Geometric (0.4)	1	1	0.053	0.051	0.051	0.050
Normal (0, 0.5)	Binomial (2, 0.7)	1	1	0.053	0.051	0.051	0.050
Uniform (−1, 1)	Geometric (0.4)	1	2	0.724	0.846	0.924	0.996
Uniform (−1, 1)	Binomial (2, 0.7)	1	2	0.734	0.813	0.942	0.952
Normal (0, 0.5)	Geometric (0.4)	1	2	0.737	0.818	0.914	0.959
Normal (0, 0.5)	Binomial (2, 0.7)	1	2	0.764	0.819	0.949	0.998
Uniform (−1, 1)	Geometric (0.4)	1	5	0.797	0.808	0.904	0.959
Uniform (−1, 1)	Binomial (2, 0.7)	1	5	0.760	0.869	0.919	0.978
Normal (0, 0.5)	Geometric (0.4)	1	5	0.793	0.843	0.917	0.988
Normal (0, 0.5)	Binomial (2, 0.7)	1	5	0.765	0.876	0.910	0.983
Uniform (−1, 1)	Geometric (0.4)	2	1	0.742	0.868	0.934	0.954
Uniform (−1, 1)	Binomial (2, 0.7)	2	1	0.730	0.810	0.925	0.966
Normal (0, 0.5)	Geometric (0.4)	2	1	0.725	0.867	0.911	0.981
Normal (0, 0.5)	Binomial (2, 0.7)	2	1	0.769	0.868	0.930	0.996
Uniform (−1, 1)	Geometric (0.4)	2	2	0.763	0.816	0.905	0.982
Uniform (−1, 1)	Binomial (2, 0.7)	2	2	0.706	0.895	0.935	0.951
Normal (0, 0.5)	Geometric (0.4)	2	2	0.723	0.866	0.909	0.981
Normal (0, 0.5)	Binomial (2, 0.7)	2	2	0.765	0.857	0.903	0.974
Uniform (−1, 1)	Geometric (0.4)	2	5	0.710	0.867	0.910	0.950
Uniform (−1, 1)	Binomial (2, 0.7)	2	5	0.764	0.837	0.904	0.981
Normal (0, 0.5)	Geometric (0.4)	2	5	0.778	0.891	0.933	0.987
Normal (0, 0.5)	Binomial (2, 0.7)	2	5	0.726	0.819	0.946	0.967

Table 4. The values of $\hat{\alpha}$ and $\hat{\pi}$ for Example 4.

X_1	X_2	$(\beta_0, \beta_1, \beta_2)$ Second	$(\beta_0, \beta_1, \beta_2)$ Third	(n_1, n_2, n_3) (10, 10, 10)	(20, 40, 60)	(50, 75, 100)	(75, 100, 150)
Uniform (0, 2)	Exponential (5)	(2, 1, 2)	(2, 1, 2)	0.052	0.052	0.050	0.049
Uniform (0, 2)	Geometric (0.3)	(2, 1, 2)	(2, 1, 2)	0.053	0.052	0.050	0.049
Binomial (3, 0.5)	Exponential (5)	(2, 1, 2)	(2, 1, 2)	0.052	0.052	0.051	0.049
Binomial (3, 0.5)	Geometric (0.3)	(2, 1, 2)	(2, 1, 2)	0.052	0.051	0.050	0.048
Uniform (0, 2)	Exponential (5)	(2, 1, 2)	(0, 2, 1)	0.734	0.893	0.923	0.961
Uniform (0, 2)	Geometric (0.3)	(2, 1, 2)	(0, 2, 1)	0.787	0.887	0.947	0.964
Binomial (3, 0.5)	Exponential (5)	(2, 1, 2)	(0, 2, 1)	0.766	0.813	0.943	0.973
Binomial (3, 0.5)	Geometric (0.3)	(2, 1, 2)	(0, 2, 1)	0.762	0.897	0.909	0.993
Uniform (0, 2)	Exponential (5)	(2, 1, 2)	(3, 2, 1)	0.706	0.866	0.936	0.966
Uniform (0, 2)	Geometric (0.3)	(2, 1, 2)	(3, 2, 1)	0.746	0.882	0.946	0.960
Binomial (3, 0.5)	Exponential (5)	(2, 1, 2)	(3, 2, 1)	0.716	0.875	0.948	0.975
Binomial (3, 0.5)	Geometric (0.3)	(2, 1, 2)	(3, 2, 1)	0.757	0.811	0.939	0.950
Uniform (0, 2)	Exponential (5)	(0, 2, 1)	(2, 1, 2)	0.792	0.866	0.936	0.985
Uniform (0, 2)	Geometric (0.3)	(0, 2, 1)	(2, 1, 2)	0.768	0.824	0.902	0.995
Binomial (3, 0.5)	Exponential (5)	(0, 2, 1)	(2, 1, 2)	0.773	0.841	0.933	0.983
Binomial (3, 0.5)	Geometric (0.3)	(0, 2, 1)	(2, 1, 2)	0.795	0.801	0.940	0.992
Uniform (0, 2)	Exponential (5)	(0, 2, 1)	(0, 2, 1)	0.790	0.891	0.912	0.953
Uniform (0, 2)	Geometric (0.3)	(0, 2, 1)	(0, 2, 1)	0.784	0.855	0.924	0.951
Binomial (3, 0.5)	Exponential (5)	(0, 2, 1)	(0, 2, 1)	0.739	0.842	0.908	0.961
Binomial (3, 0.5)	Geometric (0.3)	(0, 2, 1)	(0, 2, 1)	0.749	0.880	0.905	0.963
Uniform (0, 2)	Exponential (5)	(0, 2, 1)	(3, 2, 1)	0.745	0.854	0.918	0.956
Uniform (0, 2)	Geometric (0.3)	(0, 2, 1)	(3, 2, 1)	0.739	0.825	0.946	0.955
Binomial (3, 0.5)	Exponential (5)	(0, 2, 1)	(3, 2, 1)	0.743	0.883	0.926	0.960
Binomial (3, 0.5)	Geometric (0.3)	(0, 2, 1)	(3, 2, 1)	0.734	0.840	0.918	0.976

Table 5. The values of $\hat{\alpha}$ and $\hat{\pi}$ for Example 5.

		Y		(n₁, n₂, n₃)			
ε	X	Second	Third	(10, 10, 10)	(20, 40, 60)	(50, 75, 100)	(75, 100, 150)
Uniform (−2,2)	Normal (0, 0.5)	$e^X + \varepsilon$	$e^X + \varepsilon$	0.052	0.051	0.051	0.048
Uniform (−2,2)	Poisson (5)	$e^X + \varepsilon$	$e^X + \varepsilon$	0.053	0.051	0.051	0.049
Normal (0, 0.25)	Normal (0, 0.5)	$e^X + \varepsilon$	$e^X + \varepsilon$	0.052	0.051	0.051	0.049
Normal (0, 0.25)	Poisson (5)	$e^X + \varepsilon$	$e^X + \varepsilon$	0.052	0.051	0.050	0.048
Uniform (−2,2)	Normal (0, 0.5)	$e^X + \varepsilon$	$1 + \beta X + \varepsilon$	0.787	0.895	0.901	0.965
Uniform (−2,2)	Poisson (5)	$e^X + \varepsilon$	$1 + \beta X + \varepsilon$	0.787	0.829	0.930	0.974
Normal (0, 0.25)	Normal (0, 0.5)	$e^X + \varepsilon$	$1 + \beta X + \varepsilon$	0.725	0.848	0.912	0.991
Normal (0, 0.25)	Poisson (5)	$e^X + \varepsilon$	$1 + \beta X + \varepsilon$	0.759	0.898	0.944	0.984
Uniform (−2,2)	Normal (0, 0.5)	$e^X + \varepsilon$	$2X + \varepsilon$	0.734	0.891	0.949	0.962
Uniform (−2,2)	Poisson (5)	$e^X + \varepsilon$	$2X + \varepsilon$	0.788	0.811	0.921	0.981
Normal (0, 0.25)	Normal (0, 0.5)	$e^X + \varepsilon$	$2X + \varepsilon$	0.759	0.877	0.941	0.965
Normal (0, 0.25)	Poisson (5)	$e^X + \varepsilon$	$2X + \varepsilon$	0.704	0.868	0.948	0.989
Uniform (−2,2)	Normal (0, 0.5)	$e^X + \varepsilon$	$e^X + \varepsilon$	0.798	0.845	0.908	0.956
Uniform (−2,2)	Poisson (5)	$e^X + \varepsilon$	$e^X + \varepsilon$	0.753	0.809	0.927	0.989
Normal (0, 0.25)	Normal (0, 0.5)	$e^X + \varepsilon$	$e^X + \varepsilon$	0.731	0.865	0.910	0.990
Normal (0, 0.25)	Poisson (5)	$e^X + \varepsilon$	$e^X + \varepsilon$	0.731	0.820	0.906	0.962
Uniform (−2,2)	Normal (0, 0.5)	$1 + \beta X + \varepsilon$	$1 + \beta X + \varepsilon$	0.723	0.897	0.934	0.960
Uniform (−2,2)	Poisson (5)	$1 + \beta X + \varepsilon$	$1 + \beta X + \varepsilon$	0.799	0.807	0.949	0.982
Normal (0, 0.25)	Normal (0, 0.5)	$1 + \beta X + \varepsilon$	$1 + \beta X + \varepsilon$	0.713	0.877	0.916	0.952
Normal (0, 0.25)	Poisson (5)	$1 + \beta X + \varepsilon$	$1 + \beta X + \varepsilon$	0.743	0.872	0.925	0.965
Uniform (−2,2)	Normal (0, 0.5)	$1 + \beta X + \varepsilon$	$2X + \varepsilon$	0.725	0.892	0.901	0.996
Uniform (−2,2)	Poisson (5)	$1 + \beta X + \varepsilon$	$2X + \varepsilon$	0.795	0.886	0.944	0.959
Normal (0, 0.25)	Normal (0, 0.5)	$1 + \beta X + \varepsilon$	$2X + \varepsilon$	0.707	0.821	0.925	0.972
Normal (0, 0.25)	Poisson (5)	$1 + \beta X + \varepsilon$	$2X + \varepsilon$	0.798	0.825	0.924	0.974

4. Real Data

In this section, a practical real data is considered to study the power of the introduced approach in real world problems. Drought is a damaging natural phenomenon. To prevent this phenomenon, the hydrologists model and predict the drought datasets in a standard time period. In this research, the average monthly rainy days (1966–2010) at three Iranian synoptic stations (Fasa, Sarvestan, and Shiraz) was considered and modeled.

To model and forecast the average monthly rainy days, different polynomial regression models of orders 1 to 3 (linear, quadratic and cubic) and exponential model were fitted to datasets. The formulas of the considered models are as following:

Linear model: $Y = \beta_0 + \beta_1 X + \varepsilon$ Quadratic model:

$$Y = \beta_0 + \beta_1 X + \beta_2 X^2 + \varepsilon. \tag{10}$$

$$\text{Cubic model}: \quad Y = \beta_0 + \beta_1 X + \beta_2 X^2 + \beta_3 X^3 + \varepsilon. \tag{11}$$

$$\text{Exponential model}: \quad Y = \beta_0 + \beta_1 e^{\beta_2 X} + \varepsilon. \tag{12}$$

The numerical computations are done using the R 3.5.3 software (Library 'nlstools', lm() function for linear regression and nls() function for nonlinear regression) and Minitab 18 software.

The results of fitted regression models are summarized in Table 6. It can be observed that, for all of the stations, respectively, the polynomial regression of order 3 (cubic), and the exponential models, had the most R-square (R^2) and the least root mean square error ($RMSE$) between all fitted models.

Table 6. Indices to evaluate the fitted regression models.

Model	Station	R Square	RMSE
Linear	Fasa	0.624	1.693
	Sarvestan	0.638	1.516
	Shiraz	0.689	1.501
Quadratic	Fasa	0.734	1.350
	Sarvestan	0.743	1.285
	Shiraz	0.767	1.265
Cubic	Fasa	0.895	0.910
	Sarvestan	0.899	0.855
	Shiraz	0.976	0.529
Exponential	Fasa	0.767	0.978
	Sarvestan	0.778	0.926
	Shiraz	0.876	0.713

Now, we use the proposed approach to compare and classify these stations, for each model. The result of Friedman test is shown in Table 7. This table indicated that the fitted cubic and exponential models are significantly different in these stations ($p < 0.05$). Also, there is no significant difference between the fitted linear and quadratic models in these stations ($p > 0.05$).

Table 7. Friedman test to compare the stations.

Model	p
Linear	0.123
Quadratic	0.224
Cubic	<0.001
Exponential	<0.001

As Table 8 indicates, we can classify the stations in two clusters, for cubic and exponential models. First cluster: Fasa and Sarvestan, and second cluster: Shiraz.

Table 8. Wilcoxon test to compare and classify the stations.

Model		Stations	p
Cubic	Pair 1	Shiraz - Fasa	0.011
	Pair 2	Shiraz - Sarvestan	0.003
	Pair 3	Fasa - Sarvestan	0.144
Exponential	Pair 1	Shiraz - Fasa	0.019
	Pair 2	Shiraz - Sarvestan	<0.001
	Pair 3	Fasa - Sarvestan	0.112

5. Conclusions

In many real world problems, researchers wish to compare and classify the regression models in several datasets. In this paper, the non-parametric methods were used to construct an approach to investigate the similarity of some linear and non-linear regression models with symmetric errors. Particular approaches were evaluated using simulation and practical datasets. A simulation study indicated that the introduced approach controlled the Type I error. Also the proposed technique distinguished well between null and alternative hypotheses. The introduced approach also had many advantages. First, it was powerful. Second, it was not too computational. Third, it could be applied to compare any linear or non-linear regression models. Fourth, this method did not need the normality of errors and could be applied for all models with symmetric errors.

Author Contributions: Conceptualization, J.-J.P., M.R.M. and D.B.; Formal analysis, M.R.M., D.B. and M.M.; Investigation, J.-J.P., M.R.M. and D.B.; Methodology, J.-J.P., M.R.M., D.B. and M.M.; Project administration, J.-J.P.; Software, J.-J.P., M.R.M., D.B. and M.M.; Supervision, J.-J.P., M.R.M. and D.B.; Validation, J.-J.P., M.R.M. and D.B.; Visualization, J.-J.P., M.R.M. and D.B.; Writing—Original Draft, J.-J.P. and M.R.M.; Writing—Review and Editing, J.-J.P., M.R.M., D.B. and M.M.

Funding: This research received no external funding.

Conflicts of Interest: The authors declare no conflict of interest.

References

1. Wan, J.; Zhang, D.; Xu, W.; Guo, Q. Parameter Estimation of Multi Frequency Hopping Signals Based on Space-Time-Frequency Distribution. *Symmetry* **2019**, *11*, 648. [CrossRef]
2. Sajid, M.; Shafique, T.; Riaz, I.; Imran, M.; Jabbar Aziz Baig, M.; Baig, S.; Manzoor, S. Facial asymmetry-based anthropometric differences between gender and ethnicity. *Symmetry* **2018**, *10*, 232. [CrossRef]
3. Mahmouudi, M.R.; Maleki, M.; Pak, A. Testing the Difference between Two Independent Time Series Models. *Iran. J. Sci. Technol. Trans. A Sci.* **2017**, *41*, 665–669. [CrossRef]
4. Fisher, R.A. On the Probable Error of a Coefficient of Correlation Deduced from a Small Sample. *Metron* **1921**, *1*, 3–32.
5. Mahmoudi, M.R.; Mahmoodi, M. Inferrence on the Ratio of Correlations of Two Independent Populations. *J. Math. Ext.* **2014**, *7*, 71–82.
6. Howell, D.C. *Statistical Methods for Psychology*, 6th ed.; Thomson Wadsworth: Stamford, CT, USA, 2007.
7. Hotelling, H. The Selection of Variates for Use in Prediction with Some Comments on the General Problem of Nuisance Parameters. *Ann. Math. Stat.* **1940**, *11*, 271–283. [CrossRef]
8. Williams, E.G. The Comparison of Regression Variables. *J. R. Stat. Soc. Ser. B* **1959**, *21*, 396–399. [CrossRef]
9. Steiger, J.H. Tests for Comparing Elements of a Correlation Matrix. *Psychol. Bull.* **1980**, *87*, 245–251. [CrossRef]
10. Meng, X.; Rosenthal, R.; Rubin, D.B. Comparing Correlated Correlation Coefficients. *Psychol. Bull.* **1992**, *111*, 172–175. [CrossRef]
11. Peter, C.C.; Van Voorhis, W.R. *Statistical Procedures and Their Mathematical Bases*; McGraw-Hill: New York, NY, USA, 1940.
12. Raghunathan, T.E.; Rosenthal, R.; Rubin, D.B. Comparing Correlated but Nonoverlapping Correlations. *Psychol. Methods* **1996**, *1*, 178–183. [CrossRef]
13. Liu, W.; Jamshidian, M.; Zhang, Y. Multiple Comparison of Several Linear Regression Lines. *J. R. Stat. Soc. Ser. B* **2004**, *99*, 395–403.
14. Liu, W.; Hayter, A.J.; Wynn, H.P. Operability Region Equivalence: Simultaneous Confidence Bands for the Equivalence of Two Regression Models Over Restricted Regions. *Biom. J.* **2007**, *49*, 144–150. [CrossRef] [PubMed]
15. Liu, W.; Jamshidian, M.; Zhang, Y.; Bertz, F.; Han, X. Pooling Batches in Drug Stability Study by Using Constant-width Simultaneous Confidence Bands. *Stat. Med.* **2007**, *26*, 2759–2771. [CrossRef] [PubMed]
16. Liu, W.; Jamshidian, M.; Zhang, Y.; Bertz, F.; Han, X. Some New Methods for the Comparison of Two Linear Regression Models. *J. Stat. Plan. Inference* **2007**, *137*, 57–67. [CrossRef]
17. Hayter, A.J.; Liu, W.; Wynn, H.P. Easy-to-Construct Confidence Bands for Comparing Two Simple Linear Regression Lines. *J. Stat. Plan. Inference* **2007**, *137*, 1213–1225. [CrossRef]
18. Jamshidian, M.; Liu, W.; Bretz, F. Simultaneous Confidence Bands for all Contrasts of Three or More Simple Linear Regression Models over an Interval. *Comput. Stat Data Anal.* **2010**, *54*, 1475–1483. [CrossRef]
19. Marques, F.J.; Coelho, C.A.; Rodrigues, P.C. Testing the equality of several linear regression models. *Comput. Stat.* **2016**, *32*, 1453–1480. [CrossRef]
20. Mahmoudi, M.R.; Mahmoudi, M.; Nahavandi, E. Testing the Difference between Two Independent Regression Models. *Commun. Stat. Theory Methods* **2016**, *45*, 6284–6289. [CrossRef]
21. Mahmoudi, M.R. On Comparing Two Dependent Linear and Nonlinear Regression Models. *J. Test. Eval.* **2018**, *47*, 449–458. [CrossRef]
22. Mahmoudi, M.R.; Maleki, M.; Pak, A. Testing the Equality of Two Independent Regression Models. *Commun. Stat. Theory Methods* **2018**, *47*, 2919–2926. [CrossRef]
23. Conover, W.J. *Practical Nonparametric Statistics*, 3rd ed.; John Wiley: Hoboken, NJ, USA, 1980.

24. Friedman, M. The Use of Ranks to Avoid the Assumption of Normality Implicit in the Analysis of Variance. *J. R. Stat. Soc. Ser. B* **1937**, *32*, 675–701. [CrossRef]
25. Friedman, M. A Correction: The Use of Ranks to Avoid the Assumption of Normality Implicit in the Analysis of Variance. *J. R. Stat. Soc. Ser. B* **1939**, *34*, 109. [CrossRef]
26. Friedman, M. A Comparison of Alternative Tests of Significance for the Problem of *m* Rankings. *Ann. Math. Stat.* **1940**, *11*, 86–92. [CrossRef]

 © 2019 by the authors. Licensee MDPI, Basel, Switzerland. This article is an open access article distributed under the terms and conditions of the Creative Commons Attribution (CC BY) license (http://creativecommons.org/licenses/by/4.0/).

Article

Vibration Analysis of a Guitar considered as a Symmetrical Mechanical System

Mariana D. Stanciu [1],*, Sorin Vlase [1] and Marin Marin [2]

1. Department of Mechanical Engineering, Faculty of Mechanical Engineering, Transilvania University of Brașov, B-dul Eroilor 29, 500036 Brașov, Romania; svlase@unitbv.ro
2. Faculty of Mathematics and Computer Science, Transilvania University of Brașov, B-dul Eroilor 29, 500036 Brașov, Romania; m.marin@unitbv.ro
* Correspondence: mariana.stanciu@unitbv.ro

Received: 11 April 2019; Accepted: 27 May 2019; Published: 28 May 2019

Abstract: This paper aimed to use the symmetry that exists to the body of a guitar to ease the analysis behavior to vibrations. Symmetries can produce interesting properties when studying the dynamic and steady-state response of such systems. These properties can, in some cases, considerably decrease the effort made for dynamic analysis at the design stage. For a real guitar, these properties are used to determine the eigenvalues and eigenvectors. Finite element method (FEM) is used for a numerical modeling and to prove the theoretically determined properties in this case. In this paper, different types of guitar plates related to symmetrical reinforcement patterns were studied in terms of modal analysis performed using finite element analysis (FEA). The dynamic response differs in terms of amplitude, eigenvalues, modal shapes in accordance with number and pattern of stiffening bars. In this study, the symmetrical and asymmetric modes of modal analysis were highlighted in the case of constructive symmetrical structures.

Keywords: symmetric geometry; guitar's plate; modal analysis; skew symmetric eigenmodes

1. Introduction

The dynamical study of mechanical systems with elastic elements that present symmetries or have identical parts occurs in many engineering fields, especially in mechanical engineering and civil engineering, but also in the automotive industry and aerospace engineering. In all engineering fields, there are certain products or parts of products or components containing repetitive or identical elements where different types of symmetry can occur. The study of such systems was made by many researchers [1–3]. In civil engineering for example, most buildings, works of art or halls have, in their structure, identical parts and symmetries. This has happened since antiquity for different causes. First, as an easier, faster and a cheaper way to design, then for easy manufacturing and (less important for engineers but important to the beneficiaries) for aesthetic reasons. Systems consisting of two identical parts represent a particular case of these types of problems. Such systems exhibit interesting vibration properties which computation is presented in literature [4–6]. Depending on the types of structural elements (rod and shell), two methods were developed, namely asymptotic and variational, which are presented by Le [7] from a mathematical point of view, emphasizing the formulation of boundary-value problems. Zhou [8] presents an exact solution of free vibration of a thin rectangular plate attached with sprung masses based on differential equations.

For this type of problem, its eigenmodes can be classified in symmetrical modes and skew-symmetrical modes. The guitar is a vibrant mechanical system where we can easily identify a symmetry plan in terms of model geometry. For this reason, it can fit into the class of mechanical systems described above and the vibration study of it can be much simplified. In this paper, the symmetrical system of the guitar is analyzed and the vibration properties that are the result of considering this symmetry are revealed. An example for a real system is made in the second part of the paper. Derveaux et al. [9] developed a guitar model based on advanced numerical methods to simulate the three-dimensional sound-pressure field of a guitar in the time domain. Due to the complexity of the problem, the main parts of guitar (plates, body, strings, air from cavity, holes) that are implied in acoustic radiation and vibration were simplified in terms of mathematical models and interactions. Compare to [9], in this paper, the frequency spectrum and eigenmodes of different symmetric types of guitar plates were analyzed and in future works, the guitars bodies with different bars pattern will be numerical modeled. Elejabarrieta et al. [10] analyzed the influence of each component from acoustic box of guitar on the vibrational behavior using finite element method in a progressive manner of constructive system.

These types of problems occur frequently in practical applications; many mechanical systems exhibit different kinds of symmetry properties, resulting from a design process for constructive, simplistic, logistic or cost considerations. Knowing these properties allows to increase the precision of calculations in such issues. Properties determined by symmetries have been observed by researchers and are mainly used in the static analysis. They are presented in the classical courses of Strength of Material or Structural Analysis. From a historical point of view, the symmetries in mechanics have been studied by mathematicians [11,12]. Symmetry effects occur in the writing the motion equations, but the solutions can be symmetric and antisymmetric according to the assumed boundary conditions [7]. In accordance with [13], the methods of mechanics (Lagrangian and Hamiltonian) use symmetry to solve complex problems. Moreover, these methods and symmetry principles are based on the developments of string theory. In case of the sandwich plate, [13] present the solution of symmetric vibration and the effects of the elastic, geometric and inertia parameters on dynamic behavior of plate.

Unfortunately, there are fewer applications in practice. In January 2018, a special issue of the Symmetry magazine dedicated to applications in Structural Mechanics (Civil Engineering and Symmetry - 2018, A special issue of Symmetry -ISSN 2073-8994) was launched. A European project was also funded for the study of these types of problems (Mechanics and Symmetry in Europe: the FP5-Human Potential). The latter held courses at the Solid Mechanics Centre—International Center for Mechanical Sciences (CISM) at UDINE (similarly, the Symmetry and Group Theoretical Methods in Mechanics held lectures in the International Centre for Mechanical Sciences in September 7, 2015–September 11, 2015) [14–19]. In the vibration field, the effect of symmetries was less used, but there are papers that began to study this type of problem [20,21]. There are, however, many situations that can be studied and that is why paper aims to complete the studied case.

The aim of this study is to present the solutions of eigenvalues and eigenvectors based on symmetry principles in case of a symmetric structure as guitar plate. The effect of inertia parameters (fan struts and bars) with respect the symmetry provided in this paper can be used as benchmark for optimal soundboard strutting system of guitar during the manufacturing.

2. Guitar as a Mechanical System Subjected to Vibration

Being a complex structure, modeling of the guitar requires the identification of the component parts from the point of view of the geometric model represented by them, the system of loads acting on them, as well as the mechanical connections between the structures and the ground. Thus, in the structure of the guitar, nearly all types of geometric models can be identified by the strength of the materials, namely: wire, plates, trusses, massive blocks (Figure 1). In the paper the object of the main theoretical interest is the structure of plates with different degrees of complexity and with various stiffening systems [21–23].

Consider the guitar plate, presented in Figure 2. The density of the material is ρ. This it is in transverse bending vibration.

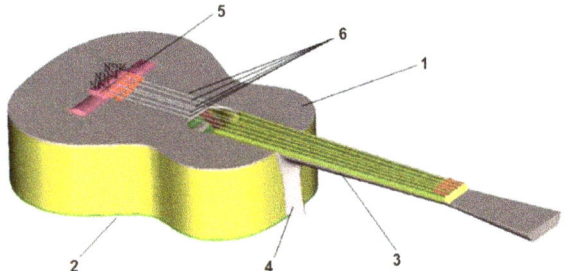

Figure 1. The main parts of a guitar: 1—top plate; 2—back plate; 3—guitar's neck; 4—heel; 5—bridge; 6—strings.

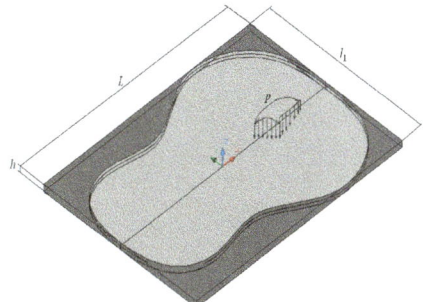

Figure 2. The original geometry of a guitar plate.

Based on Kirchhoff's hypotheses and applying the d'Alembert principle or another energy method, it is obtained the dynamic response of the plate [23,24]:

$$\Delta\Delta w(x,y,t) + \frac{\rho h}{D}\frac{\partial^2 w(x,y,t)}{\partial t^2} = \frac{p(x,y,t)}{D}, \qquad (1)$$

where $w(x,y,t)$—the normal instantaneous displacement on the median surface of the plate; $D = \frac{Eh^3}{12(1-\nu^2)}$—bending modulus of the plate, h—thickness of plate, E—the modulus of elasticity and ν—Poisson's coefficient; $p(x,y,t)$—the distributed load acting perpendicular to the median surface of the plate; $\Delta = \frac{\partial^2}{\partial x^2} + \frac{\partial^2}{\partial y^2}$—Laplacian differential operator.

If $p(x,y,t) = 0$, from (1) it results in the free transverse vibration Equations (2):

$$\frac{\partial^4 w(x,y,t)}{\partial x^4} + 2\frac{\partial^4 w(x,y,t)}{\partial x^2 \partial y^2} + \frac{\partial^4 w(x,y,t)}{\partial y^4} = -\rho\frac{h}{D}\frac{\partial^2 w(x,y,t)}{\partial t^2}, \qquad (2)$$

To solve (1), the Fourier-Bernoulli method is used, finding a solution under the form:

$$w(x,y,t) = w(x,y) \cdot w(t), \qquad (3)$$

Using (3) in (2) results in:

$$\frac{D}{\rho h}\frac{1}{w(x,y)}\left[\frac{d^4 w(x,y)}{dx^4} + 2\frac{d^4 w(x,y)}{dx^2 dy^2} + \frac{d^4 w(x,y)}{dy^4}\right] = -\frac{1}{w(t)}\frac{d^2 w_t(t)}{dt^2} = \omega^2, \qquad (4)$$

where ω is an arbitrary constant.

From (4), we have:

$$\begin{cases} \frac{d^2 w_t(t)}{dt^2} + \omega^2 w_t(t) = 0 \\ \frac{d^4 w(x,y)}{dx^4} + 2\frac{d^4 w(x,y)}{dx^2 dy^2} + \frac{d^4 w(x,y)}{dy^4} = \omega^2 \frac{\rho h}{d} w(x,y) \end{cases}, \qquad (5)$$

For Equation (5), the solution is [25–27]:

$$w_t(t) = A^{(t)} \cos(\omega t - \phi), \qquad (6)$$

$A^{(t)}$ and ϕ are integration constants depending on boundary conditions.

A good approximation for the guitar domain is a rectangular domain (Figure 2). In this case, the second Equation (5) should satisfy the boundary conditions. Some common boundary conditions are:

a. In case of plate with one free edge (where $x = x_0$), we assume that the bending moment M_y and shear force T_{zx} are zero. Thus, the condition where the bending moment along the free edge is zero, is:

$$\begin{aligned} \left(\frac{\partial^2 w(x,y)}{\partial x^2} + \nu \frac{\partial^2 w(x,y)}{\partial y^2}\right)\bigg|_{x=x_0} &= 0, \\ \left(\frac{\partial^3 w(x,y)}{\partial x^3} + (2-\nu)\frac{\partial^3 w(x,y)}{\partial x \partial y^2}\right)\bigg|_{x=x_0} &= 0, \end{aligned} \qquad (7)$$

b. In case of the plate being supported at all edges ($x = x_0$), the displacement are null and along the all edges, the bending moment is zero, too. In Equation (8), the boundary conditions are readily seen to be satisfied exactly:

$$\begin{aligned} w(x,y)\big|_{x=x_0} &= 0 \\ \left(\frac{\partial^2 w(x,y)}{\partial x^2} + \nu \frac{\partial^2 w(x,y)}{\partial y^2}\right)\bigg|_{x=x_0} &= 0 \end{aligned} \qquad (8)$$

c. For a plate with one clamped edge ($x = x_0$), the displacement and rotation in a perpendicular plan to the side are null:

$$\begin{aligned} w(x,y)\big|_{x=x_0} &= 0 \\ \left(\frac{\partial w}{\partial x}\right)\bigg|_{x=x_0} &= 0 \end{aligned} \qquad (9)$$

In [7], Le makes the difference between the clamped edge and the fixed edge of a structure considering based on the variational principle.

When the contour is not rectangular or circular and when the support is continuous, as is the case with the guitar plates, the boundary conditions become more complicated. In the particular case of the guitar plate supported at all edges, the boundary conditions (8) result in:

$$\begin{cases} w(x,y)\big|_{\substack{x=0 \\ x=a}} = 0, \left(\frac{\partial^2 w(x,y)}{\partial x^2} + \nu \frac{\partial^2 w(x,y)}{\partial y^2}\right)\bigg|_{\substack{x=0 \\ x=a}} = 0 \\ w(x,y)\big|_{\substack{y=0 \\ y=b}} = 0, \left(\frac{\partial^2 w(x,y)}{\partial y^2} + \nu \frac{\partial^2 w(x,y)}{\partial x^2}\right)\bigg|_{\substack{y=0 \\ y=b}} = 0 \end{cases}, \qquad (10)$$

It can be shown that the function $w(x,y)$ in the conditions (10) is expressed by (11):

$$w(x,y) = A^{(x,y)} \sin\frac{\pi m}{L} x \sin\frac{\pi n}{l_1} y, (m, n = 1, 2, \ldots), \qquad (11)$$

where m and n are the numbers of vibration nodes.

The eigenvalues corresponding of the plate with the eigenmode "*mn*" are:

$$\omega_{m,n} = \pi^2\left(\frac{m^2}{L^2} + \frac{n^2}{l_1^2}\right)\sqrt{\frac{D}{\rho h}}, (m,n = 1,2,\ldots), \quad (12)$$

According to Equation (12) the considered rectangular plate has duplicate eigenfrequencies due to an arbitrary linear combination of the two modes [28]. Thus, one of the two eigenfrequencies will hold the line, while the other varies with the parameters of the system [9].

3. Properties of a System Consisting of Two Identical Parts

From the big class of systems that have different types of symmetry, we will study a particular one: a system made by two identical parts. Two example are presented in Figure 3a,b, [12,19].

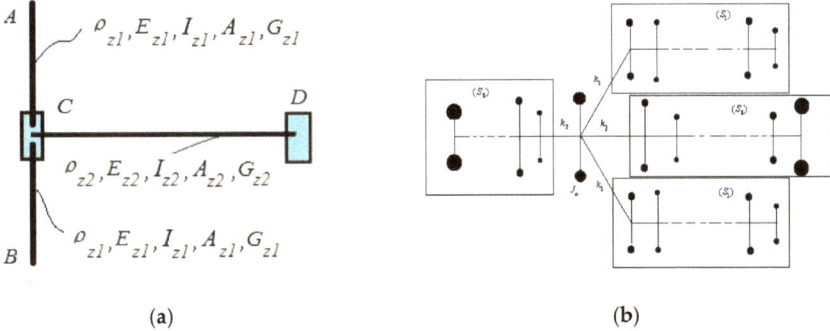

Figure 3. Schemes of systems with two identical parts: (**a**) a continuous mechanical system with two identical bars [19]; (**b**) model with wheels of a two engine acting a transmission (a truck used in the oil industry) [12].

From a mechanical point of view, the guitar can be reduced to a structure consisting of two parts: a bar represented by the neck of the guitar that is clamped on the body and the sound box (guitar body), which is made of coupled plates. The strings system that loads the two components is compounded by simple and complex wires which give an asymmetry of the entire acoustical structure.

If the strings are neglected, from the geometrical point of view, the guitar plate has a symmetry plan *(xz)* and can be considered as being composed of two identical parts connected to each other, as can be seen in Figure 4 [29]. This symmetry is also respected as material so that the wood used in the structure of the guitar plates has a symmetrical structure. In the following, we will highlight the vibration properties of such a system.

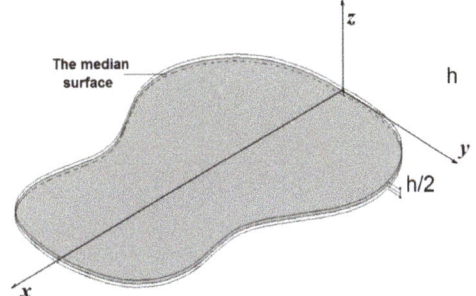

Figure 4. The symmetric guitar plate related to longitudinal axis (x).

To determine the mechanical response of such type of structure we try to model the system. The whole structure (S) can be considered as being composed by two identical sub-structure, denoted with (S_1) and (S_r) (Figures 3 and 4). We denote by Δ_a the common nodes of the two structures (S_1), with Δ_l the nodes of the left structure (S_1), different from Δ_a and with Δ_r the nodes of the right structure (S_r), different from Δ_a. The equations of the undamped free vibrations of the entire structure (S), the left substructure (S_1) (Equation (13)) and the right substructure (S_r) (Equation (14)) are, respectively [4,28–30]:

$$\begin{bmatrix} m_a & m_{ab} \\ m_{ab} & m_b \end{bmatrix} \begin{Bmatrix} \ddot{\Delta}_l \\ \ddot{\Delta}_a \end{Bmatrix} + \begin{bmatrix} k_a & k_{ab} \\ k_{ab}^T & k_b \end{bmatrix} \begin{Bmatrix} \Delta_l \\ \Delta_a \end{Bmatrix} = 0, \qquad (13)$$

$$\begin{bmatrix} m_a & m_{ab} \\ m_{ab} & m_b \end{bmatrix} \begin{Bmatrix} \ddot{\Delta}_r \\ \ddot{\Delta}_a \end{Bmatrix} + \begin{bmatrix} k_a & k_{ab} \\ k_{ab}^T & k_b \end{bmatrix} \begin{Bmatrix} \Delta_r \\ \Delta_a \end{Bmatrix} = 0, \qquad (14)$$

$$\begin{bmatrix} m_a & 0 & m_{ab} \\ 0 & m_a & m_{ab} \\ m_{ab} & m_{ab} & m_{bb} \end{bmatrix} \begin{Bmatrix} \ddot{\Delta}_l \\ \ddot{\Delta}_r \\ \ddot{\Delta}_a \end{Bmatrix} + \begin{bmatrix} k_a & 0 & k_{ab} \\ 0 & k_a & k_{ab} \\ k_{ab}^T & k_{ab}^T & k_{bb} \end{bmatrix} \begin{Bmatrix} \Delta_l \\ \Delta_r \\ \Delta_a \end{Bmatrix} = 0, \qquad (15)$$

It is obvious that the matrices involved in the Equations (13) and (14) are the same. This means that the problem of eigenvalues and eigenvectors offer the same solution in the both cases. For the full system (Equation (15)), we have the following property (proved for a more general context in [26]):
P1—the eigenvalues for the subsystem (S_l) (for the differential, Equations (1) or (2)) are eigenvalues for the system (S)).

This property expresses the fact that the eigenvalues corresponding to the subsystem (S_l) verified the eigenvalues problem for the whole system (S). That means that the solutions of Equations (16) are also solutions of the algebraic Equation (17):

$$\begin{vmatrix} k_a - \omega^2 m_a & k_{ab}^T - \omega^2 m_{ab} \\ k_{ab}^T - \omega^2 m_{ab} & k_b - \omega^2 m_b \end{vmatrix} = 0, \qquad (16)$$

$$\begin{vmatrix} k_a - \omega^2 m_a & 0 & k_{ab} - \omega^2 m_{ab} \\ 0 & k_a - \omega^2 m_a & k_{ab} - \omega^2 m_{ab} \\ k_{ab}^T - \omega^2 m_{ab}^T & k_{ab}^T - \omega^2 m_{ab}^T & k_b - \omega^2 m_b \end{vmatrix} = 0, \qquad (17)$$

which implies that the polynomial in ω^2 expressed by the Equation (17) is divided by the polynomial in ω^2 expresses by the Equation (16).

If we denote:

$$A_{11} = \begin{bmatrix} k_a - \omega^2 m_a \end{bmatrix}; A_{13} = A_{31} = \begin{bmatrix} k_{ab} - \omega^2 m_{ab} \end{bmatrix}; A_{33} = \begin{bmatrix} k_b - \omega^2 m_b \end{bmatrix}, \qquad (18)$$

and:

$$S = \begin{bmatrix} A_{11} & 0 & A_{13} \\ 0 & A_{11} & A_{13} \\ A_{31} & A_{31} & A_{33} \end{bmatrix}, \qquad (19)$$

the characteristic Equation of (17) can be written:

$$\det \begin{bmatrix} A_{11} & 0 & A_{13} \\ 0 & A_{11} & A_{13} \\ A_{31} & A_{31} & A_{33} \end{bmatrix} = \det[S] = 0 \qquad (20)$$

To determine the eigenvectors, once the eigenvalues are known, it must to solve the linear system [17,19,31]:

$$[S]\{\Phi\} = 0, \quad (21)$$

where we noted with $\{\Phi\}$ as the eigenvectors. Any eigenvalue corresponds to an eigenvector. Concerning these, the following two properties are interesting for our research: **P2**—for the common eigenvalues of the system presented in Figure 3a and of the system presented in Figure 3b, the eigenvectors are of the form:

$$\Phi = \begin{Bmatrix} \Phi_1 \\ -\Phi_1 \\ 0 \end{Bmatrix}, \quad (22)$$

The components of the eigenmodes, corresponding to the two identical parts are skew symmetric, the other components are zero-we call these skewsymmetric eigenmodes.

Proof: the existence of common eigenvalues for the system presented in Figure 3a,b is proven by the previous presented theorem P1. With the eigenvalues obtained from (20), the linear homogeneous system offers the eigenvectors:

$$\begin{bmatrix} A_{11} & 0 & A_{13} \\ 0 & A_{11} & A_{13} \\ A_{31} & A_{31} & A_{33} \end{bmatrix} \begin{Bmatrix} \Phi_l \\ \Phi_r \\ \Phi_a \end{Bmatrix} = \begin{Bmatrix} 0 \\ 0 \\ 0 \end{Bmatrix}, \quad (23)$$

with:

$$\det A_{11} = 0, \quad (24)$$

Equation (24) implies that we can find a vector Φ_l, so that:

$$A_{11}\Phi_l = 0, \quad (25)$$

and then (23) becomes:

$$A_{13}\Phi_a = 0, \quad (26)$$

$$A_{11}\Phi_r + A_{13}\Phi_a = 0, \quad (27)$$

$$A_{31}(\Phi_l + \Phi_r) + A_{33}\Phi_a = 0, \quad (28)$$

From (26), we immediately have:

$$\Phi_a = 0, \quad (29)$$

and by replacing in (28), we obtain $\Phi_l = \Phi_r$, which also verifies (27) if we consider (25). If we note: $\Phi_l = \Phi_1$, we easily obtain (22).

P3—for the other eigenvalues, (not obtained from (S_l)), the eigenvectors are of the form (symmetric eigenmodes) [32,33]:

$$\Phi = \begin{Bmatrix} \Phi_1 \\ \Phi_1 \\ \Phi_3 \end{Bmatrix}, \quad (30)$$

The components of the eigenmodes corresponding to the two identical beams are the same—we call this symmetric eigenmodes [34]. □

Proof: for these eigenvalues we have to solve the system (23), considering $\det A_{11} \neq 0$, or:

$$A_{11}\Phi_l + A_{13}\Phi_a = 0, \quad (31)$$

$$A_{11}\Phi_r + A_{13}\Phi_a = 0, \tag{32}$$

$$A_{31}(\Phi_l + \Phi_r) + A_{33}\Phi_a = 0 \tag{33}$$

By subtracting (31) from (32), the following can be obtained:

$$A_{11}(\Phi_l - \Phi_r) = 0 \tag{34}$$

However, $\det A_{11} \neq 0$, then we have $\Phi_l - \Phi_r = 0$ so $\Phi_l = \Phi_r = \Phi_1$. □

4. Finite Element Analysis (FEA) of Guitar Plates

The behavior of wood (a commonly used material for the guitar plate) under the action of sound is influenced by the sound energy that comes into contact with the wood and, on the other hand, by the quality of the wood material. It is known that wood is an orthotropic material characterized by three planes of symmetry. The macro and microscopic structure of the wood: cell membrane construction, dimensions and cohesion fibers, chemicals structure, the humidity and temperature of the wood, the elastic properties and orientation of the structure with respect to the dynamic force influences the dynamic response of plates [35]. As a result of vibrations, the internal friction occurring in the wood transforms the original sound energy into a modified sound energy, obtaining a resonance phenomenon as well as a caloric energy due to intermolecular energy exchanges.

The most important acoustic property of wood is the ability to receive sounds with a frequency close to or identical to the frequency of its membranes. This produces the resonance phenomenon that leads to the amplification of sounds and their timber enriched with overtones due to its complex and anisotropic structure. Depending on the external excitation frequency period, the resonance occurs whenever the pulse of forced vibration passes in the vicinity of eigenfrequency of the wood [28,29]. The complex phenomena that develop during cyclical stresses are structural and acoustic in nature, interdependent with each other. From this point of view, the analytical theories underlying the dynamic guitars' response mainly deal with one aspect, which is why a series of simplifying hypotheses are introduced. However, the calculations that use these theories are numerous and laborious. At present, the mathematical and analytical modeling of the guitar is replaced by numerical methods: the finite element method (FEM) or finite difference method (DFM) underlying the numerical modeling software [30,31]. For the modal analysis of the guitar plates, nine symmetric variants of plates were geometrically shaped respecting the dimensions practiced for a real guitar: a simple plate with an acoustic hole (denoted PSah) without stiffening elements (Figure 5a); a plate with three transversal reinforcing bars (denoted P3BT) (Figure 5b); a plate with three radial resonance bars, symmetric disposed related to longitudinal axis of plate (denoted P3BR) (Figure 5c); a plate with five radial resonance bars, symmetric disposed related to longitudinal axis of plate radial (denoted P5BR) (Figure 5d); a plate with three radial and two transversal bars (denoted P3BR2T) (Figure 5e); a plate with five radial and two transverse bars (denoted P5BR2T) (Figure 6f); a plate with three radial bars, two transverse and two oblique (denoted P3BR2V) (Figure 5g); a plate with five radial bars, two transverse and two oblique bars (denoted P5BR2V) (Figure 5h); a plate with seven radial bars two transverse and two oblique bars (denoted P7BR2V) (Figure 5i) [35].

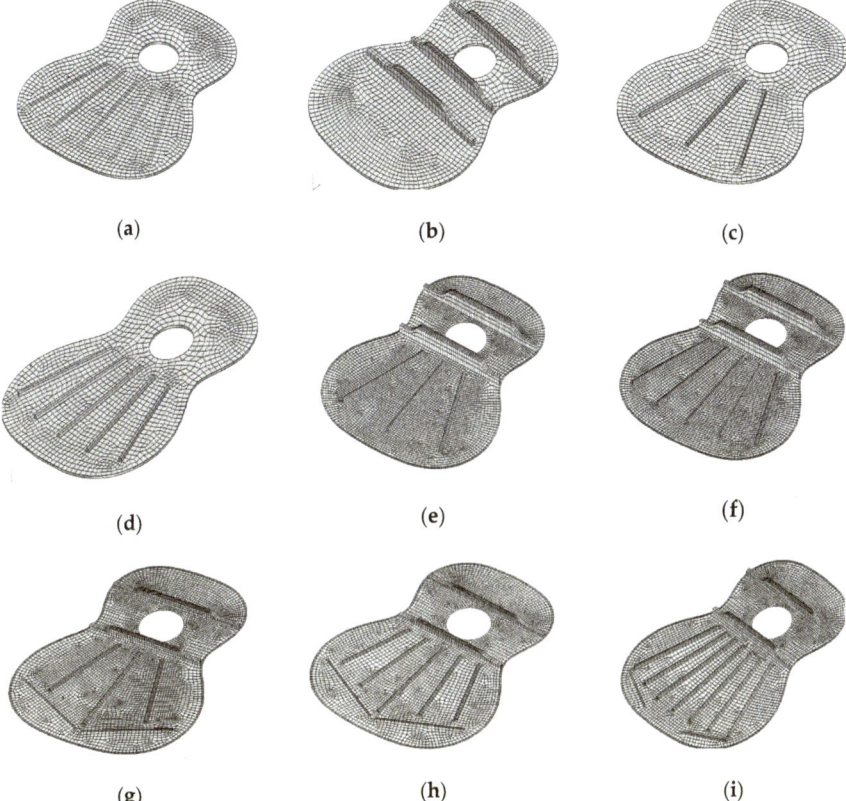

Figure 5. Different reinforcement solutions for the guitar plates: (**a**) simple plate with acoustic hole (denoted PSah) without stiffening elements; (**b**) plate with three transversal reinforcing bars (denoted P3BT); (**c**) plate with three radial resonance bars, symmetric disposed related to longitudinal axis of plate (denoted P3BR); (**d**) plate with five radial resonance bars, symmetric disposed related to longitudinal axis of plate radial (denoted P5BR); (**e**) plate with three radial and two transversal bars (denoted P3BR2T); (**f**) plate with five radial and two transverse bars (denoted P5BR2T); (**g**) plate with three radial bars, two transverse and two oblique (denoted P3BR2V); (**h**) plate with five radial bars, two transverse and two oblique bars (denoted P5BR2V); (**i**) plate with seven radial bars two transverse and two oblique bars (denoted P7BR2V) [35].

Modeling and simulation were done using the Patran Nastran 2004 package. For the meshing in finite elements, elements shell type (QUAD4) were used; the number of degrees of freedom on the element node being 6. For all the nodes on the contour, all the degrees of freedom were fixed. In the preprocessing stage, the following parameters specific to the material and the geometry of the plate were introduced: thickness (h = 1.5, 2, 2.5, 3, 3.5 mm), Young's modulus (E = 10,000, 12,000, 14,000 MPa), density (ρ = 350, 400, 450, 500 kg/m^3), system of beams; keeping Poisson's coefficient constant (ν = 0.36) and shear's modulus (G = 5000 MPa). The values of the Young's modulus E and the density ρ were taken from the literature and based on the results of the analytical modeling [35–37].

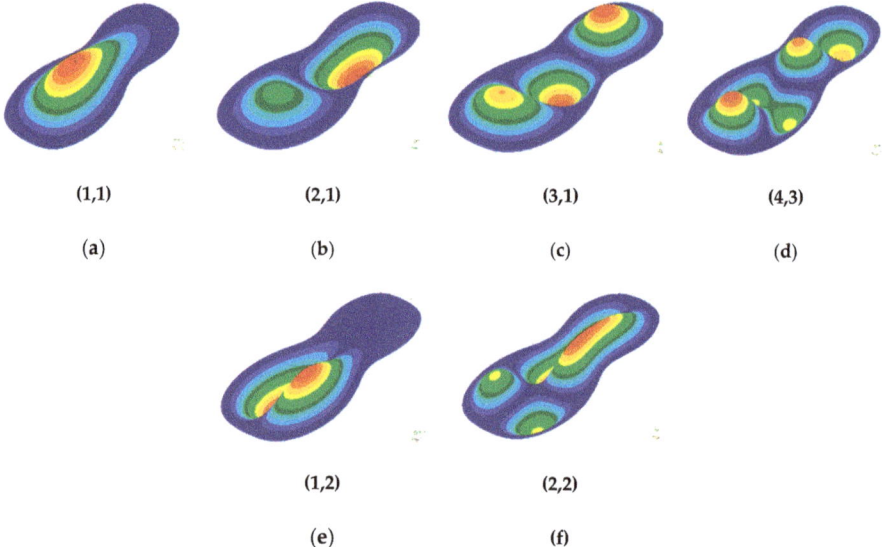

Figure 6. Guitar plate without acoustic hole PS: (**a**–**d**) symmetrical eigenmodes; (**e**,**f**) skew symmetrical eigenmodes.

5. Discussion: Use of the Symmetry in the Calculus of the Eigenvalues of the Guitar

In the following, we will present the procedure that facilitates the calculation, taking into account the mentioned symmetry properties the case of free vibrations of such a system. First of all, we will examine an example to see if these properties are met in the case of guitars. The calculation is done for the entire structure and then only for half of the structure, with a numerical example; for a guitar that is considered in two cases: once without the acoustic hole and the second time with the acoustic hole. The results are shown in Table 1. It is noted that the first four frequencies for the guitar half are the 2, 5, 8, 9 skew-symmetrical eigenmodes of the guitar taken in its entirety. It follows that the properties listed above are met by the guitar structure. This means that we can ease the calculation in this case using the following method: first, we calculate eigenvalues for a half guitar. Then, the polynomial characteristic equation for the entire structure is divided into the characteristic equation for half structure. The result obtained, which is a polynomial, provides the rest of the structure's eigenvalues, which will define the symmetrical eigenmodes.

In the studied case, after the reduction of the dimension of the characteristic equation, the calculated values are obtained with a very good approximation. In the following study, this method is used to determine the eigenvalues, which reduces the time required to study. The properties P2 and P3 (from previous chapter) offer the eigenmodes of the structure. In order to determine how the existence of the acoustic hole influences the eigenmodes and the eigenfrequencies, the element of the guitar plate, without stiffening elements and without the acoustic hole was analyzed using FEA and the same plate provided with a rosette. In the preprocessing step, the same characteristics of the material were introduced, so as to obtain the eigenfrequencies determined by the change of the plate structure in the presence of the sound hole. In Figures 6 and 7 are shown the eigenmodes, and in Table 1 the eigenfrequencies. This study was previously used to verify the properties presented in the Section 2 of this paper.

Table 1. Influence of the acoustic hole on the eigenvalues of the guitar plates for: E = 13,000 MPa, G = 2300 MPa, v = 0.4, ρ = 500 kg/m^3, h = 2.5 mm.

All Eigen-Mode	Eigen Frequency of Plate without Acoustic Hole, (Hz)	Eigen Frequency of Plate with Acoustic Hole, (Hz)	Difference between the Two Cases (%)	Symmetrical/ Skew Symmetrical Modes	Eigen Frequency of Plate without Acoustic Hole, (Hz)	Eigen Frequency of Plate with Acoustic Hole, (Hz)
1	191.64	191.65	+0.00052	Symmetric		
2	295.79	310.84	+4.5	Symmetric	296.21	310.23
3	405.56	404.99	−0.14	Skew (1)		
4	437.14	428.00	−2.09	Symmetric		
5	638.70	598.35	−6.8	Skew (2)	638.22	598.43
6	646.94	619.39	−4.258	Symmetric		
7	702.26	695.79	−0.921	Symmetric		
8	727.18	698.61	−3.92	Skew (3)	727.67	698.35
9	931.33	914.99	−1.754	Skew (4)	931.73	914.46
10	943.61	944.98	+0.145	Symmetric		

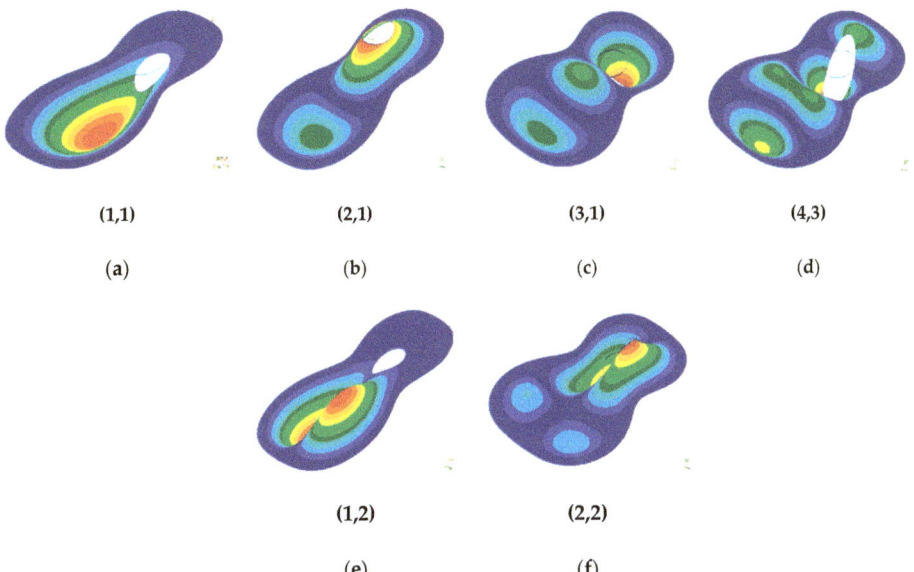

(1,1) (2,1) (3,1) (4,3)
(a)   (b)   (c)   (d)

(1,2) (2,2)
(e)   (f)

Figure 7. Guitar plate with acoustic hole: (a–d) symmetrical eigenmodes PSah; (e,f) skew symmetrical eigenmodes.

It has been found that the presence of the acoustic hole does not produce major differences in the eigenvalues of eigenmodes as can be seen in Figures 6 and 7 and in Table 1. From the point of view of eigenmodes, there are no differences between the first six eigenmodes between the single bezel plate and the acoustic hole; starting with the seventh mode, there are differences in the distribution of nodal lines. We highlighted symmetrical and skewsymmetrical modes. Figures 8–10 show eigenmodes of the plates with different stiffening systems. We have presented a selection from the multitude of the model studied. It has been observed that the eigenmodes are not influenced by the shape, size and order by the material parameters used in the preprocessing stage, namely: density, thickness, Young's modulus, etc. [38]. However, the structure and stiffening elements of the plate has a significant influence on eigenmodes. In Figure 8, the modal shapes of plate with three symmetrical radial bars are presented. If the bars are setting perpendicular to the longitudinal axis, the skew eigenmodes change in shape and orientation as can be seen in Figure 9.

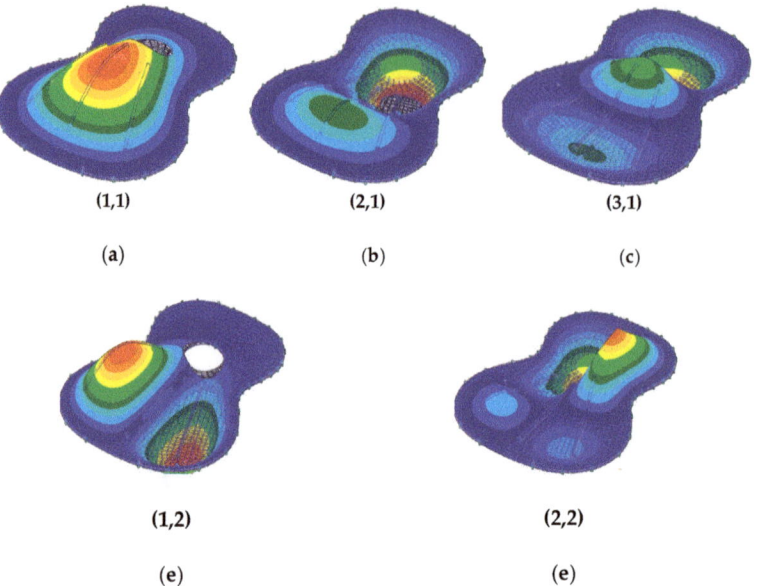

Figure 8. Guitar plate with three radial bars: (**a**–**c**), symmetrical eigenmodes P3BR; (**e**,**f**) skew symmetrical eigenmodes.

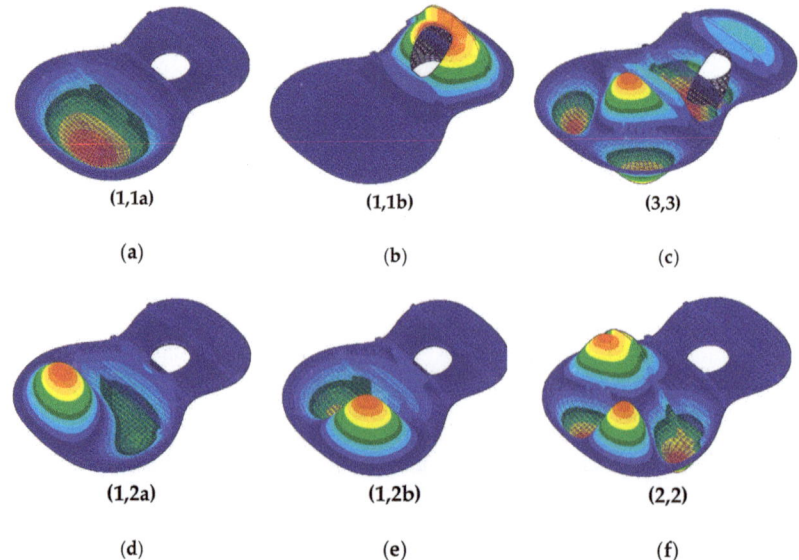

Figure 9. Guitar plate with acoustic hole and three transversal bars, P3BT: (**a**–**c**), symmetrical eigenmodes; (**d**–**f**) skew symmetrical eigenmodes.

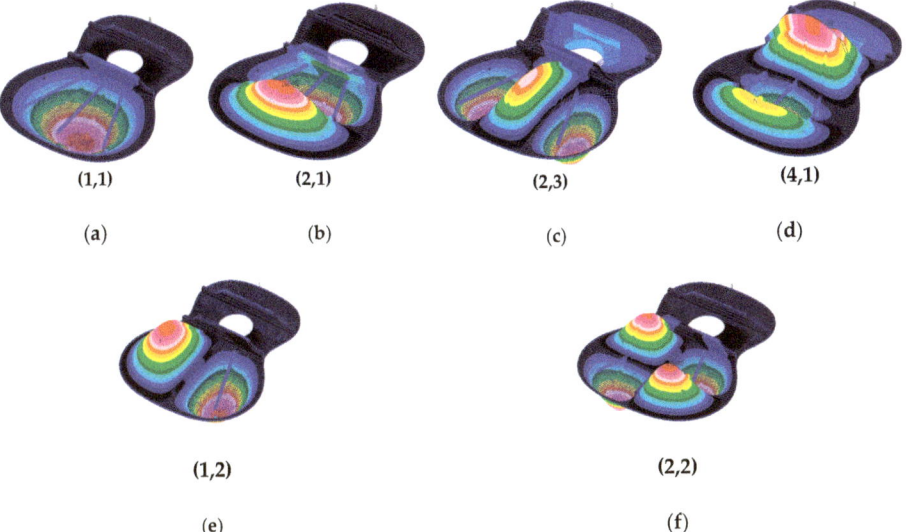

Figure 10. Guitar plate with acoustic hole, three symmetrical radial bars and two transversal bars, P3BR2T: (**a–d**) symmetrical eigenmodes; (**e,f**) skew symmetrical eigenmodes.

In Figure 10, the symmetrical eigenmodes correspond to breathing mode or synclastic (1,1), to anticlastic bending of first order (2,1) and second order (4,1). The skew symmetrical eigenmodes reported to plane of symmetry are the center bout rotation (1,2) and transverse dipole in plane (2,2).

Analyzing the eigenmodes obtained in the case of studied plates, it was observed that:

- The first eigenmode of vibration (1,1) presents the same shape, regardless the stiffening pattern. Compare to plates with transversal bars, the plates without them (Figures 6–8) are characterized by a more extended shape along the longitudinal axis. Additionally, the plates without transverse stiffening elements have the second eigenmode of the type (1,2) compared to the other plates whose eigenmode is of the shape (2,1);
- The stiffening bars change the eigenmodes and the order on the plates. Additionally, the increases of number of bars leads to the natural frequency also increased (Table 2);
- The eigenmodes for the plates with the complete stiffening bars: radial, oblique and transverse bars, regardless of their number, are similar;
- An interesting situation is recorded for plate with only three transversal bars (Figure 9), where the antisymmetric modes (1,2a) and (1,2b) are obtained related to a different plane from longitudinal symmetry plane, which is inclined to about 40-45 degrees (Figure 9d,e).

In Table 2, the values of the first natural frequency obtained by means FEA, for the different elasticity moduli are shown.

For plates without stiffening bars, the eigenmodes are formed over the entire surface of the plate; the plate's eigenmodes are not significantly influenced by thickness and material (density, Young's modulus, Poisson's ratio).

Table 2. The eigenvalues of the guitar plates with different stiffening pattern, for: $G = 2300$ MPa, $\nu = 3.6$, $\rho = 450$ kg/m³, $h = 2.5$ mm (Legend: PS – simple plate without acoustic hole; PSah - simple plate with acoustic hole, without stiffening bars; P3BR – plate with three radial resonance bars, symmetric disposed related to longitudinal axis of plate; P5BR – plate with five radial resonance bars, symmetric disposed related to longitudinal axis of plate radial; P3BT – plate with three transversal reinforcing bars; P3BR2T – plate with three radial and two transversal bars; P5BR2T – plate with five radial and two transverse bars; P3BR2V – plate with three radial bars, two transverse and two oblique; P5BR2V – plate with five radial bars, two transverse and two oblique bars; P7BR2V - plate with seven radial bars two transverse and two oblique bars.

Plates	Frequency (Hz)		
	14,000 MPa	12,000 MPa	10,000 MPa
PS	101	92	87
PSah	106	98	89
P3BR	207	175	155
P5BR	209	193	176
P3BT	210	194	178
P3BR2T	262	242	221
P5BR2T	264	243	223
P3BR2V	261	241	220
P5BR2V	264	243	224
P7BR2V	265	245.72	224.43

6. Conclusions

For some practical reasons related to material costs, execution and design, identical parts in the engineering are commonly used. Using identical parts in a project a device or a machine can be designed and can be executed cheaper and faster. The paper made an analysis of such a structure to determine if these repetitive parts can induce characteristic properties and can bring some advantages in design and calculus. The paper studied a particular case of a structure made by two identical parts with application to real guitar plates. Although the modal analysis of the guitar plates does not provide a comprehensive eigenmodes map of the whole guitar, the natural modes of soundboards presented in this paper play an important role on the vibrational behavior of coupled plates on the entire guitar structure.

In this way, the computational time decreases significantly. The dimension of the system decrease and it's easier to make a vibration analysis. The FEA allowed verification of the stated properties and was used for the numerical study of the problem.

Author Contributions: Conceptualization, M.D.S., M.M. and S.V.; methodology, S.V., M.M. and M.D.S.; investigation and software M.D.S.; validation, M.M.; formal analysis, S.V.; resources, M.D.S., S.V.; data curation, M.D.S.; writing—original draft preparation, S.V.; writing—review and editing, M.D.S. and M.M.; visualization, M.D.S.; supervision, S.V.

Funding: This research received no external funding.

Acknowledgments: The authors would like to thank company SC Hora S.A., Reghin, Romania for providing the patterns and geometry of the guitar plates involved in this study.

Conflicts of Interest: The authors declare no conflict of interest.

References

1. Celep, Z. On the axially symmetric vibration of thick circular plates. *Ing. Arch.* **1978**, *47*, 411–420. [CrossRef]
2. Zingoni, A. Group-theoretic exploitations of symmetry in computational solid and structural mechanics. *Int. J. Numer. Meth. Eng.* **2009**, *79*, 253–289. [CrossRef]
3. Zingoni, A. On the symmetries and vibration modes of layered space grids. *J. Eng. Struct.* **2005**, *27*, 629–638. [CrossRef]

4. Mangeron, D.; Goia, I.; Vlase, S. Symmetrical Branched Systems Vibrations. *Sci. Mem. Rom. Acad. Buchar.* **1991**, *12*, 232–236.
5. Meirovitch, L. *Principles and Techniques of Vibrations*, 1st ed.; Prentice Hall: Upper Saddle River, NJ, USA, 1997; pp. 455–570.
6. Le, K.C.; Nguyen, L.T.K. *Energy Methods in Dynamics*, 2nd ed.; Springer: Basel, Switzerland, 2014.
7. Le, K.C. *Vibrations of Shells and Rods*; Springer Science & Business Media: Berlin, Germany, 2012.
8. Zhou, D.; Ji, T. Free vibration of rectangular plates with attached discrete sprung masses. *Shock Vib.* **2012**, *19*, 101–118. [CrossRef]
9. Derveaux, G.; Chaigne, A.; Joly, P.; Becache, E. Time-domain simulation of a guitar. Model and method. *J. Acoust. Soc. Am.* **2003**, *114*, 3368–3383. [CrossRef]
10. Elejabarrieta, M.J.; Ezcurra, A.; Santamarıa, C. Coupled modes of the resonance box of the guitar. *J. Acoust. Soc. Am.* **2002**, *111*, 2282–2292. [CrossRef]
11. Chen, Y.; Feng, J. Generalized Eigenvalue Analysis of Symmetric Prestressed Structures using Group Theory. *J Comput. Civil Eng.* **2012**, *26*, 488–497. [CrossRef]
12. Holm, D.D.; Stoica, C.; Ellis, D.C.P. Group Actions, Symmetries and reduction. In *Geometric Mechanics and Symmetry*, 1st ed.; Oxford University Press: Oxford, UK, 2009; pp. 184–211.
13. Marsden, J.E.; Ratiu, T.S. The Free Rigid Body. In *Introduction to Mechanics and Symmetry: A Basic Exposition of Classical Mechanical Systems*, 2nd ed.; Springer Science & Business Media: Berlin, Germany, 2002; pp. 483–586.
14. Lopatin, A.V.; Morozov, E.V. Symmetrical vibration modes of composite sandwich plates. *J. Sandw. Struct. Mater.* **2010**, *13*, 189–211. [CrossRef]
15. Shi, C.Z.; Parker, R.G. Modal structure of centrifugal pendulum vibration absorber systems with multiple cyclically symmetric groups of absorbers. *J. Sound Vib.* **2013**, *332*, 4339–4353. [CrossRef]
16. Singer, S.F. *Symmetry in Mechanics*, 2nd ed.; Birkhäuser: Basel, Switzerland, 2004; pp. 83–112.
17. Vlase, S.; Năstac, D.C.; Marin, M.; Mihălcică, M. A method for the study of the vibration of mechanical bars systems with symmetries. *VLASEV* **2017**, *60*, 539.
18. Zavadskas, E.K.; Bausys, R.; Antucheviciene, J. Civil Engineering and Symmetry. *Symmetry* **2019**, *11*, 501. [CrossRef]
19. Ganghoffer, J.F.; Mladenov, I. *Similarity, Symmetry and Group Theoretical Methods in Mechanics*; Lectures at the International Centre for Mechanical Sciences, CISM: Udine, Italy, 2015.
20. Vlase, S.; Paun, M. Vibration Analysis of a Mechanical System consisting of Two Identical Parts. *Ro. J. Techn. Sci. Appl. Mech.* **2015**, *60*, 216–230.
21. Vlase, S.; Marin, M.; Scutaru, M.L.; Munteanu, R. Coupled transverse and torsional vibrations in a mechanical system with two identical beams. *AIP Adv.* **2017**, *7*, 065301. [CrossRef]
22. Curtu, I.; Stanciu, M.D.; Grimberg, R. Correlations between the Plates' Vibrations from the Guitar's Structure and the Physical, Mechanical and Elastically Characteristics of the Composite Materials. In Proceedings of the 9th International Conference on Acoustics & Music: Theory & Applications (Amta '08), Bucharest, Romania, 24–26 June 2008.
23. Curtu, I.; Stanciu, M.D.; Itu, C.; Grimberg, R. Numerical Modelling of the Acoustic Plates as Constituents of Stringed Instruments. In Proceedings of the 6th International Conference Baltic Industrial Engineering, Tallinn, Estonia, 24–26th April 2008; pp. 53–58.
24. Lee, M.K.; Fouladi, M.H.; Namasivayam, S.N. Mathematical Modelling and Acoustical Analysis of Classical Guitars and Their Soundboards. *Adv. Acoust. Vib.* **2016**. Article ID 6084230. [CrossRef]
25. Douglas, T. Eigenvalues and Eigenvectors. In *Structural Dynamics and Vibrations in Practice: An Engineering Handbook*, 1st ed.; Butterworth-Heinemann: Oxford, UK; pp. 159–180.
26. Marin, M. Cesaro means in thermoelasticity of dipolar bodies. *Acta. Mech.* **1997**, *122*, 155–168. [CrossRef]
27. Marin, M.; Öchsner, A. The effect of a dipolar structure on the Holder stability in Green-Naghdi thermoelasticity. *Contin. Mech. Thermodyn.* **2017**, *29*, 1365–1374. [CrossRef]
28. Senjanović, I.; Tomić, M.; Vladimir, N.; Hadžić, N. An Analytical Solution to Free Rectangular Plate Natural Vibrations by Beam Modes—Ordinary and Missing Plate Modes. *Trans. FAMENA* **2016**, *40*, 1–18. [CrossRef]
29. Senjanović, I.; Tomić, M.; Vladimir, N.; Hadžić, N. An approximate analytical procedure for natural vibration analysis of free rectangular plates. *Thin-Walled Struct.* **2015**, *95*, 101–114. [CrossRef]
30. Rades, M. *Mechanical Vibrations*; Printech Press: Bucharest, Romania, 2006.
31. Vlase, S. Elimination of Lagrangean Multipliers. *Mech. Res. Commun.* **1987**, *14*, 17–20. [CrossRef]

32. Negrean, I. New Formulations in Analytical Dynamics of Systems. *Acta Tech. Napocensis Ser. Appl. Math. Mech. Eng.* **2017**, *60*, 49–56.
33. Shepherd, M.R.; Hambric, S.A.; Wess, D.B. The effects of wood variability on the free vibration of an acoustic guitar top plate. *J. Acous. Soc. Am.* **2014**, *136*, EL357. [CrossRef] [PubMed]
34. Öchsner, A.; Öchsner, M. Classical Plate Elements. In *Computational Statics and Dynamics: An Introduction Based on the Finite Element Method*, 1st ed.; Springer: Singapore, 2016; pp. 279–310.
35. Stanciu, M.D.; Curtu, I. Analytical and numerical simulation of structures of classical guitar. In *Dinamica structurii chitarei clasice (Ro)*; Printhouse of Transilvania University of Brasov: Brasov, Romania, 2012; pp. 51–120.
36. Curtu, I.; Ghelmeziu, N. *Mechanics of Wood and Wood Based Materials*, 1st ed.; Editura Tehnica: Bucharest, Romania, 1984; pp. 165–204.
37. Bucur, V. *Acoustic of Wood*, 2nd ed.; Springer: New York, NY, USA, 2006; pp. 173–216.
38. Rossing, T.D.; Fletcher, N.H. Two Dimensional Systems: Membranes and Plates. In *Principle of Vibration and Sound*, 2nd ed.; Springer: New York, NY, USA, 2004; pp. 65–92.

© 2019 by the authors. Licensee MDPI, Basel, Switzerland. This article is an open access article distributed under the terms and conditions of the Creative Commons Attribution (CC BY) license (http://creativecommons.org/licenses/by/4.0/).

Article

A Hybrid Model Based on a Two-Layer Decomposition Approach and an Optimized Neural Network for Chaotic Time Series Prediction

Xinghan Xu [1,*] and Weijie Ren [2,*]

1. Department of Environmental Engineering, Kyoto University, Kyoto 615-8540, Japan
2. Faculty of Electronic Information and Electrical Engineering, Dalian University of Technology, Dalian 116024, China
* Correspondence: xu.xinghan.57u@st.kyoto-u.ac.jp (X.X.); renweijie@mail.dlut.edu.cn (W.R.); Tel.: +86-0411-84706002-2607 (W.R.)

Received: 2 April 2019; Accepted: 26 April 2019; Published: 1 May 2019

Abstract: The prediction of chaotic time series has been a popular research field in recent years. Due to the strong non-stationary and high complexity of the chaotic time series, it is difficult to directly analyze and predict depending on a single model, so the hybrid prediction model has become a promising and favorable alternative. In this paper, we put forward a novel hybrid model based on a two-layer decomposition approach and an optimized back propagation neural network (BPNN). The two-layer decomposition approach is proposed to obtain comprehensive information of the chaotic time series, which is composed of complete ensemble empirical mode decomposition with adaptive noise (CEEMDAN) and variational mode decomposition (VMD). The VMD algorithm is used for further decomposition of the high frequency subsequences obtained by CEEMDAN, after which the prediction performance is significantly improved. We then use the BPNN optimized by a firefly algorithm (FA) for prediction. The experimental results indicate that the two-layer decomposition approach is superior to other competing approaches in terms of four evaluation indexes in one-step and multi-step ahead predictions. The proposed hybrid model has a good prospect in the prediction of chaotic time series.

Keywords: chaotic time series prediction; neural network; firefly algorithm; CEEMDAN; VMD

---

1. Introduction

Chaotic time series exist in a wide range of areas, such as nature, the economy, society, and industry. They contain many important and valuable information, useful for complex system modeling and prediction. Time series data mining is an important means of the control and decision-making of practical problems in various fields [1,2]. In recent years, scholars have carried out a great amount of research work on chaotic time series analysis and prediction. Han et al. [3] proposed an improved extreme learning machine combined with a hybrid variable selection algorithm for the prediction of multivariate chaotic time series, which can achieve high predictive accuracy and reliable performance. Chandra [4] put forward a competitive cooperative coevolution algorithm to train recurrent neural networks (RNNs) for chaotic time series prediction. Yaslan et al. [5] presented a hybrid model based on empirical mode decomposition (EMD) and support vector regression (SVR) for electricity load forecasting. Chen [6] proposed a prediction model of a radial basis function (RBF) neural network optimized by an artificial bee colony algorithm for prediction of traffic flow time series. A multilayered echo state machine with the addition of multiple layers of reservoirs was introduced in [7], and it could be more robust than the echo state network with a conventional reservoir in dealing with chaotic time series prediction.

The time series prediction models proposed in recent years are usually divided into three types: statistical models, artificial intelligence models, and hybrid models. The statistical models mainly include autoregressive (AR) models [8], the autoregressive moving average (ARMA), the autoregressive integrated moving average (ARIMA) [9], multivariate linear regression, the Gaussian process [10], and so on. Since statistical models require time series to be subject to certain a priori assumptions, such as stationarity, they are not ideal for practical systems with many uncertainties. With the development of computational intelligence, artificial intelligence models have obtained widespread attention. They are data-driven methods that do not require any a priori assumptions and thus have a wide range of applications. Commonly used artificial intelligence models include support vector regression [11], RBF neural networks [12], Elman neural networks [4], echo state networks [7], deep neural networks [13], extended Kalman filters [14], adaptive neuro-fuzzy inference systems (ANFISs) [15–17], etc. So as to improve the performance of single-model-based prediction models, a novel framework based on decomposition algorithm has been introduced for time series prediction [18]. Multiple decomposition methods have been put forward to analyze time series, thus forming the hybrid prediction models [19]. The subsequences obtained by the decomposition algorithms are much easier to predict than the original time series, which brings forward a new means of predicting nonlinear and non-stationary time series [20]. Ren et al. [21] introduced a hybrid model, EMD combined with kNN, for wind speed prediction. The model generated a set of feature vectors from the components obtained by EMD, and kNN was then employed for prediction. The suggested hybrid model performed well for long-term wind speed forecasting. An ensemble EMD (EEMD)–ARIMA model has been proposed to predict annual runoff time series [22]. According to the experimental results, it was concluded that the introduction of EEMD could observably improve prediction performance, and the EEMD–ARIMA model was superior to the ARIMA. It is confirmed that hybrid models perform better than their corresponding single models in chaotic time series prediction.

Though these existing hybrid prediction models have indeed increased the performance of chaotic time series prediction, they still cannot handle the chaotic time series with strong non-stationary and nonlinear very well. Hence, the hybrid models can be further improved to obtain more accurate predictions. For the sake of enhancing the accuracy of actual chaotic time series prediction, we put forward a novel hybrid model based on a two-layer decomposition technique and an optimized back propagation neural network (BPNN). The main contents and contributions of this paper are summarized as follows.

1. A hybrid model based on a two-layer decomposition technique is proposed in this paper. For the sake of solving the problem that the prediction model based on single decomposition technique cannot completely deal with the nonlinear and non-stationary of chaotic time series, this paper puts forward a two-layer decomposition technique based on CEEMDAN and VMD, which is able to fully extract the complex characteristics of time series and improve prediction accuracy.
2. A firefly algorithm (FA) is applied to optimize the weights between input and hidden layer, the weights between the hidden and output layer and the thresholds of neuron nodes, which can reduce the human interference of parameter settings and improve the function approximation ability of the neural network. A BPNN optimized by the FA is applied to predict the subsequences obtained by two-layer decomposition.
3. The real world chaotic time series, daily maximum temperature time series in Melbourne, is used to assess the validity of the proposed hybrid model. The experimental results indicate that our hybrid model has a significant improvement in prediction accuracy compared to the existing single-model-based approaches and hybrid models based on the single layer decomposition technique.

The remainder of the paper is organized as follows. Preliminaries and related works are introduced in Section 2. In Section 3, we introduce the principles and the implementation steps of the proposed method. Section 4 presents the experiments illustrating the availability of the proposed model. Finally, conclusions and future directions are demonstrated in Section 5. For the convenience of reading, the notations used in this paper are shown in Table 1.

Table 1. The notations used in this paper.

Notation	Meaning
$\varepsilon(t)$	independent Gaussian white noise with unit variance
ω_0	a noise coefficient
$r_1(t)$	the first residue
$E_j(\cdot)$	a function to extract the j-th intrinsic mode function (IMF) decomposed by EMD
$u_k(t)$	the k-th mode of decomposition
ω_k	the center frequency of mode k
$\partial_t(\cdot)$	partial derivative
σ	the Dirac distribution
$*$	convolution computation
α	the balancing parameter of the data-fidelity constraint
λ	the Lagrangian multiplier
$\hat{f}(\omega)$	the Fourier transforms of $f(t)$
I_0	the intensity of the light source
γ	the light absorption coefficient
r_{ij}	the distance between firefly i and j
β_0	the attractiveness at the light source ($r = 0$)
s_i	the space positions of firefly i

2. Preliminaries and Related Works

In this section, we will firstly introduce the basic methods, including CEEMDAN, VMD, and the FA. Furthermore, related works of the hybrid prediction model will be described.

2.1. Complete Ensemble Empirical Mode Decomposition with Adaptive Noise

Ensemble EMD (EEMD) adds white noise to the original signal to solve the mode mixing problem of EMD. However, the white noise sequence cannot be absolutely canceled through the finite average, and the magnitude of reconstruction error depends on the times of integration. In addition, increasing the times of integration would lead to an increase computational burden. To solve these problems, CEEMDAN was proposed as an improved version of EEMD [23]. By adding a finite number of adaptive white noise at each stage, the CEEMDAN method is able to obtain the reconstruction error close to zero through a small average number of times of integration. Thus, CEEMDAN avoids the mode mixing problem in EMD, while reducing computational complexity compared to EEMD.

2.2. Variational Mode Decomposition

VMD is a novel non-recursive signal processing approach [24], which decomposes the original signal into a group of subsequences. Each subsequence is called a mode and is concentrated near a specific central pulsation frequency. For assessing the bandwidth of each mode, there are three main schemes: Firstly, apply the Hilbert transform to each mode separately, calculate the associated analytic signals, and then obtain a number of unilateral frequency spectrums. Then, adjust each mode to its estimated center frequency by adding an exponential term, so as to shift the frequency spectrums of these modes to the respective baseband. Finally, estimate the bandwidth of every mode by means of the Gaussian smoothness of the demodulated signal, for example, the L2-norm of the gradient.

2.3. Firefly Algorithm

The firefly algorithm (FA) [25] is a heuristic optimization algorithm based on firefly behavior, whereby flash signals are used to attract potential mates for positional movement. It is a swarm intelligence optimization algorithm and therefore has the advantages of a swarm intelligence algorithm.

In addition, compared with other similar algorithms, it has the following two advantages. Firstly, it can realize automatic segmentation, which is well suited for solving highly nonlinear optimization problems. Secondly, the FA has multi-modal characteristics and can deal with multi-modal problems quickly and efficiently. The basic idea of the FA is to treat every point in space as a firefly, which is attracted to the brighter fireflies and moves in this direction. During the movement of the weakly glowing firefly, the position is updated until the optimal position is found. Since it is proposed, FA has been widely used to solve various practical problems.

2.4. Related Works

Aiming at the analysis of nonlinear and non-stationary time series, many hybrid models have been proposed. In [26], the EMD–BPNN is presented and applied to forecast tourism demand, namely the number of tourists. The hybrid model showed better performance than a single BPNN or ARIMA model. Zhou et al. [27] put forward a hybrid model of EEMD–GRNN (general regression neural network) for PM2.5 forecasting. Simulation results indicated that the EEMD–GRNN model outperformed the GRNN, multiple linear regression, and the ARIMA. This research was significant for the development of air quality warning systems. Moreover, a comparison work of hybrid prediction models based on wavelet decomposition, wavelet packet decomposition, EMD, and fast EEMD was implemented in [28]. The study investigated the decomposition and prediction performance of multiple hybrid models. In [29], VMD was adopted to decompose wind power time series into multiple modes, and Gram–Schmidt orthogonalization was used to eliminate redundant attributes. Next, the hybrid model based on VMD and extreme learning machines was proposed for short-term wind power forecasting. Similarly, Lahmiri [30] proposed a hybrid model for economic and financial time series forecasting, called VMD–GRNN. Jianwei E et al. [31] raised a hybrid model VMD–ICA–ARIMA for crude oil price forecasting. Wang et al. [32] suggested a hybrid model based on VMD, phase space reconstruction, and a wavelet neural network, which is reliable for multi-step prediction of wind speed time series.

As mentioned before, hybrid models based on decomposition methods, such as EMD, EEMD, and VMD, have been extensively used for time series modeling and have exhibited satisfactory performance.

3. Methodology

3.1. The Structure of CEEMDAN–VMD–FABP Model

Subsequences obtained by an effective decomposition technique are much easier to analyze than an original time series, and a single-layer decomposition pattern is one of the most commonly adopted methods in existing hybrid models. Models based on a single-layer decomposition technique are able to increase the predictive performance of various chaotic time series to some extent, but they are difficult in completely reflecting the non-stationarity and irregularity of original signals. In view of this, we propose a novel two-layer decomposition approach using CEEMDAN and VMD for chaotic time series forecasting, which is shown in Figure 1. In addition, we adopt a BPNN optimized by an FA (FABP) to predict subsequences obtained by a two-layer decomposition operation. FABP has the ability to automatically optimize weights and thresholds, which is able to reduce the randomness of parameter selection and ultimately strengthen the function approximation ability of a neural network. Next, we will comprehensively introduce the structure of the proposed model.

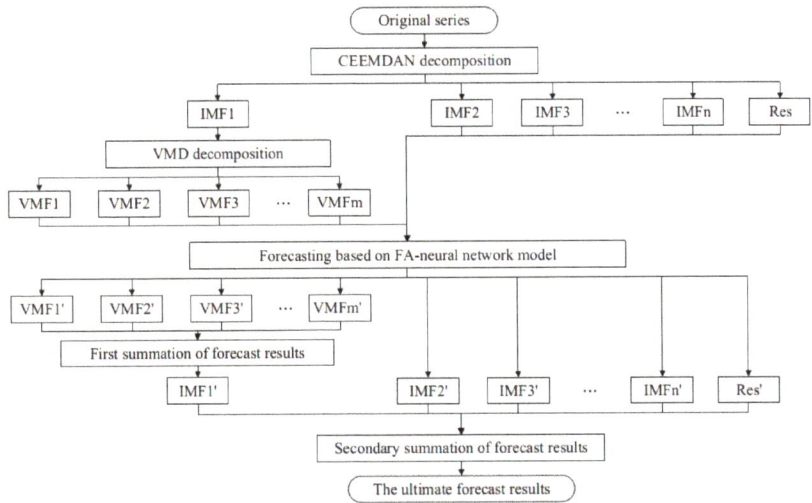

Figure 1. The structure of the two-layer decomposition prediction model.

The CEEMDAN method with noise robustness is first used to decompose the original signal. CEEMDAN has good performance in anti-noise, so the original signal is not denoised in this paper. At present, many scholars are committed to improving the CEEMDAN algorithm by increasing its robustness and improving its adaptive anti-noise performance so that it omits the process of denoising the original signal. Therefore, the original signal is automatically decomposed into a group of intrinsic mode functions (IMFs), namely IMF1, IMF2, ... , IMFn, and a residue.

IMF1 has the highest frequency, which is the most difficult to track and predict. To improve the overall prediction accuracy of the original signal, the VMD algorithm is introduced here for the secondary decomposition of IMF1. IMF1 is decomposed by the VMD algorithm into VMF1, VMF2, ... , VMFm. Next, they are each predicted by FABP. Afterwards, the prediction results of these subsequences are combined into the final result of IMF1. Furthermore, the subsequences IMF2, ... , IMFn and the residue are each predicted using FABP. The summation of the prediction results of IMF2, ... , IMFn and the residue is superimposed with the prediction result of the IMF1. The final forecast result of the original chaotic signal is obtained. At this point, the entire forecast process is complete.

3.2. Algorithm Design

3.2.1. CEEMDAN for Original Time Series

First, the CEEMDAN decomposition algorithm [23] is adopted to obtain a group of IMFs adaptively. The computational process of CEEMDAN is as follows.

Step 1. A collection of noise-added original time series is created: $x^i(t) = x(t) + \omega_0 \varepsilon^i(t)$, $i \in \{1, \ldots, I\}$, where $\varepsilon(t)$ is independent Gaussian white noise with unit variance, and ω_0 is a noise coefficient.

Step 2. For each $x^i(t)$, EMD is used to obtain the first IMF and take the average:

$$c_1(t) = \frac{1}{I}\sum_{i=1}^{I} c_1^i(t). \tag{1}$$

The first residue is then $r_1(t) = x(t) - c_1(t)$.

Essentially, the procedure of the EMD algorithm is a sifting process from which IMFs can be obtained [33]. The specific computation process of the signal $x^i(t)$ is described as follows. Firstly, we obtain every local maxima and local minima of $x^i(t)$. Next, the upper envelope $u(t)$ and the lower envelope $l(t)$ are structured by cubic spline interpolation. Finally, we calculate the average of $u(t)$ and $l(t)$ and record the average as $m(t)$.

$$m(t) = (u(t) + l(t))/2. \tag{2}$$

We subtract $m(t)$ from the original signal $x^i(t)$ and record the result as $c_1^i(t)$.

$$c_1^i(t) = x^i(t) - m(t). \tag{3}$$

It is then determined whether $c_1^i(t)$ is an integral part of the IMF by checking whether $c_1^i(t)$ satisfies the above two conditions. The IMFs should conform to the following two conditions: (1) the absolute value of the difference between the number of extreme points and zero crossing must be less than or equal to 1 in the whole data series; (2) the average of the upper envelope and lower envelope must be zero at any point. If $c_1^i(t)$ meets the above two conditions, $c_1^i(t)$ is recognized as IMF1. In the meantime, $r(t) = x(t) - c(t)$ is used as the residue instead of $x(t)$. If $c_1^i(t)$ is not an IMF, $c_1^i(t)$ is used instead of $x^i(t)$, and the above process is repeated until we IMF1 is obtained.

Step 3. The second IMFs is obtained by decomposing the noise-added residue $r_1 + \omega_1 E_1(\varepsilon_1(t))$:

$$c_2(t) = \frac{1}{I} \sum_{i=1}^{I} E_1\big(r_1 + \omega_1 E_1(\varepsilon^i(t))\big), \tag{4}$$

where $E_j(\cdot)$ represents the j-th IMF obtained by EMD.

Step 4. The remaining IMFs are repeated until the number of extreme points of the residual signal does not exceed two.

So far, the original signal has been decomposed into a series of IMF components, namely IMF1, IMF2, ..., IMFn. Due to the highest frequency and unpredictability of IMF1, this research introduces the second decomposition of the IMF1 based on the VMD.

3.2.2. VMD for IMF1

The main task of the VMD algorithm [34] is to solve the following constrained optimization problem:

$$\min_{\{u_k\},\{\omega_k\}} \sum_k \left\| \partial_t \left[\left(\sigma(t) + \frac{j}{\pi t} \right) * u_k(t) \right] e^{-j\omega_k t} \right\|_2^2 \\ s.t. \sum_k u_k = f \tag{5}$$

where $u_k(t)$ denotes the k-th mode of decomposition, ω_k is the center frequency of mode k, $\partial_t(\cdot)$ denotes partial derivative, σ indicates the Dirac distribution, $*$ denotes convolution computation, and f is the original signal to be decomposed.

In order to solve the above constrained optimization problem, a quadratic penalty term and a Lagrangian multiplier are shown in Equation (6). The former is conducive to enforcing the constraint, and the latter helps to improve convergence. Hence, the translated unconstrained form is shown as

$$L(\{u_k\}, \{\omega_k\}, \lambda) = \alpha \sum_k \left\| \partial_t \left[\left(\sigma(t) + \frac{j}{\pi t} \right) * u_k(t) \right] e^{-j\omega_k t} \right\|_2^2 + \left\| f - \sum_k u_k(t) \right\|_2^2 + \left\langle \lambda(t), f(t) - \sum_k u_k(t) \right\rangle, \tag{6}$$

where α represents the balancing parameter, and λ is the Lagrangian multiplier.

An alternate direction method of multipliers (ADMM) [35] is applied to settle the optimization problem of (6). The saddle point of the obtained augmented Lagrangian L in iterative optimization process is determined by AMDD. The suboptimized solution is embedded into ADMM and optimized in the Fourier domain. The detailed calculation process can be found in [24]. To implement the

VMD algorithm and update the modes, u_k and ω_k are iterated in two directions according to the following solutions:

$$u_k^{n+1} = \operatorname{argmin}\left\{\alpha \sum_k \left\|\partial_t\left[\left(\sigma(t) + \frac{j}{\pi t}\right) * u_k(t)\right]e^{-j\omega_k t}\right\|_2^2 + \left\|f - \sum_k u_i(t) - \frac{\lambda(t)}{2}\right\|_2^2\right\}, \tag{7}$$

$$\omega_k^{n+1} = \operatorname{argmin}\left\|\partial_t\left[\left(\sigma(t) + \frac{j}{\pi t}\right) * u_k(t)\right]e^{-j\omega_k t}\right\|_2^2. \tag{8}$$

Subsequently, the solutions are represented as follows:

$$\hat{u}_k^{n+1}(\omega) = \frac{\hat{f}(\omega) - \sum_{i \neq k} \hat{u}_i(\omega) + \frac{\hat{\lambda}(\omega)}{2}}{1 + 2\alpha(\omega - \omega_k)^2} \tag{9}$$

$$\omega_k^{n+1} = \frac{\int_0^\infty \omega |\hat{u}_k(\omega)|^2 d\omega}{\int_0^\infty |\hat{u}_k(\omega)|^2 d\omega} \tag{10}$$

where $\hat{f}(\omega)$, $\hat{u}_k(\omega)$, and $\hat{\lambda}(\omega)$ represent the Fourier transforms of $f(t)$, $u_k(t)$, and $\lambda(t)$, and n denotes the number of iterations.

The complete calculation process of VMD is organized in Algorithm 1. IMF1 is decomposed by VMD algorithm into VMF1, VMF2, ..., VMFm.

Algorithm 1: ADMM Optimization Process for VMD

Initialize $\{\hat{u}_k^1\}, \{\omega_k^1\}, \hat{\lambda}^1, n = 0$
repeat
$\quad n = n + 1$
$\quad$ **for** $k = 1 : K$ **do**
$\quad\quad$ Update $\hat{u}_k^1$ for all $\omega \geq 0$
$\quad\quad \hat{u}_k^{n+1}(\omega) = \frac{\hat{f}(\omega) - \sum_{i \neq k} \hat{u}_i(\omega) + \frac{\hat{\lambda}(\omega)}{2}}{1 + 2\alpha(\omega - \omega_k)^2}$
$\quad\quad$ Update ω_k
$\quad\quad \omega_k^{n+1} = \frac{\int_0^\infty \omega |\hat{u}_k(\omega)|^2 d\omega}{\int_0^\infty |\hat{u}_k(\omega)|^2 d\omega}$
$\quad$ **end for**
$\quad$ for all $\omega \geq 0$
$\quad \hat{\lambda}^{n+1}(\omega) \leftarrow \hat{\lambda}^n(\omega) + \gamma\left[\hat{f}(\omega) - \sum_k \hat{u}_k^{n+1}(\omega)\right]$
until convergence $\sum_k \|\hat{u}_k^{n+1} - \hat{u}_k^n\|_2^2 / \|\hat{u}_k^n\|_2^2 < \varepsilon$

3.2.3. BPNN Optimized by a Firefly Algorithm

After two-layer decomposition, a BPNN is used to predict all subsequences obtained by decomposition. The FA is adopted to optimize the parameters of the BPNN and is capable of improving the BPNN's function approximation ability.

The search process is related to two important parameters of fireflies: the brightness and mutual attraction of the fireflies. Bright fireflies will attract the weak fireflies to move to them. The brighter the lights are, the better their positions are, and the brightest fireflies represent the optimal solution of the function. The higher a firefly's brightness is, the more attractive it is to other fireflies. If the luminance is the same, the fireflies will engage in a random motion, and the two important parameters are inversely proportional to the distance. The greater the distance is, the smaller the attraction is.

The relative fluorescent brightness of a firefly is defined as

$$I = I_0 e^{-\gamma r_{ij}^2},\qquad(11)$$

where I_0 denotes the intensity of the light source. The better the objective function value is, the higher the firefly's brightness is. γ represents the light absorption coefficient. As the distance increases and the transmission medium weakens the light intensity, the fluorescence gradually decreases. The light absorption coefficient is set to reflect this characteristic. The parameter r_{ij} denotes the distance between firefly i and j, which is calculated according to the following formula:

$$r_{ij} = \sqrt{\sum_{k=1}^{n}(s_{ik} - s_{jk})^2},\qquad(12)$$

where n represents the dimension of the problem.

Once a firefly is attracted by the flash of other fireflies, the attractiveness β is updated based on

$$\beta = \beta_0 e^{-\gamma r_{ij}^2},\qquad(13)$$

where β_0 denotes the attractiveness at the light source ($r = 0$).

The formula for updating the location of fireflies is as follows:

$$s_i(t+1) = s_i(t) + \beta_0 e^{-\gamma r_{ij}^2}(s_j(t) - s_i(t)) + \alpha\varepsilon,\qquad(14)$$

where s_i and s_j denotes the space positions of firefly i and j, separately. α denotes the step factor, and ε represents a random value.

The main algorithm process of the FA can be described as Algorithm 2.

Algorithm 2: Process of the Firefly Algorithm

Initialize $n, \beta_0, \gamma, \alpha, \varepsilon, t = 0$
Define the maximum number of iterations (MaxGeneration).
while $t <$ MaxGeneration
  $t = t + 1$
  for $i = 1 : n$
    for $j = 1 : i$
      Calculate light intensity I_i at s_i position.
      If $I_j > I_i$
        Move firefly i towards j.
      end if
      Update the attractiveness values.
      Evaluate the new solutions and update the light intensity.
    end for
  end for
  Rank the fireflies and find the current best.
end while
Output the global optimal value.

We then obtain all prediction results of all subsequence, and the final prediction results of all sub-signals obtained by decomposition are superimposed. The advantages of this approach are verified in the next section.

4. Experimental Results

In this section, the daily maximum temperature time series in Melbourne is applied to analyze and verify the availability of the presented method. The experimental data is collected from the real world. The evaluation criteria are selected as root-mean-squared error (RMSE), normalized root-mean-square error (NRMSE), mean absolute percentage error (MAPE), and symmetric mean absolute percentage error (SMAPE). All prediction errors shown next are the mean values of 50 experimental results. The expressions are as follows:

$$\text{RMSE} = \left[\frac{1}{n-1} \sum_{k=1}^{n} [\hat{y}(k) - y(k)]^2 \right]^{1/2}, \quad (15)$$

$$\text{NRMSE} = \frac{\left[\frac{1}{n-1} \sum_{k=1}^{n} [\hat{y}(k) - y(k)]^2 \right]^{1/2}}{y_{\max} - y_{\min}}, \quad (16)$$

$$\text{MAPE} = \frac{100}{n} \sum_{k=1}^{n} \left| \frac{\hat{y}(k) - y(k)}{y(k)} \right|, \quad (17)$$

$$\text{SMAPE} = \frac{1}{n} \sum_{k=1}^{n} \frac{|y(k) - \hat{y}(k)|}{|y(k)| + |\hat{y}(k)|}, \quad (18)$$

where $y(k)$ denotes the real value, $\hat{y}(k)$ denotes the predicted value, and n denotes the number of samples.

The daily maximum temperature time series in Melbourne is used in the experiment. The dataset contains the daily maximum temperature from 3 January 1981 to 31 December 31 1990, a total of 3650 samples, which is shown in Figure 2. The training set is made up of the first 3000 samples, and the testing set is composed of the remaining 650. In order to compare the effectiveness of the model, six other methods were used in the comparative experiments: an RBF neural network, [36], an ANFIS [37], the original BP model, FABP, FABP with CEEMDAN decomposition (CEEMDAN–FABP), and FABP with VMD decomposition (VMD–FABP). The experimental environment involved the Windows 7 operating system, and all experiments were carried out using MATLAB R2016a on a 3.50 GHz, Intel(R) Core i3-4150M CPU with 6 GB RAM.

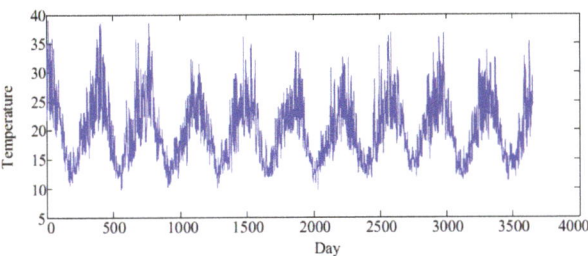

Figure 2. The daily maximum temperature time series in Melbourne.

Figure 3 shows the decomposition results of the original time series based on CEEMDAN, which adaptively obtains 13 IMFs. CEEMDAN has good anti-noise performance, so this paper leaves out the de-noising process of the original time series.

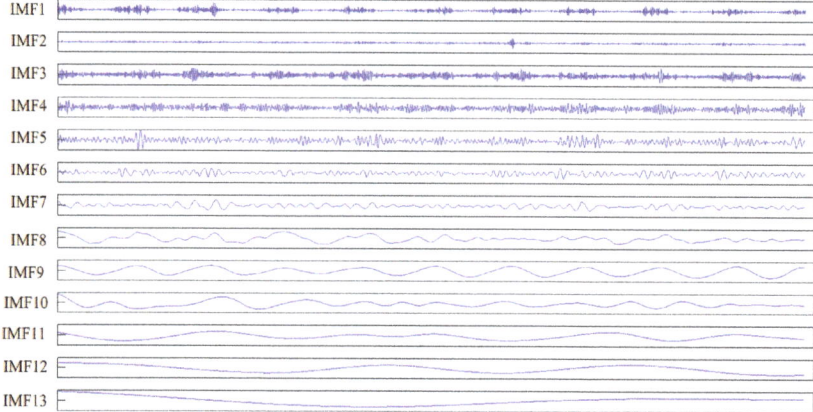

Figure 3. Decomposition results of the daily maximum temperature time series in Melbourne based on complete ensemble empirical mode decomposition with adaptive noise (CEEMDAN).

In this paper, sample entropy is an indicator for judging the way IMF predicts. Figure 4 shows the sample entropies of the IMFs. The entropy of the original time series is 0.81, which is indicated by the dotted line. In this research, the IMF components, whose entropy is greater than the entropy of the original time series, are predicted separately, because these IMFs are highly complex. The IMFs whose entropies are smaller than the original time series are combined and superimposed. After the combined operation, the model complexity and computational time are significantly reduced, while ensuring accuracy of prediction. Based on the results in Figure 4, the first four IMF components should be predicted separately, while the IMF components from 5 to 13 should be combined for prediction.

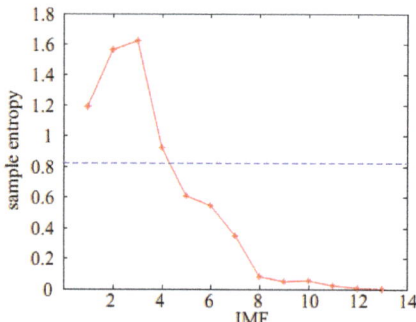

Figure 4. Sample entropies of intrinsic mode functions (IMFs).

We obtained the prediction results of five IMF components, as shown in Figure 5. It can be observed that the prediction curve of each IMF is able to track the actual values, and the prediction trend is basically consistent. Due to the high frequency characteristics of IMF1, accurate prediction and tracking is more difficult, resulting in larger prediction errors for IMF1.

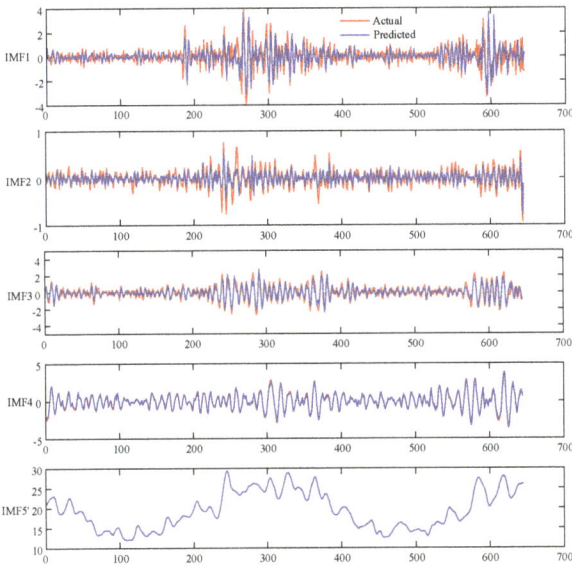

Figure 5. Prediction results of IMFs.

Figure 6 shows the final prediction results of the hybrid model of CEEMDAN and FABP. It can be seen that the prediction curve can track the real curve well and that the prediction error is small. The prediction result can be obtained in the case of the original time series without denoising.

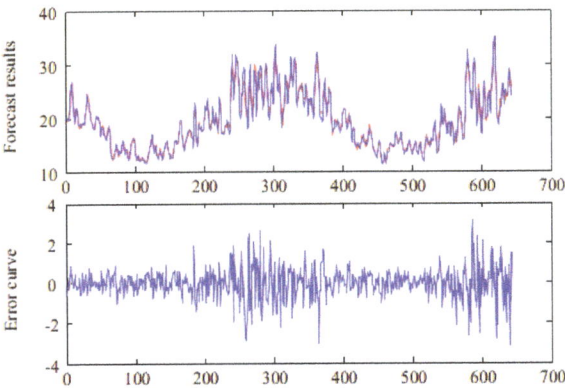

Figure 6. Prediction results and prediction errors of the daily maximum temperature in Melbourne based on the hybrid model of CEEMDAN and FABP.

The prediction errors presented in Figures 5 and 6 are shown in Table 2. As can be seen from Table 2, the overall prediction error of the time series is 0.7763, while the prediction error of IMF1 is 0.6361. Therefore, if the prediction accuracy of IMF1 can be improved, the overall prediction accuracy will be greatly improved. Based on the above analysis, we propose the application of a two-layer decomposition strategy to further decompose the IMF1 component.

Table 2. Overall prediction error and prediction error of IMF1.

Prediction Error	RMSE	NRMSE	MAPE	SMAPE
Overall prediction error	0.7763	0.0325	0.0277	0.0138
IMF1 prediction error	0.6361	0.0808	1.8515	0.5829

The IMF1 component is decomposed into six subsequences based on the VMD algorithm. The decomposition results of the IMF1 component are shown in Figure 7. We use FABP to predict VMF1, ..., VMF6, and we combine the prediction results to obtain the prediction results of the IMF1 component. Next, we combine the prediction results of IMF1 with IMF2, ..., IMF5' components, thereby obtaining the prediction results of daily maximum temperature time series in Melbourne, which are shown in Figure 8. The error indexes of different prediction models are shown in Table 3. It can be seen that the proposed two-layer decomposition algorithm can significantly reduce the prediction error and improve prediction accuracy compared with the direct prediction. This is mainly because the decomposition algorithm converts complex original signals into several simple and easy-to-analyze sub-signals, which is conducive to analysis and prediction. Based on the two-layer decomposition model, the prediction error is smaller than the single-layer approach, and the prediction performance is improved.

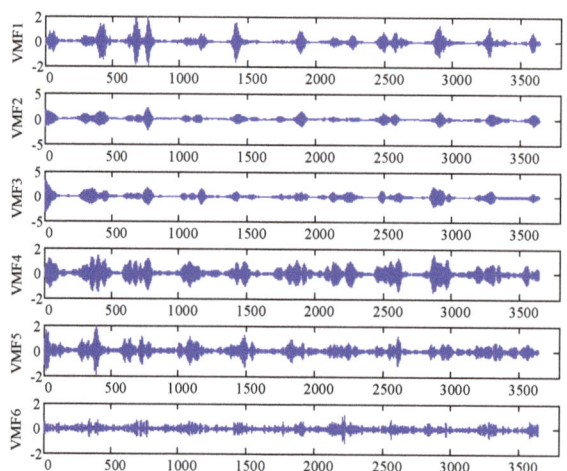

Figure 7. Decomposition results of IMF1 based on the variational mode decomposition (VMD) algorithm.

We conducted one-step-, two-step-, three-step-, and five-step-ahead prediction experiments on the time series. The corresponding prediction errors, including RMSE, NRMSE, MAPE, and SMAPE, are shown in Tables 3–6. According to the results presented in the tables, the proposed model has a minimum prediction error in multiple prediction experiments, which indicates that the hybrid model of CEEMDAN–VMD–FABP has the best prediction performance. We can also say that the hybrid model based on the two-layer decomposition approach is better than the hybrid models based on a single decomposition approach. Surprisingly, although the ANFIS performed poorly in one-step prediction, its one-step prediction and multi-step prediction have similar effects, especially in the five-step ahead prediction experiment, and its performance is better than the RBF in multi-step prediction. The ANFIS method shows stability ability in terms of prediction, although the overall effect was not satisfactory. The prediction results of other models except for the ANFIS basically conform to the actual law. The more advanced the steps are, the more difficult it is to predict the chaotic time series.

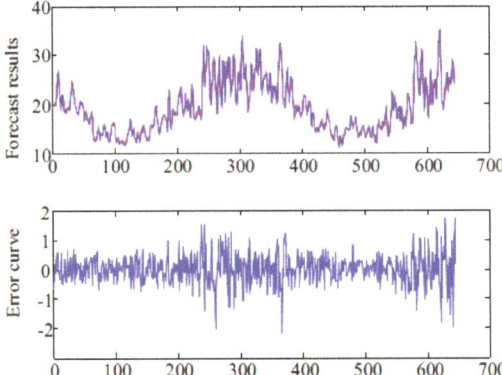

Figure 8. Prediction results and prediction errors of the daily maximum temperature in Melbourne based on the two-layer decomposition algorithm and a BPNN optimized by a firefly algorithm (FABP).

Table 3. Prediction errors of daily maximum temperature time series based on different algorithms (one step ahead).

Model	RMSE	NRMSE	MAPE	SMAPE	Training Time	Testing Time
RBF	1.7241	0.0721	0.0695	0.0345	0.9984	0.1092
ANFIS	3.2410	0.1356	0.1224	0.0598	22.3393	0.0936
BP	1.3818	0.0578	0.0511	0.0255	0.4368	0.0468
FABP	1.3618	0.0570	0.0505	0.0251	34.4294	0.0780
CEEMDAN–FABP	0.7763	0.0325	0.0277	0.0138	151.3834	0.1404
VMD–FABP	0.7026	0.0294	0.0266	0.0132	197.1852	0.2340
CEEMDAN–VMD–FABP	0.5131	0.0215	0.0198	0.0099	307.5092	0.2964

Table 4. Prediction errors of daily maximum temperature time series based on different algorithms (two steps ahead).

Model	RMSE	NRMSE	MAPE	SMAPE
RBF	2.6741	0.1119	0.1051	0.0517
ANFIS	3.4725	0.1453	0.1317	0.0644
BP	2.4105	0.1009	0.0912	0.0454
FABP	2.4032	0.1006	0.0924	0.0456
CEEMDAN–FABP	0.9292	0.0389	0.0346	0.0172
VMD–FABP	0.7240	0.0303	0.0276	0.0138
CEEMDAN–VMD–FABP	0.6910	0.0289	0.0262	0.0130

Table 5. Prediction errors of daily maximum temperature time series based on different algorithms (three steps ahead).

Model	RMSE	NRMSE	MAPE	SMAPE
RBF	3.3242	0.1391	0.1286	0.0628
ANFIS	3.4715	0.1453	0.1330	0.0651
BP	3.1497	0.1318	0.1211	0.0589
FABP	3.1435	0.1315	0.1194	0.0587
CEEMDAN–FABP	1.1266	0.0471	0.0420	0.0209
VMD–FABP	0.9105	0.0381	0.0345	0.0172
CEEMDAN–VMD–FABP	0.8692	0.0364	0.0333	0.0166

Table 6. Prediction errors of daily maximum temperature time series based on different algorithms (five steps ahead).

Model	RMSE	NRMSE	MAPE	SMAPE
RBF	3.4960	0.1463	0.1353	0.0662
ANFIS	3.4408	0.1440	0.1333	0.0654
BP	3.3497	0.1402	0.1295	0.0633
FABP	3.3595	0.1406	0.1285	0.0632
CEEMDAN–FABP	1.5822	0.0662	0.0598	0.0297
VMD–FABP	1.2500	0.0523	0.0487	0.0242
CEEMDAN–VMD–FABP	0.9864	0.0413	0.0370	0.0183

To show the above experimental results more intuitively, we transform the error values, i.e. the RMSE, NRMSE, MAPE, and SMAPE shown in Tables 3–6, into a column chart, presented in Figure 9. It can be seen that the proposed hybrid model of CEEMDAN–VMD–FABP has the best performance. The proposed two-layer decomposition model has minimum prediction error in one-step-, two-step-, three-step-, and five-step-ahead prediction.

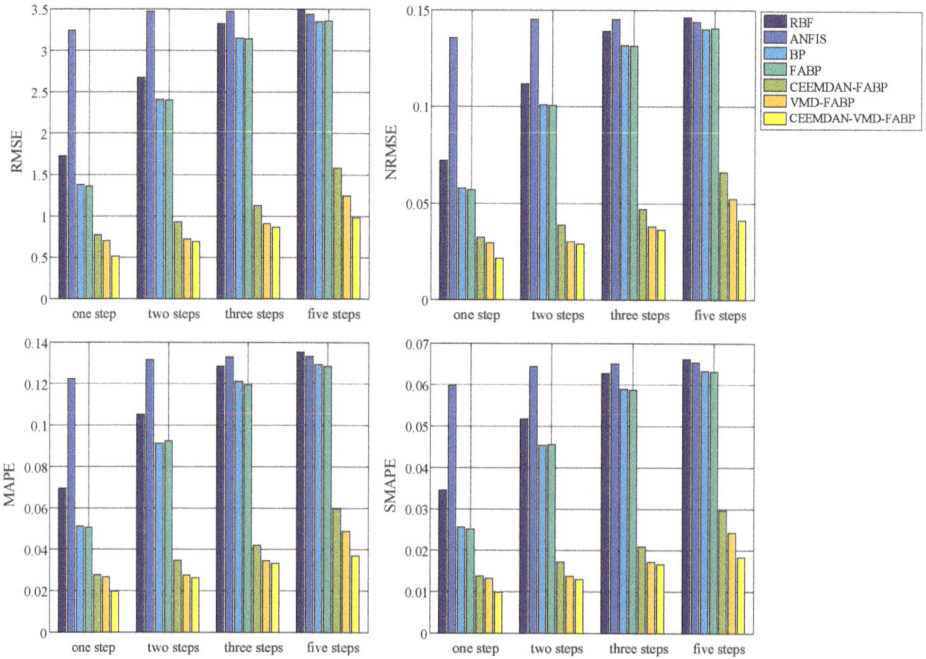

Figure 9. Performance comparison of different models with different prediction steps.

Moreover, we compared the training time and testing time in one-step-ahead prediction. Under different prediction steps, the running time is not much different, so we only list the running time of one-step prediction in Table 3. As can be seen in the table, the testing time of each method is not much different. Although the method proposed in this paper has the longest training time, the experimental results prove that the proposed two-layer decomposition model can obtain the best prediction accuracy, which shows the effectiveness of the proposed method. Moreover, even if the training time is long, the overall running time is only a few minutes (not long), completely within the acceptable range.

5. Conclusions

In this paper, we propose a two-layer decomposition technique consisting of CEEMEAN and VMD. We obtained a group of subsequences by a two-layer decomposition method. These subsequences were separately predicted, and the prediction results were combined to obtain the final result. For the prediction, we used a BPNN optimized by a firefly algorithm. From this work, the following can be concluded:

1. The actual time series is usually non-stationary and noisy. It is generally difficult to analyze the original time series. CEEMDAN is an anti-noise decomposition method, and VMD can handle non-stationary signals very well. Therefore, subsequences decomposed by CEEMDAN and VMD are easy to analyze and predict.
2. After decomposition of the original signal, the BPNN was used for prediction. At this stage, the parameters in the BPNN greatly influenced prediction accuracy. Therefore, in order to reasonably select the model parameters, the FA algorithm was introduced to optimize the parameters of BP.

In general, in order to improve prediction accuracy, the following can be considered for further study: Firstly, original input variables were analyzed to eliminate factors that are not conducive to analysis and prediction. Secondly, we optimized the prediction model at the prediction stage. We studied the two aspects simultaneously, and the experimental results demonstrate the effectiveness of the proposed method.

Author Contributions: Conceptualization: X.X. and W.R.; methodology: X.X. and W.R.; data curation: X.X. and W.R.; writing—original draft preparation: X.X. and W.R.

Funding: This work was supported by the National Natural Science Foundation of China (61773087).

Conflicts of Interest: The authors declare no conflict of interest.

References

1. Shah, H.; Tairan, N.; Garg, H.; Ghazali, R. A Quick Gbest Guided Artificial Bee Colony Algorithm for Stock Market Prices Prediction. *Symmetry* **2018**, *10*, 292. [CrossRef]
2. Zhai, H.; Cui, L.; Nie, Y.; Xu, X.; Zhang, W. A Comprehensive Comparative Analysis of the Basic Theory of the Short Term Bus Passenger Flow Prediction. *Symmetry* **2018**, *10*, 369. [CrossRef]
3. Han, M.; Zhang, R.; Xu, M. Multivariate Chaotic Time Series Prediction Based on ELM–PLSR and Hybrid Variable Selection Algorithm. *Neural Process. Lett.* **2017**, *46*, 705–717. [CrossRef]
4. Chandra, R. Competition and collaboration in cooperative coevolution of Elman recurrent neural networks for time-series prediction. *IEEE Trans. Neural Netw. Learn. Syst.* **2015**, *26*, 3123–3136. [CrossRef] [PubMed]
5. Yaslan, Y.; Bican, B. Empirical mode decomposition based denoising method with support vector regression for time series prediction: A case study for electricity load forecasting. *Measurement* **2017**, *103*, 52–61. [CrossRef]
6. Chen, D. Research on traffic flow prediction in the big data environment based on the improved RBF neural network. *IEEE Trans. Ind. Inform.* **2017**, *13*, 2000–2008. [CrossRef]
7. Malik, Z.K.; Hussain, A.; Wu, Q.J. Multilayered echo state machine: A novel architecture and algorithm. *IEEE Trans. Cybern.* **2017**, *47*, 946–959. [CrossRef]
8. Gibson, J. Entropy Power, Autoregressive Models, and Mutual Information. *Entropy* **2018**, *20*, 750. [CrossRef]
9. Alsharif, M.H.; Younes, M.K.; Kim, J. Time Series ARIMA Model for Prediction of Daily and Monthly Average Global Solar Radiation: The Case Study of Seoul, South Korea. *Symmetry* **2019**, *11*, 240. [CrossRef]
10. Yan, J.; Li, K.; Bai, E.; Yang, Z.; Foley, A. Time series wind power forecasting based on variant Gaussian Process and TLBO. *Neurocomputing* **2016**, *189*, 135–144. [CrossRef]
11. Nava, N.; Di Matteo, T.; Aste, T. Financial Time Series Forecasting Using Empirical Mode Decomposition and Support Vector Regression. *Risks* **2018**, *6*, 7. [CrossRef]
12. Baghaee, H.R.; Mirsalim, M.; Gharehpetian, G.B. Power calculation using RBF neural networks to improve power sharing of hierarchical control scheme in multi-DER microgrids. *IEEE J. Emerg. Sel. Top. Power Electron.* **2016**, *4*, 1217–1225. [CrossRef]

13. Ahn, J.; Shin, D.; Kim, K.; Yang, J. Indoor Air Quality Analysis Using Deep Learning with Sensor Data. *Sensors* **2017**, *17*, 2476. [CrossRef] [PubMed]
14. Takeda, H.; Tamura, Y.; Sato, S. Using the ensemble Kalman filter for electricity load forecasting and analysis. *Energy* **2016**, *104*, 184–198. [CrossRef]
15. Bogiatzis, A.; Papadopoulos, B. Global Image Thresholding Adaptive Neuro-Fuzzy Inference System Trained with Fuzzy Inclusion and Entropy Measures. *Symmetry* **2019**, *11*, 286. [CrossRef]
16. Mlakić, D.; Baghaee, H.R.; Nikolovski, S. A novel ANFIS-based islanding detection for inverter–interfaced microgrids. *IEEE Trans. Smart Grid* **2018**, in press.
17. Alhasa, K.M.; Mohd Nadzir, M.S.; Olalekan, P.; Latif, M.T.; Yusup, Y.; Iqbal Faruque, M.R.; Ahamad, F.; Abd Hamid, H.H.; Aiyub, K.; Md Ali, S.H.; et al. Calibration Model of a Low-Cost Air Quality Sensor Using an Adaptive Neuro-Fuzzy Inference System. *Sensors* **2018**, *18*, 4380. [CrossRef]
18. Zhou, J.; Yu, X.; Jin, B. Short-Term Wind Power Forecasting: A New Hybrid Model Combined Extreme-Point Symmetric Mode Decomposition, Extreme Learning Machine and Particle Swarm Optimization. *Sustainability* **2018**, *10*, 3202. [CrossRef]
19. Fan, G.-F.; Qing, S.; Wang, H.; Hong, W.-C.; Li, H.-J. Support Vector Regression Model Based on Empirical Mode Decomposition and Auto Regression for Electric Load Forecasting. *Energies* **2013**, *6*, 1887–1901. [CrossRef]
20. Liu, H.; Tian, H.Q.; Liang, X.F.; Li, Y.F. Wind speed forecasting approach using secondary decomposition algorithm and Elman neural networks. *Appl. Energy* **2015**, *157*, 183–194. [CrossRef]
21. Ren, Y.; Suganthan, P.N. Empirical mode decomposition-k nearest neighbor models for wind speed forecasting. *J. Power Energy Eng.* **2014**, *2*, 176–185. [CrossRef]
22. Wang, W.C.; Chau, K.W.; Xu, D.M.; Chen, X.Y. Improving forecasting accuracy of annual runoff time series using ARIMA based on EEMD decomposition. *Water Resour. Manag.* **2015**, *29*, 2655–2675. [CrossRef]
23. Torres, M.E.; Colominas, M.A.; Schlotthauer, G.; Flandrin, P. A complete ensemble empirical mode decomposition with adaptive noise. In Proceedings of the IEEE International Conference on Acoustics, Speech, and Signal Processing (ICASSP), Prague, Czech Republic, 22–27 May 2011.
24. Dragomiretskiy, K.; Zosso, D. Variational mode decomposition. *IEEE Trans. Signal Process.* **2014**, *62*, 531–544. [CrossRef]
25. Yang, X.S. Firefly algorithm, stochastic test functions and design optimization. *Int. J. Bio-Inspired Comput.* **2010**, *2*, 78–84. [CrossRef]
26. Chen, C.F.; Lai, M.C.; Yeh, C.C. Forecasting tourism demand based on empirical mode decomposition and neural network. *Knowl.-Based Syst.* **2012**, *26*, 281–287. [CrossRef]
27. Zhou, Q.; Jiang, H.; Wang, J.; Zhou, J. A hybrid model for PM2. 5 forecasting based on ensemble empirical mode decomposition and a general regression neural network. *Sci. Total Environ.* **2014**, *496*, 264–274. [CrossRef]
28. Liu, H.; Tian, H.Q.; Li, Y.F. Comparison of new hybrid FEEMD-MLP, FEEMD-ANFIS, Wavelet Packet-MLP and Wavelet Packet-ANFIS for wind speed predictions. *Energy Convers. Manag.* **2015**, *89*, 1–11. [CrossRef]
29. Abdoos, A.A. A new intelligent method based on combination of VMD and ELM for short term wind power forecasting. *Neurocomputing* **2016**, *203*, 111–120. [CrossRef]
30. Lahmiri, S. A variational mode decompoisition approach for analysis and forecasting of economic and financial time series. *Expert Syst. Appl.* **2016**, *55*, 268–273. [CrossRef]
31. Jianwei, E.; Bao, Y.; Ye, J. Crude oil price analysis and forecasting based on variational mode decomposition and independent component analysis. *Phys. A Stat. Mech. Appl.* **2017**, *484*, 412–427.
32. Wang, D.; Luo, H.; Grunder, O.; Lin, Y. Multi-step ahead wind speed forecasting using an improved wavelet neural network combining variational mode decomposition and phase space reconstruction. *Renew. Energy* **2017**, *113*, 1345–1358. [CrossRef]
33. Huang, N.E.; Shen, Z.; Long, S.R.; Wu, M.C.; Shih, H.H.; Zheng, Q.; Yen, N.; Tung, C.C.; Liu, H.H. The empirical mode decomposition and the Hilbert spectrum for nonlinear and non-stationary time series analysis. *Proc. R. Soc. Lond. Ser. A Math. Phys. Eng. Sci.* **1998**, *454*, 903–995. [CrossRef]
34. Li, Y.; Li, Y.; Chen, X.; Yu, J. Denoising and Feature Extraction Algorithms Using NPE Combined with VMD and Their Applications in Ship-Radiated Noise. *Symmetry* **2017**, *9*, 256. [CrossRef]

35. Ghadimi, E.; Teixeira, A.; Shames, I.; Johansson, M. Optimal parameter selection for the alternating direction method of multipliers (ADMM): Quadratic problems. *IEEE Trans. Autom. Control* **2015**, *60*, 644–658. [CrossRef]
36. Baghaee, H.R.; Mirsalim, M.; Gharehpetan, G.B.; Talebi, H.A. Nonlinear load sharing and voltage compensation of microgrids based on harmonic power-flow calculations using radial basis function neural networks. *IEEE Syst. J.* **2018**, *12*, 2749–2759. [CrossRef]
37. Nikolovski, S.; Reza Baghaee, H.; Mlakić, D. ANFIS-based peak power shaving/curtailment in microgrids including PV units and besss. *Energies* **2018**, *11*, 2953. [CrossRef]

© 2019 by the authors. Licensee MDPI, Basel, Switzerland. This article is an open access article distributed under the terms and conditions of the Creative Commons Attribution (CC BY) license (http://creativecommons.org/licenses/by/4.0/).

Article

Multiparametric Analytical Solution for the Eigenvalue Problem of FGM Porous Circular Plates

Krzysztof Kamil Żur * and Piotr Jankowski

Faculty of Mechanical Engineering, Bialystok University of Technology, 45C, 15-351 Bialystok, Poland; piotrjankowski1995@gmail.com
* Correspondence: k.zur@pb.edu.pl

Received: 11 March 2019; Accepted: 20 March 2019; Published: 22 March 2019

Abstract: Free vibration analysis of the porous functionally graded circular plates has been presented on the basis of classical plate theory. The three defined coupled equations of motion of the porous functionally graded circular/annular plate were decoupled to one differential equation of free transverse vibrations of plate. The one universal general solution was obtained as a linear combination of the multiparametric special functions for the functionally graded circular and annular plates with even and uneven porosity distributions. The multiparametric frequency equations of functionally graded porous circular plate with diverse boundary conditions were obtained in the exact closed-form. The influences of the even and uneven distributions of porosity, power-law index, diverse boundary conditions and the neglected effect of the coupling in-plane and transverse displacements on the dimensionless frequencies of the circular plate were comprehensively studied for the first time. The formulated boundary value problem, the exact method of solution and the numerical results for the perfect and imperfect functionally graded circular plates have not yet been reported.

Keywords: eigenvalue problem; axisymmetric and non-axisymmetric vibrations; multiparametric special functions; circular plate; functionally graded porous material

1. Introduction

Functionally graded materials (FGMs) are a class of composite materials, which are made of the ceramic and metal mixture such that the material properties vary continuously in appropriate directions of structural components. In the processes of preparing functionally graded material, micro-voids and porosities may appear inside material in view of the technical issues. Zhu et al. [1] reported that many porosities appear in material during the functionally graded material preparation process by the non-pressure sintering technique. Wattanasakulpong et al. [2] reported that many porosities exist in the intermediate area of the functionally graded material fabricated by utilizing a multi-step sequential infiltration technique because of the problem with infiltration of the secondary material into the middle area. In that case, less porosities appear in the top and bottom area of material because infiltration of the material is easier in these zones.

In recent years, a significant number of articles about the free vibrations of porous functionally graded (FGM) plates have appeared in the literature due to their wide applications in many fields of engineering such as aeronautical, civil, mechanical, automotive, and ocean engineering. The gradation of properties in functionally graded materials and the diverse distributions of porosity have a significant effect on distributions of the mass and the stiffness of plates and therefore their natural frequencies. The knowledge about influence of distribution of the material properties on dynamics of plates is very important because it allows us to predict the frequency of plates and find their optimal parameters. Additionally, the comprehensive investigation of the effect of functionally graded material with porosities and diverse boundary conditions on the natural frequencies of plates is the first important step to designing their safe and rational active vibration control system.

We note that, in most engineering applications, the classical plate theory is often used to analyze the dynamic behavior of thin lightweight plates. It is impossible to review all works focused on mechanical behavior of porous FGM structures; then, we limit ourselves to chronological review of some of the works focused on mechanical behavior of porous and porous FGM plates that are closely related to our work.

Jabbari et al. [3] studied the buckling of thin saturated porous circular plate with the layers of piezoelectric actuators. Buckling load was obtained for clamped circular plate under uniform radial compressive loading. The same authors presented the buckling analysis of clamped thin saturated porous circular plate with sensor–actuator layers under uniform radial compression [4,5] investigated thermal and mechanical stability of clamped thin saturated and unsaturated porous circular plates with piezoelectric actuators. Rad and Shariyat [6] solved the three-dimensional magneto-elastic problem for asymmetric variable thickness porous FGM circular supported on the Kerr elastic foundation using the differential quadrature method and the state space vector technique. Barati et al. [7] studied buckling of functionally graded piezoelectric rectangular plates with porosities based on the four-variable plate theory. Mechab et al. [8] studied free vibration of the FGM nanoplate with porosities resting on Winkler and Pasternak elastic foundation based on the two-variable plate theory. Mojahedin et al. [9] analyzed buckling of radially loaded clamped saturated porous circular plates based on higher order shear deformation theory. Wang and Zu [10] analyzed vibration behaviors of thin FGM rectangular plates with porosities and moving in the thermal environment using the method of harmonic balance and the Runge–Kutta technique. Gupta and Talha [11] analyzed flexural and vibration response of porous FGM rectangular plates using nonpolynomial higher-order shear and the normal deformation theory. Wang and Zu [12] analyzed vibration characteristics of longitudinally moving sigmoid porous FGM plates based on the von Kármán nonlinear plate theory. Ebrahimi et al. [13] studied free vibration of smart shear deformable rectangular plates made of porous magneto-electro-elastic functionally graded materials. Feyzi and Khorshidvand [14] studied axisymmetric post-buckling behavior of a saturated porous circular plate with simply supported and clamped boundary conditions. Wang and Zu [15] studied large-amplitude vibration of thin sigmoid functionally graded plates with porosities. Wang et al. [16] studied vibrations of longitudinally travelling FGM porous thin rectangular plates using the Galerkin method and the four-order Runge–Kutta method. Ebrahimi et al. [17] used a four-variable shear deformation refined plate theory for free vibration analysis of embedded smart rectangular plates made of magneto-electro-elastic porous functionally graded materials. Shahverdi and Barati [18] developed nonlocal strain-gradient elasticity model for vibration analysis of porous FGM nano-scale rectangular plates. Shojaeefard et al. [19] studied free vibration and thermal buckling of micro temperature-dependent FGM porous circular plate using the generalized differential quadrature method. Barati and Shahverdi [20] presented a new solution to examine large amplitude vibration of a porous nanoplate resting on a nonlinear elastic foundation modeled based on the four-variable plate theory. Kiran et al. [21] studied free vibration of porous FGM magneto-electro-elastic skew plates using the finite element formulation. Cong et al. [22] presented an analytical approach to buckling and post-buckling behavior analysis of FGM rectangular plates with porosities under thermal and thermomechanical loads based on the Reddy's higher-order shear deformation theory. Kiran and Kattimani [23] studied free vibration and static behavior of porous FGM magneto-electro-elastic rectangular plates using the finite element method. Arshid and Khorshidvand [24] analyzed free vibration of saturated porous FGM circular plates integrated with piezoelectric actuators using the differential quadrature method. Shahsavari et al. [25] used the quasi-3D hyperbolic theory for free vibration of porous FGM rectangular plates resting on Winkler, Pasternak and Kerr foundations.

2. Contribution of Current Study

The aim of the paper is to formulate and solve the boundary value problem for the free axisymmetric and non-axisymmetric vibrations of FGM circular plate with even and uneven porosity distributions and diverse boundary conditions. The defined coupled equations of motion for the porous

FGM circular plate were decoupled based on the properties of physical neutral surface. The general solution of the decoupled equation of motion of a porous FGM circular plate was defined as the linear combination of the Bessel functions functionally dependent on the material parameters. The obtained characteristic equations allow us to comprehensively study the effect of the distribution of material parameters and the formulated boundary conditions on the natural frequencies of axisymmetric and non-axisymmetric vibrations of the circular plates without the necessity to solve a new eigenvalue problem for plates with a steady distribution of parameters.

Authors of many previous papers (e.g., [26–30]) presented the free transverse vibration analysis of the perfect (without porosity) FGM circular plates using the equation of motion including only the coefficient of the pure bending stiffness varying in the thickness direction of the plate. The coefficients of the extensional stiffness and the bending-extensional coupling stiffness were neglected because the effect of the coupled in-plane and transverse displacements was omitted for obtaining simplified solution to the eigenvalue problem.

In the present paper, the obtained equation of motion of the perfect and imperfect FGM circular plates includes the coefficients of extensional stiffness, bending-extensional coupling stiffness and bending stiffness, which appeared by decoupling the in-plane and transverse displacements using the properties of the physical neutral surface. The differences between the values of numerical results for the eigenfrequencies of the perfect FGM circular plate with and without the coupling effect are shown for diverse boundary conditions.

To the best knowledge of authors, there are no studies which focus on the free axisymmetric and non-axisymmetric vibrations of FGM and porous FGM circular plates. In particular, the obtained exact solution, the multiparametric frequency equations and the calculated eigenfrequencies for the free vibrations of perfect and imperfect FGM circular plates with clamped, simply supported, sliding and free edges have not yet been reported. The present paper fills this void in the literature.

3. FGM Circular Plate with Porosities

Consider a porous FGM thin circular plate with radius R and thickness h presented in the cylindrical coordinate (r, θ, z) with the z-axis along the longitudinal direction. The geometry and the coordinate system of the considered circular plate are shown in Figure 1. The FGM plate contains evenly (*e*) and unevenly (*u*) distributed porosities along the plate's thickness direction. The cross-sections of the FGM circular plates with the two various types of distribution of porosities are shown in Figure 2.

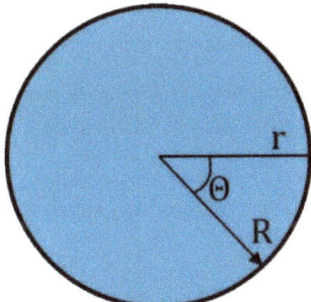

Figure 1. The geometry and the coordinate system of the porous FGM circular plate.

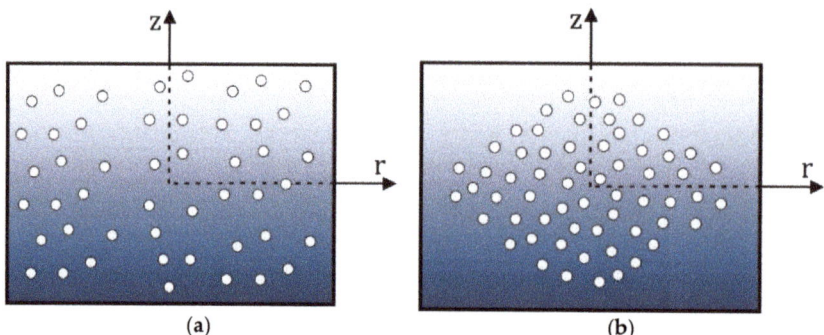

Figure 2. The cross-sections of the porous FGM circular plate: (**a**) even distribution; (**b**) uneven distribution.

The functionally graded material is a mixture of a ceramic (*c*) and a metal (*m*). If the volume fraction of the ceramic part is V_c and the metallic part is V_m, we have the well-known dependence:

$$V_c(z) + V_m(z) = 1. \tag{1}$$

Based on the modified rule of mixtures [16] with the porosity volume fraction ψ ($\psi \ll 1$), the Young's modulus, the density and the Poisson's ratio for evenly (*e*) distributed porosities over the cross-section of the plate have the general forms:

$$E^e(z, \psi) = E_c \left[V_c(z) - \frac{\psi}{2} \right] + E_m \left[V_m(z) - \frac{\psi}{2} \right], \tag{2a}$$

$$\rho^e(z, \psi) = \rho_c \left[V_c(z) - \frac{\psi}{2} \right] + \rho_m \left[V_m(z) - \frac{\psi}{2} \right], \tag{2b}$$

$$\nu^e(z, \psi) = \nu_c \left[V_c(z) - \frac{\psi}{2} \right] + \nu_m \left[V_m(z) - \frac{\psi}{2} \right]. \tag{2c}$$

The volume fraction of the ceramic part changes continually along the thickness and can be defined as [31]

$$V_c(z, g) = \left(\frac{z}{h} + \frac{1}{2} \right)^g, \quad g \geq 0, \tag{3}$$

where *g* is the power-law index of the material. A change in the power *g* of functionally graded material results in a change in the portion of the ceramic and metal components in the circular plate. We assume that the composition is varied from the bottom surface ($z = -h/2$) to the top surface ($z = h/2$) of the circular plate. After substituting the variation of the ceramic part $V_c(z, g)$ from Equation (3) into Equation (2), the material properties of the functionally graded circular plate with evenly distributed porosities are defined in the final form:

$$E^e(z, g, \psi) = (E_c - E_m) \left(\frac{z}{h} + \frac{1}{2} \right)^g + E_m - \frac{\psi}{2}(E_c + E_m), \tag{4a}$$

$$\rho^e(z, g, \psi) = (\rho_c - \rho_m) \left(\frac{z}{h} + \frac{1}{2} \right)^g + \rho_m - \frac{\psi}{2}(\rho_c + \rho_m), \tag{4b}$$

$$\nu^e(z, g, \psi) = (\nu_c - \nu_m) \left(\frac{z}{h} + \frac{1}{2} \right)^g + \nu_m - \frac{\psi}{2}(\nu_c + \nu_m). \tag{4c}$$

For the functionally graded circular plate with unevenly (u) distributed porosities [16], the material properties in Equations (4) can be replaced by the following forms:

$$E^u(z,g,\psi) = (E_c - E_m)\left(\frac{z}{h} + \frac{1}{2}\right)^g + E_m - \frac{\psi}{2}(E_c + E_m)\left(1 - \frac{2|z|}{h}\right), \quad (5a)$$

$$\rho^u(z,g,\psi) = (\rho_c - \rho_m)\left(\frac{z}{h} + \frac{1}{2}\right)^g + \rho_m - \frac{\psi}{2}(\rho_c + \rho_m)\left(1 - \frac{2|z|}{h}\right), \quad (5b)$$

$$\nu^u(z,g,\psi) = (\nu_c - \nu_m)\left(\frac{z}{h} + \frac{1}{2}\right)^g + \nu_m - \frac{\psi}{2}(\nu_c + \nu_m)\left(1 - \frac{2|z|}{h}\right). \quad (5c)$$

In this case, the porosity linearly decreases to zero at the top and the bottom of the cross-section of the plate. The effect of Poisson's ratio is much less on the mechanical behavior of FGM plates than the Young's modulus [32,33], thus the Poisson's ratio will assume to be constant $\nu^e = \nu^u = \nu$ in the whole volume of the porous FGM circular plate.

4. Constitutive Relations and Governing Equations

In most practical applications, the ratio of the radius R to the thickness h of the plate is more than 10; then, the assumptions of classical plate theory (CPT) are applicable and rotary inertia and shear deformation can be successfully omitted.

For a thin circular plate, the displacement field has the form:

$$u_r(r,\theta,z,t) = u(r,\theta,t) - z\frac{\partial w(r,\theta,t)}{\partial r}, \quad (6a)$$

$$u_\theta(r,\theta,z,t) = v(r,\theta,t) - \frac{z}{r}\frac{\partial w(r,\theta,t)}{\partial \theta}, \quad (6b)$$

$$w(r,\theta,z,t) = w(r,\theta,t), \quad (6c)$$

where u, v and w are the radial, circumferential and transverse displacements of the midplane ($z = 0$) of the plate at time t. Based on the linear strain–displacement relations and Hook's law, the resultant forces and the moments for porous FGM circular plate ($i = \{e, u\}$) can be expressed in the following form [34]:

$$\begin{pmatrix} N_{rr}^i \\ N_{\theta\theta}^i \\ N_{r\theta}^i \end{pmatrix} = \begin{bmatrix} A_{11}^i & A_{12}^i & 0 \\ A_{12}^i & A_{11}^i & 0 \\ 0 & 0 & A_{33}^i \end{bmatrix} \begin{pmatrix} \varepsilon_{rr}^0 \\ \varepsilon_{\theta\theta}^0 \\ \gamma_{r\theta}^0 \end{pmatrix} + \begin{bmatrix} B_{11}^i & B_{12}^i & 0 \\ B_{12}^i & B_{11}^i & 0 \\ 0 & 0 & B_{33}^i \end{bmatrix} \begin{pmatrix} \kappa_{rr} \\ \kappa_{\theta\theta} \\ \kappa_{r\theta} \end{pmatrix}, \quad (7a)$$

$$\begin{pmatrix} M_{rr}^i \\ M_{\theta\theta}^i \\ M_{r\theta}^i \end{pmatrix} = \begin{bmatrix} B_{11}^i & B_{12}^i & 0 \\ B_{12}^i & B_{11}^i & 0 \\ 0 & 0 & B_{33}^i \end{bmatrix} \begin{pmatrix} \varepsilon_{rr}^0 \\ \varepsilon_{\theta\theta}^0 \\ \gamma_{r\theta}^0 \end{pmatrix} + \begin{bmatrix} D_{11}^i & D_{12}^i & 0 \\ D_{12}^i & D_{11}^i & 0 \\ 0 & 0 & D_{33}^i \end{bmatrix} \begin{pmatrix} \kappa_{rr} \\ \kappa_{\theta\theta} \\ \kappa_{r\theta} \end{pmatrix}, \quad (7b)$$

where

$$\left(\varepsilon_{rr}^0, \varepsilon_{\theta\theta}^0, \gamma_{r\theta}^0\right) = \left(\frac{\partial u}{\partial r}, \frac{1}{r}\frac{\partial v}{\partial \theta} + \frac{u}{r}, \frac{1}{r}\frac{\partial u}{\partial \theta} + \frac{\partial v}{\partial r} - \frac{v}{r}\right), \quad (8a)$$

$$(\kappa_{rr}, \kappa_{\theta\theta}, \kappa_{r\theta}) = \left(-\frac{\partial^2 w}{\partial r^2}, -\frac{1}{r^2}\frac{\partial^2 w}{\partial \theta^2} - \frac{1}{r}\frac{\partial w}{\partial r}, -\frac{2}{r}\frac{\partial^2 w}{\partial r \partial \theta} + \frac{2}{r^2}\frac{\partial w}{\partial \theta}\right) \quad (8b)$$

are the in-plane strains and curvatures of midplane, respectively.

We assume that the material properties are varied from the bottom surface ($z = -h/2$) to the top surface ($z = h/2$) of the plate; then, the coefficients of extensional stiffness A_{kl}^i, bending-extensional coupling stiffness B^i_{kl} and bending stiffness D^i_{kl} can be defined for FGM circular plate with i-th distribution of porosities in the general forms:

$$\left(A^i_{11}, B^i_{11}, D^i_{11}\right) = \int_{-h/2}^{h/2} \frac{E^i(z, g, \psi)}{1 - \nu^2}\left(1, z, z^2\right) dz, \tag{9a}$$

$$\left(A^i_{12}, B^i_{12}, D^i_{12}\right) = \int_{-h/2}^{h/2} \frac{\nu E^i(z, g, \psi)}{1 - \nu^2}\left(1, z, z^2\right) dz, \tag{9b}$$

$$\left(A^i_{33}, B^i_{33}, D^i_{33}\right) = \int_{-h/2}^{h/2} \frac{E^i(z, g, \psi)}{2(1 + \nu)}\left(1, z, z^2\right) dz. \tag{9c}$$

Additionally, the stiffness coefficients from Equation (9) satisfy the equations

$$A^i_{12} + 2A^i_{33} = A^i_{11}, \quad B^i_{12} + 2B^i_{33} = B^i_{11}, \quad D^i_{12} + 2D^i_{33} = D^i_{11}. \tag{10}$$

The resultant forces and the moments can be also defined by

$$\left(N^i_{rr}, N^i_{\theta\theta}, N^i_{r\theta}\right) = \int_{-h/2}^{h/2} \left(\sigma^i_{rr}, \sigma^i_{\theta\theta}, \tau^i_{r\theta}\right) dz, \tag{11a}$$

$$\left(M^i_{rr}, M^i_{\theta\theta}, M^i_{r\theta}\right) = \int_{-h/2}^{h/2} \left(\sigma^i_{rr} z, \sigma^i_{\theta\theta} z, \tau^i_{r\theta} z\right) dz, \tag{11b}$$

where the stress components and the strain components have the form:

$$\begin{pmatrix} \sigma^i_{rr} \\ \sigma^i_{\theta\theta} \\ \tau^i_{r\theta} \end{pmatrix} = \begin{pmatrix} \frac{E^i(z,g,\psi)}{1-\nu^2}(\varepsilon_{rr} + \nu\varepsilon_{\theta\theta}) \\ \frac{E^i(z,g,\psi)}{1-\nu^2}(\varepsilon_{\theta\theta} + \nu\varepsilon_{rr}) \\ \frac{E^i(z,g,\psi)}{2(1+\nu)}(2\gamma_{r\theta}) \end{pmatrix}, \tag{12}$$

$$\begin{pmatrix} \varepsilon_{rr} \\ \varepsilon_{\theta\theta} \\ 2\gamma_{r\theta} \end{pmatrix} = \begin{pmatrix} \varepsilon^0_{rr} + z\kappa_{rr} \\ \varepsilon^0_{\theta\theta} + z\kappa_{\theta\theta} \\ \gamma^0_{r\theta} + z\kappa_{r\theta} \end{pmatrix}. \tag{13}$$

4.1. Coupled Equations of Motion

Using the Hamilton's principle [34] and ignoring in-plane inertia forces, the equilibrium equations of motion of the porous FGM thin circular plate have the forms:

$$\frac{\partial N^i_{rr}}{\partial r} + \frac{1}{r}\left(\frac{\partial N^i_{r\theta}}{\partial \theta} + N^i_{rr} - N^i_{\theta\theta}\right) = 0, \tag{14a}$$

$$\frac{\partial N^i_{r\theta}}{\partial r} + \frac{1}{r}\frac{\partial N^i_{\theta\theta}}{\partial \theta} + \frac{2}{r}N^i_{r\theta} = 0, \tag{14b}$$

$$\frac{\partial^2 M^i_{rr}}{\partial r^2} + \frac{2}{r}\frac{\partial M^i_{rr}}{\partial r} + \frac{1}{r^2}\frac{\partial^2 M^i_{\theta\theta}}{\partial \theta^2} - \frac{1}{r}\frac{\partial M^i_{\theta\theta}}{\partial r} + \frac{2}{r}\frac{\partial^2 M^i_{r\theta}}{\partial r \partial \theta} + \frac{2}{r^2}\frac{\partial M^i_{r\theta}}{\partial \theta} = \rho^i h \frac{\partial^2 w}{\partial t^2}, \tag{14c}$$

where the resultants forces and the moments can be obtained using Equations (7) and (8), and can be presented in the following form:

$$N^i_{rr} = A^i_{11}\frac{\partial u}{\partial r} + A^i_{12}\left(\frac{1}{r}\frac{\partial v}{\partial \theta} + \frac{u}{r}\right) - B^i_{11}\frac{\partial^2 w}{\partial r^2} - B^i_{12}\left(\frac{1}{r^2}\frac{\partial^2 w}{\partial \theta^2} + \frac{1}{r}\frac{\partial w}{\partial r}\right), \tag{15a}$$

$$N^i_{\theta\theta} = A^i_{12}\frac{\partial u}{\partial r} + A^i_{11}\left(\frac{1}{r}\frac{\partial v}{\partial \theta} + \frac{u}{r}\right) - B^i_{12}\frac{\partial^2 w}{\partial r^2} - B^i_{11}\left(\frac{1}{r^2}\frac{\partial^2 w}{\partial \theta^2} + \frac{1}{r}\frac{\partial w}{\partial r}\right), \tag{15b}$$

$$N_{r\theta}^i = A_{33}^i \left(\frac{1}{r} \frac{\partial u}{\partial \theta} + \frac{\partial v}{\partial r} - \frac{v}{r} \right) - B_{33}^i \left(\frac{2}{r} \frac{\partial^2 w}{\partial r \partial \theta} - \frac{2}{r^2} \frac{\partial w}{\partial \theta} \right), \quad (15c)$$

$$M_{rr}^i = B_{11}^i \frac{\partial u}{\partial r} + B_{12}^i \left(\frac{1}{r} \frac{\partial v}{\partial \theta} + \frac{u}{r} \right) - D_{11}^i \frac{\partial^2 w}{\partial r^2} - D_{12}^i \left(\frac{1}{r^2} \frac{\partial^2 w}{\partial \theta^2} + \frac{1}{r} \frac{\partial w}{\partial r} \right), \quad (16a)$$

$$M_{\theta\theta}^i = B_{12}^i \frac{\partial u}{\partial r} + B_{11}^i \left(\frac{1}{r} \frac{\partial v}{\partial \theta} + \frac{u}{r} \right) - D_{12}^i \frac{\partial^2 w}{\partial r^2} - D_{11}^i \left(\frac{1}{r^2} \frac{\partial^2 w}{\partial \theta^2} + \frac{1}{r} \frac{\partial w}{\partial r} \right), \quad (16b)$$

$$M_{r\theta}^i = B_{33}^i \left(\frac{1}{r} \frac{\partial u}{\partial \theta} + \frac{\partial v}{\partial r} - \frac{v}{r} \right) - D_{33}^i \left(\frac{2}{r} \frac{\partial^2 w}{\partial r \partial \theta} - \frac{2}{r^2} \frac{\partial w}{\partial \theta} \right). \quad (16c)$$

In Equation (14c), ρ^i is the averaged material density of the FGM circular plate for the *i*-th distribution of porosities presented in the general form:

$$\rho^i \equiv \rho^i(g, \psi) = \frac{1}{h} \int_{-h/2}^{h/2} \rho^i(z, g, \psi) dz, \quad i = \{e, u\}. \quad (17)$$

Substituting Equations (15) and (16) into Equation (14), and using relations given in Equation (10), we get the coupled equilibrium equations of motion of the porous FGM circular plate presented in terms of displacement components:

$$A_{11}^i \left(\frac{\partial^2 u}{\partial r^2} + \frac{1}{r} \frac{\partial u}{\partial r} - \frac{u}{r^2} - \frac{1}{r^2} \frac{\partial v}{\partial \theta} + \frac{1}{r} \frac{\partial^2 v}{\partial r \partial \theta} \right) + A_{33}^i \left(\frac{1}{r^2} \frac{\partial^2 u}{\partial \theta^2} - \frac{1}{r} \frac{\partial^2 v}{\partial r \partial \theta} - \frac{1}{r^2} \frac{\partial v}{\partial \theta} \right) - B_{11}^i \frac{\partial \nabla^2 w}{\partial r} = 0, \quad (18a)$$

$$A_{11}^i \left(\frac{1}{r^2} \frac{\partial u}{\partial \theta} + \frac{1}{r} \frac{\partial^2 u}{\partial r \partial \theta} + \frac{1}{r^2} \frac{\partial^2 v}{\partial \theta^2} \right) + A_{33}^i \left(\frac{1}{r^2} \frac{\partial u}{\partial \theta} - \frac{1}{r} \frac{\partial^2 u}{\partial r \partial \theta} + \frac{\partial^2 v}{\partial r^2} + \frac{1}{r} \frac{\partial v}{\partial r} - \frac{v}{r^2} \right) - B_{11}^i \frac{1}{r} \frac{\partial \nabla^2 w}{\partial \theta} = 0, \quad (18b)$$

$$D_{11}^i \nabla^2 \nabla^2 w - B_{11}^i \nabla^2 \varepsilon = -\rho^i h \frac{\partial^2 w}{\partial t^2}, \quad (18c)$$

where $\nabla^2 = \frac{\partial^2}{\partial r^2} + \frac{1}{r} \frac{\partial}{\partial r} + \frac{1}{r^2} \frac{\partial^2}{\partial \theta^2}$ is the Laplace operator presented in polar coordinates and

$$\varepsilon = \frac{\partial u}{\partial r} + \frac{1}{r} \frac{\partial v}{\partial \theta} + \frac{u}{r}. \quad (19)$$

4.2. Decoupled Equation of Motion

Equation (18) show that the in-plane stretching and bending are coupled because the reference surface is a geometrical midplane. We can eliminate this coupling by introducing the physical neutral surface, where the in-plane displacements will be omitted. The in-plane displacements of the midplane can be expressed in terms of the slopes of deflection in the following form:

$$u(r, \theta, t) = z_0 \frac{\partial w(r, \theta, t)}{\partial r}, \quad (20a)$$

$$v(r, \theta, t) = z_0 \frac{1}{r} \frac{\partial w(r, \theta, t)}{\partial \theta}, \quad (20b)$$

where z_0 is the distance between the midplane and the physical neutral surface. By substituting Equation (20) into Equations (6) and (15) and introducing $z = z_0$, the in-plane displacements u, v and the in-plane forces N_{rr}^i, $N_{\theta\theta}^i$, $N_{r\theta}^i$ must equal zero based on properties of the physical neutral surface. By substituting Equation (20) into Equation (15)

$$N_{rr}^i = \left(z_0 A_{11}^i - B_{11}^i \right) \frac{\partial^2 w}{\partial r^2} + \left(z_0 A_{12}^i - B_{12}^i \right) \left(\frac{1}{r^2} \frac{\partial^2 w}{\partial \theta^2} + \frac{1}{r} \frac{\partial w}{\partial r} \right) = 0, \quad (21a)$$

$$N_{\theta\theta}^i = \left(z_0 A_{12}^i - B_{12}^i \right) \frac{\partial^2 w}{\partial r^2} + \left(z_0 A_{11}^i - B_{11}^i \right) \left(\frac{1}{r^2} \frac{\partial^2 w}{\partial \theta^2} + \frac{1}{r} \frac{\partial w}{\partial r} \right) = 0, \quad (21b)$$

$$N_{r\theta}^{i} = \left(z_{0}A_{33}^{i} - B_{33}^{i}\right)\left(\frac{2}{r}\frac{\partial^{2}w}{\partial r\partial\theta} - \frac{2}{r^{2}}\frac{\partial w}{\partial\theta}\right) = 0 \tag{21c}$$

and assuming that the Poisson's ratio is constant, distance z_0 can be obtained from relations:

$$z_{0}A_{11}^{i} - B_{11}^{i} = z_{0}A_{12}^{i} - B_{12}^{i} = z_{0}A_{33}^{i} - B_{33}^{i} = 0, \tag{22}$$

where

$$z_{0} = \frac{B_{11}^{i}}{A_{11}^{i}} = \frac{B_{12}^{i}}{A_{12}^{i}} = \frac{B_{33}^{i}}{A_{33}^{i}} = \frac{\int_{-h/2}^{h/2} E^{i}(z,g,\psi)z\,dz}{\int_{-h/2}^{h/2} E^{i}(z,g,\psi)dz}. \tag{23}$$

By substituting Equations (20) and (23) into Equations (18c) and (19), we obtain the decoupled equation of transverse vibration of the porous FGM thin circular plate in the form:

$$\mathfrak{D}^{i}\nabla^{2}\nabla^{2}w = -\rho^{i}h\frac{\partial^{2}w}{\partial t^{2}}, \tag{24}$$

where

$$\mathfrak{D}^{i} = D_{11}^{i} - \frac{\left(B_{11}^{i}\right)^{2}}{A_{11}^{i}}. \tag{25}$$

5. Solution of the Problem

Taking into account a harmonic solution, the small vibration of the porous FGM circular plate may be expressed as follows:

$$w(r,\theta,t) = W(r)\cos(n\theta)\cos(\omega t), \tag{26}$$

where $W(r)$ is the radial mode function as the small deflection compared with the thickness h of the plate, n is the integer number of diagonal nodal lines, θ is the angular coordinate, and ω is the natural frequency. By substituting Equation (26) into Equation (24) using the dimensionless coordinate $\xi = r/R$ ($0 < \xi \leq 1$), the general governing differential equation assumes the following form:

$$\mathcal{L}_{n}(W) = \rho^{i}h\omega^{2}W, \tag{27}$$

where $\mathcal{L}_n(\cdot)$ is the differential operator defined by

$$\mathcal{L}_{n}(\cdot) \equiv \mathfrak{D}^{i}\frac{d^{4}}{d\xi^{4}} + \frac{2\mathfrak{D}^{i}}{\xi}\frac{d^{3}}{d\xi^{3}} - \frac{(1+2n^{2})\mathfrak{D}^{i}}{\xi^{2}}\frac{d^{2}}{d\xi^{2}} + \frac{(1+2n^{2})\mathfrak{D}^{i}}{\xi^{3}}\frac{d}{d\xi} + \frac{(n^{4}-4n^{2})\mathfrak{D}^{i}}{\xi^{4}}. \tag{28}$$

The calculated general forms of material density ρ^i and the coefficients of extensional stiffness $\left(A_{11}^{i}\right)$, extensional-bending coupling stiffness $\left(B_{11}^{i}\right)$ and bending stiffness $\left(D_{11}^{i}\right)$ for the porous FGM circular plate are presented in the following general forms:

$$\rho^{i} = \frac{\rho_{c}(2x - \psi - g\psi) + \rho_{m}(2xg - \psi - \psi g)}{2x(1+g)}, \tag{29a}$$

$$A_{11}^{i} = \frac{E_{c}h}{1-\nu^{2}}\left[\frac{(2x - \psi - g\psi) + \frac{E_{m}}{E_{c}}(2xg - \psi - g\psi)}{2x(1+g)}\right], \tag{29b}$$

$$B_{11}^{e} = B_{11}^{u} = \frac{E_{c}h^{2}}{(1-\nu^{2})}\left[\frac{g\left(1 - \frac{E_{m}}{E_{c}}\right)}{2(1+g)(2+g)}\right], \tag{29c}$$

$$D^{i} = \frac{E_{c}h^{3}}{12(1-\nu^{2})}\left[\frac{y(6g^{2}+6g+12)-\psi(1+g)(2+g)(3+g)+\frac{E_{m}}{E_{c}}\left[y(2g^{3}+6g^{2}+16g)-\psi(1+g)(2+g)(3+g)\right]}{2y(1+g)(2+g)(3+g)}\right], \tag{29d}$$

where $x = y = 1$ for the even distribution $(i = e)$ of porosities and $x = 2$, $y = 4$ for the uneven $(i = u)$ distribution of porosities. The extensional-bending coupling stiffness B_{11}^i has the same form for both types of porosities.

By substituting the obtained forms from Equation (29) into Equation (27), the generalized ordinary differential equation with variable coefficients is obtained as:

$$\mathcal{L}_n(W)_\chi = \lambda^2 \mu^i W, \tag{30}$$

where

$$\mathcal{L}_n(\cdot)_\chi \equiv (\chi_1^i + \chi_2^i)\frac{d^4}{d\xi^4} + \frac{2(\chi_1^i + \chi_2^i)}{\xi}\frac{d^3}{d\xi^3} - \frac{(1+2n^2)(\chi_1^i + \chi_2^i)}{\xi^2}\frac{d^2}{d\xi^2} + \frac{(1+2n^2)(\chi_1^i + \chi_2^i)}{\xi^3}\frac{d}{d\xi} + \frac{(n^4 - 4n^2)(\chi_1^i + \chi_2^i)}{\xi^4}, \tag{31}$$

$$\chi_1^i = \frac{6xg^2(E_c - E_m)^2}{E_c(1+g)(2+g)^2[E_c(\psi + g\psi - 2x) + E_m(\psi + g\psi - 2xg)]}, \tag{32}$$

$$\chi_2^i = \frac{E_c[y(12 + 6g + 6g^2) - \psi(1+g)(2+g)(3+g)] + E_m[y(16g + 6g^2 + 2g^3) - \psi(1+g)(2+g)(3+g)]}{2yE_c(1+g)(2+g)(3+g)}, \tag{33}$$

$$\mu^i = \frac{(-g\psi - \psi + 2x) - \frac{\rho_m}{\rho_c}(g\psi + \psi - 2xg)}{2x(1+g)}, \tag{34}$$

$$\lambda = \omega R^2 \sqrt{\rho_c h / D_c}, \tag{35}$$

$$D_c = \frac{E_c h^3}{12(1-\nu^2)}. \tag{36}$$

The boundary conditions on the outer edge ($\xi = 1$) of the porous FGM circular plate may be one of the following: clamped, simply supported, sliding supported and free. These conditions may be written in terms of the radial mode function $W(\xi)$ in the following form:

- Clamped:

$$W(\xi)|_{\xi=1} = 0, \tag{37a}$$

$$\frac{dW}{d\xi}\bigg|_{\xi=1} = 0. \tag{37b}$$

- Simply supported:

$$W(\xi)|_{\xi=1} = 0, \tag{38a}$$

$$M(W)|_{\xi=1} = \left[\frac{d^2 W}{d\xi^2} + \frac{\nu}{\xi}\frac{dW}{d\xi} - \frac{\nu n^2}{\xi^2}W\right]_{\xi=1} = 0. \tag{38b}$$

- Sliding supported:

$$\frac{dW}{d\xi}\bigg|_{\xi=1} = 0, \tag{39a}$$

$$V(W)|_{\xi=1} = \left[\frac{d^3 W}{d\xi^3} + \frac{1}{\xi}\frac{d^2 W}{d\xi^2} - \left(\frac{1+2n^2 - \nu n^2}{\xi^2}\right)\frac{dW}{d\xi} + \left(\frac{3n^2 - \nu n^2}{\xi^3}\right)W\right]_{\xi=1} = 0, \tag{39b}$$

- Free:

$$M(W)|_{\xi=1} = 0, \tag{40a}$$

$$V(W)|_{\xi=1} = 0. \tag{40b}$$

The static forces $M(W)$ and $V(W)$ are the normalized radial bending moment and the normalized effective shear force, respectively.

The one multiparametric general solution of the defined differential Equation (30) for FGM circular/annular plates with the two various types of distribution of porosities ($i = \{e, u\}$) is obtained in the following form:

$$W_n^i(\xi, \lambda, g, \psi) = C_1 J_n\left[\left(\lambda\sqrt{\mathfrak{M}_i}\right)^{1/2}\xi\right] + C_2 I_n\left[\left(\lambda\sqrt{\mathfrak{M}_i}\right)^{1/2}\xi\right] + C_3 Y_n\left[\left(\lambda\sqrt{\mathfrak{M}_i}\right)^{1/2}\xi\right] + C_4 K_n\left[\left(\lambda\sqrt{\mathfrak{M}_i}\right)^{1/2}\xi\right], \quad (41)$$

where n ($n \in \mathbb{N}^+$) is the number of nodal lines, C_1, C_2, C_3, C_4 are the constants of integration, $J_n\left[\left(\lambda\sqrt{\mathfrak{M}_i}\right)^{1/2}\xi\right]$, $I_n\left[\left(\lambda\sqrt{\mathfrak{M}_i}\right)^{1/2}\xi\right]$, $Y_n\left[\left(\lambda\sqrt{\mathfrak{M}_i}\right)^{1/2}\xi\right]$, $K_n\left[\left(\lambda\sqrt{\mathfrak{M}_i}\right)^{1/2}\xi\right]$ are the Bessel functions as particular solutions of Equation (30), and $\mathfrak{M}_i$ is the generalized multiparametric function defined as:

$$\mathfrak{M}_i \equiv \mathfrak{M}_i(x, y, g, \psi, E_m, E_c, \rho_m, \rho_c) = \frac{\Omega_1^i}{\Omega_2^i + \Omega_3^i}, \quad \mathfrak{M}_i \geq 1 \forall g \in [0, \infty] \wedge \forall \psi \in [0, 1), \quad (42)$$

where

$$\Omega_1^i = -E_c x(2 + g)^2 [\rho_c(g\psi + \psi - 2x) + \rho_m(g\psi + \psi - 2xg)], \quad (43a)$$

$$\Omega_2^i = \frac{12xyg^2(E_c - E_m)^2 \rho_c}{E_c(g\psi + \psi - 2x) + E_m(g\psi + \psi - 2xg)}, \quad (43b)$$

$$\Omega_3^i = \frac{(2+g)\rho_c\left[E_c\left[y\left(12+6g+6g^2\right)-\psi(1+g)(2+g)(3+g)\right]+E_m\left[y\left(16g+6g^2+2g^3\right)-\psi(1+g)(2+g)(3+g)\right]\right]}{3+g}. \quad (43c)$$

The functions $J_n\left[\left(\lambda\sqrt{\mathfrak{M}_i}\right)^{1/2}\xi\right]$ and $I_n\left[\left(\lambda\sqrt{\mathfrak{M}_i}\right)^{1/2}\xi\right]$ are the limited linear independent solutions $\left(\lim_{\xi \to 0} J_n\left[\left(\lambda\sqrt{\mathfrak{M}_i}\right)^{1/2}\xi\right] < \infty, \lim_{\xi \to 0} I_n\left[\left(\lambda\sqrt{\mathfrak{M}_i}\right)^{1/2}\xi\right] < \infty\right)$ of Equation (30) for the axisymmetric and non-axisymmetric deflections at center ($\xi = 0$) of the porous FGM circular plate and diverse values of the physically justified parameters λ, g and ψ. The particular solutions $Y_n\left[\left(\lambda\sqrt{\mathfrak{M}_i}\right)^{1/2}\xi\right]$ and $K_n\left[\left(\lambda\sqrt{\mathfrak{M}_i}\right)^{1/2}\xi\right]$ are unlimited $\left(\lim_{\xi \to 0} Y_n\left[\left(\lambda\sqrt{\mathfrak{M}_i}\right)^{1/2}\xi\right] = -\infty, \lim_{\xi \to 0} K_n\left[\left(\lambda\sqrt{\mathfrak{M}_i}\right)^{1/2}\xi\right] = \infty\right)$ for the deflection at the center of the plate, then, the general solution (41) for the porous FGM circular plate can be presented in the new form:

$$W_n^i(\xi, \lambda, g, \psi) = C_1 \Psi_1 + C_2 \Psi_2, \quad (44)$$

where

$$\Psi_1 \equiv J_n\left[\left(\lambda\sqrt{\mathfrak{M}_i}\right)^{1/2}\xi\right], \quad (45a)$$

$$\Psi_2 \equiv I_n\left[\left(\lambda\sqrt{\mathfrak{M}_i}\right)^{1/2}\xi\right]. \quad (45b)$$

By applying the general solution (44) and the boundary conditions (37–40) as well as assuming the existence of the non-trivial constants C_1 and C_2, the general nonlinear multiparametric characteristic equations of the FGM circular plate with the two various types of distribution of porosities were obtained in the form:

- Clamped (C):

$$\Delta_C^i(\lambda, g, \psi, n, x, y) \equiv \left. \begin{vmatrix} \Psi_1 & \Psi_2 \\ \frac{\partial \Psi_1}{\partial \xi} & \frac{\partial \Psi_2}{\partial \xi} \end{vmatrix} \right|_{\xi=1} = 0; \quad (46a)$$

- Simply supported (SS):

$$\Delta_{SS}^i(\lambda, g, \psi, n, x, y) \equiv \left. \begin{vmatrix} \Psi_1 & \Psi_2 \\ M[\Psi_1] & M[\Psi_2] \end{vmatrix} \right|_{\xi=1} = 0; \quad (46b)$$

- Sliding supported (S):

$$\Delta^i_S(\lambda, g, \psi, n, x, y) \equiv \left| \begin{array}{cc} \frac{\partial \Psi_1}{\partial \xi} & \frac{\partial \Psi_2}{\partial \xi} \\ V[\Psi_1] & V[\Psi_2] \end{array} \right|_{\xi=1} = 0; \qquad (46c)$$

- Free (F):

$$\Delta^i_F(\lambda, g, \psi, n, x, y) \equiv \left| \begin{array}{cc} M[\Psi_1] & M[\Psi_2] \\ V[\Psi_1] & V[\Psi_2] \end{array} \right|_{\xi=1} = 0. \qquad (46d)$$

If $x = y = 1$ is introduced to Equations (42) and (45), then the obtained characteristic Equation (46) will be valid for the FGM circular plates with even ($i = e$) distribution of porosities. If $x = 2, y = 4$ is introduced to Equations (42) and (45), then the obtained characteristic equations (46) will be valid for the FGM circular plates with uneven ($i = u$) distribution of porosities.

The general solution for the perfect (without porosity) FGM circular plate can be obtained from Equation (44) and presented in the following form:

$$W_n(\xi, \lambda, g) \equiv \lim_{\psi \to 0} W_n^i(\xi, \lambda, g, \psi) = C_1 \lim_{\psi \to 0} J_n\left[\left(\lambda \sqrt{\mathfrak{M}_i}\right)^{1/2} \xi\right] + C_2 \lim_{\psi \to 0} I_n\left[\left(\lambda \sqrt{\mathfrak{M}_i}\right)^{1/2} \xi\right]. \qquad (47)$$

After calculations, the final form of general solution for the perfect FGM circular plate is expressed as

$$W_n(\xi, \lambda, g) = J_n\left[\left(\lambda \sqrt{\mathfrak{Y}}\right)^{1/2} \xi\right] + I_n\left[\left(\lambda \sqrt{\mathfrak{Y}}\right)^{1/2} \xi\right], \qquad (48)$$

where

$$\mathfrak{Y} = \frac{E_c(2+g)^2(3+g)(E_c + gE_m)(\rho_c + g\rho_m)}{\rho_c(1+g)[12E_c^2 + (28g + 16g^2 + 4g^3)E_cE_m + (7g^2 + 4g^3 + g^4)E_m^2]}. \qquad (49)$$

The general solution for the perfect FGM circular plate with negligible effect of the coupling in-plane and transverse displacements ($A^i_{11} \to 0$, $B^i_{11} \to 0$) has the form:

$$W_n(\xi, \lambda, g) = C_1 J_n\left[\left(\lambda \sqrt{\mathfrak{P}}\right)^{1/2} \xi\right] + C_2 I_n\left[\left(\lambda \sqrt{\mathfrak{P}}\right)^{1/2} \xi\right], \qquad (50)$$

where

$$\mathfrak{P} = \frac{E_c(2+g)(3+g)(\rho_c + g\rho_m)}{\rho_c[3E_c(2+g+g^2) + E_m(8g+3g^2+g^3)]}. \qquad (51)$$

6. Parametric Study

The every single fundamental and lower dimensionless frequencies of the free axisymmetric and non-axisymmetric vibrations of porous FGM circular plate were calculated for diverse values of the power-law index g, the porosity volume fraction ψ and different boundary conditions using the Newton method aided by a calculation software.

The Poisson's ratio is taken as $\nu = 0.3$ and its variation is assumed to be negligible. In the present study, aluminum is taken as the metal and alumina is taken as the ceramic material. The values of Young's modulus and densities are taken as follows: $E_m = 70$ GPa, $E_c = 380$ GPa, $\rho_m = 2702$ kg/m^3, $\rho_c = 3800$ kg/m^3.

6.1. Imperfect FGM Circular Plate

The obtained numerical results for the first three dimensionless frequencies $\lambda = \omega R^2 \sqrt{\rho_c h / D_c}$ of the axisymmetric ($n = 0$) and non-axisymmetric ($n = 1$) vibrations of the perfect ($\psi \to 0$) homogeneous ($g \to 0$) circular plate with various boundary conditions are presented in Table 1 and compared with the results obtained by Wu and Liu [35], Yalcin et al. [36], Zhou et al. [37] and

Duan et al. [38]. The obtained numerical results for the perfect homogeneous circular plate are in excellent agreement with those available in the literature.

Table 1. The dimensionless frequencies of the perfect homogeneous circular plate.

λ		Clamped		Simply Supported		Sliding Supported		Free	
					n				
		0	1	0	1	0	1	0	1
λ_0	Present	10.215	21.260	4.935	13.898	14.682	3.082	9.003	20.474
	[35]	10.216	21.260	4.935	13.898	14.682	3.082	9.003	20.475
	[36]	10.215	21.260	4.935	13.898	-	-	9.003	20.474
	[37]	10.215	21.260	4.935	13.898	-	-	9.003	20.474
	[38]	10.215	21.260	4.935	13.898	-	-	9.003	20.474
λ_1	Present	39.771	60.828	29.720	48.478	49.218	28.398	38.443	59.812
	[35]	39.771	60.829	29.720	48.478	49.218	28.399	38.443	59.812
	[36]	39.771	60.828	29.720	48.479	-	-	38.443	59.811
	[37]	39.771	60.828	29.720	48.478	-	-	38.443	59.811
	[38]	39.771	60.828	29.719	48.478	-	-	38.443	59.812
λ_2	Present	89.104	120.079	74.156	102.773	103.499	72.859	87.750	118.957
	[35]	89.104	120.079	74.156	102.772	103.499	72.859	87.750	118.957
	[36]	89.104	120.079	74.156	102.773	-	-	87.705	118.957
	[37]	89.104	120.080	74.156	102.773	-	-	87.750	118.957
	[38]	89.104	120.079	74.156	102.773	-	-	87.750	118.957

The calculated fundamental dimensionless frequencies λ_0 of the axisymmetric ($n = 0$) and non-axisymmetric ($n = 1$) vibrations of the FGM circular plate with evenly ($i = e$) and unevenly ($i = u$) distributed porosity are presented in Tables 2–5. In the parametric study, values of the power-law index of FGMs is taken as $g = \{0, 0.2, 0.4, 0.6, 1, 2, 3, 4, 5\}$ and values of the porosity volume fraction is taken as $\psi = \{0, 0.05, 0.1, 0.2, 0.3\}$.

Table 2. The dimensionless fundamental frequencies of the clamped porous FGM circular plate.

i	n	ψ	g								
			0	0.2	0.4	0.6	1	2	3	4	5
							λ_0				
e	0	0	10.215	9.481	8.896	8.436	7.797	7.090	6.867	6.777	6.724
		0.05	10.286	9.522	8.905	8.414	7.718	6.920	6.661	6.559	6.503
		0.1	10.362	9.566	8.914	8.387	7.623	6.712	6.401	6.280	6.219
		0.2	10.535	9.668	8.932	8.315	7.374	6.113	5.612	5.402	5.305
		0.3	10.745	9.792	8.948	8.207	6.993	5.034	3.949	3.312	2.923
	1	0	21.260	19.731	18.514	17.557	16.228	14.756	14.292	14.105	13.993
		0.05	21.406	19.816	18.533	17.510	16.062	14.402	13.863	13.650	13.533
		0.1	21.564	19.909	18.552	17.454	15.866	13.968	13.222	13.069	12.942
		0.2	21.925	20.121	18.590	17.304	15.346	12.723	11.680	11.242	11.041
		0.3	22.362	20.380	18.622	17.081	14.554	10.478	8.220	6.894	6.084
u	0	0	10.215	9.481	8.896	8.436	7.797	7.090	6.867	6.777	6.724
		0.05	10.288	9.544	8.949	8.478	7.819	7.079	6.844	6.751	6.698
		0.1	10.364	9.611	9.004	8.521	7.840	7.065	6.816	6.719	6.666
		0.2	10.523	9.751	9.120	8.612	7.882	7.023	6.738	6.630	6.577
		0.3	10.696	9.903	9.246	8.710	7.923	6.959	6.622	6.495	6.438
	1	0	21.260	19.731	18.514	17.557	16.228	14.756	14.292	14.105	13.993
		0.05	21.411	19.864	18.624	17.644	16.272	14.733	14.244	14.050	13.940
		0.1	21.568	20.001	18.738	17.734	16.316	14.703	14.182	13.983	13.874
		0.2	21.901	20.293	18.980	17.923	16.404	14.617	14.023	13.798	13.688
		0.3	22.260	20.610	19.243	18.127	16.490	14.483	13.782	13.517	13.399

Table 3. The dimensionless fundamental frequencies of the simply supported porous FGM circular plate.

i	n	ψ	\multicolumn{9}{c	}{g}
			0	0.2	0.4	0.6	1	2	3	4	5
			\multicolumn{9}{c	}{λ_0}
e	0	0	4.935	4.580	4.297	4.075	3.767	3.425	3.317	3.274	3.248
		0.05	4.969	4.600	4.302	4.064	3.728	3.343	3.218	3.168	3.141
		0.1	5.005	4.621	4.306	4.051	3.683	3.242	3.092	3.033	3.004
		0.2	5.089	4.670	4.315	4.017	3.562	2.953	2.711	2.609	2.563
		0.3	5.190	4.730	4.322	3.965	3.378	2.432	1.908	1.600	1.412
	1	0	13.898	12.898	12.103	11.477	10.608	9.646	9.343	9.220	9.147
		0.05	13.993	12.954	12.115	11.446	10.500	9.415	9.062	8.923	8.847
		0.1	14.097	13.015	12.127	11.410	10.372	9.131	8.708	8.543	8.460
		0.2	14.333	13.153	12.152	11.312	10.032	8.317	7.635	7.349	7.218
		0.3	14.618	13.322	12.173	11.166	9.514	6.849	5.373	4.506	3.977
u	0	0	4.935	4.580	4.297	4.075	3.767	3.425	3.317	3.274	3.248
		0.05	4.970	4.611	4.323	4.095	3.777	3.420	3.306	3.261	3.236
		0.1	5.006	4.643	4.349	4.116	3.787	3.413	3.292	3.246	3.220
		0.2	5.083	4.710	4.406	4.160	3.808	3.393	3.255	3.203	3.177
		0.3	5.167	4.784	4.467	4.207	3.828	3.362	3.199	3.137	3.110
	1	0	13.898	12.898	12.103	11.477	10.608	9.646	9.343	9.220	9.147
		0.05	13.997	12.985	12.174	11.534	10.637	9.631	9.311	9.185	9.113
		0.1	14.099	13.075	12.249	11.592	10.666	9.611	9.273	9.141	9.069
		0.2	14.317	13.266	12.408	11.716	10.724	9.555	9.167	9.020	8.948
		0.3	14.551	13.473	12.580	11.850	10.780	9.468	9.009	8.836	8.759

Table 4. The dimensionless fundamental frequencies of the porous FGM circular plate with sliding support.

i	n	ψ	\multicolumn{9}{c	}{g}
			0	0.2	0.4	0.6	1	2	3	4	5
			\multicolumn{9}{c	}{λ_0}
e	0	0	14.682	13.626	12.785	12.124	11.206	10.190	9.870	9.740	9.663
		0.05	14.782	13.685	12.798	12.092	11.092	9.946	9.573	9.426	9.346
		0.1	14.892	13.749	12.811	12.053	10.956	9.646	9.199	9.025	8.938
		0.2	15.141	13.895	12.837	11.950	10.597	8.786	8.066	7.764	7.625
		0.3	15.442	14.074	12.860	11.796	10.051	7.236	5.676	4.761	4.201
	1	0	3.082	2.860	2.684	2.545	2.352	2.139	2.072	2.045	2.029
		0.05	3.103	2.873	2.687	2.538	2.328	2.088	2.010	1.980	1.962
		0.1	3.126	2.886	2.690	2.530	2.300	2.025	1.931	1.894	1.876
		0.2	3.178	2.917	2.695	2.509	2.225	1.844	1.693	1.630	1.600
		0.3	3.242	2.954	2.700	2.476	2.110	1.519	1.191	0.999	0.882
u	0	0	14.682	13.626	12.785	12.124	11.206	10.190	9.870	9.740	9.663
		0.05	14.786	13.717	12.861	12.184	11.237	10.174	9.836	9.703	9.627
		0.1	14.895	13.812	12.940	12.246	11.268	10.154	9.795	9.656	9.581
		0.2	15.124	14.014	13.107	12.377	11.328	10.094	9.684	9.529	9.453
		0.3	15.372	14.233	13.289	12.518	11.388	10.002	9.518	9.334	9.253
	1	0	3.082	2.860	2.684	2.545	2.352	2.139	2.072	2.045	2.029
		0.05	3.104	2.880	2.700	2.558	2.359	2.136	2.065	2.037	2.021
		0.1	3.127	2.890	2.716	2.571	2.365	2.131	2.056	2.027	2.011
		0.2	3.175	2.942	2.752	2.598	2.378	2.119	2.033	2.000	1.984
		0.3	3.227	2.988	2.790	2.628	2.391	2.100	1.998	1.960	1.942

Table 5. The dimensionless fundamental frequencies of the free porous FGM circular plate.

i	n	ψ	\multicolumn{9}{c}{g}
			0	0.2	0.4	0.6	1	2	3	4	5
			\multicolumn{9}{c}{λ_0}
e	0	0	9.003	8.355	7.840	7.435	6.872	6.248	6.052	5.973	5.926
		0.05	9.064	8.391	7.848	7.415	6.802	6.099	5.870	5.780	5.731
		0.1	9.132	8.431	7.856	7.391	6.718	5.915	5.641	5.534	5.480
		0.2	9.284	8.521	7.872	7.328	6.498	5.388	4.946	4.761	4.675
		0.3	9.469	8.630	7.886	7.233	6.163	4.437	3.481	2.919	2.576
	1	0	20.474	19.002	17.830	16.908	15.628	14.211	13.764	13.584	13.476
		0.05	20.615	19.084	17.848	16.863	15.468	13.870	13.350	13.145	13.033
		0.1	20.767	19.173	17.866	16.809	15.280	13.452	12.829	12.586	12.464
		0.2	21.115	19.378	17.902	16.665	14.779	12.253	11.248	10.827	10.633
		0.3	21.535	19.628	17.934	16.450	14.016	10.091	7.916	6.639	5.859
u	0	0	9.003	8.355	7.840	7.435	6.872	6.248	6.052	5.973	5.926
		0.05	9.067	8.411	7.886	7.471	6.890	6.239	6.032	5.950	5.903
		0.1	9.133	8.470	7.935	7.509	6.909	6.226	6.007	5.921	5.875
		0.2	9.274	8.593	8.037	7.590	6.947	6.190	5.938	5.843	5.796
		0.3	9.426	8.728	8.149	7.676	6.983	6.133	5.836	5.724	5.674
	1	0	20.474	19.002	17.830	16.908	15.628	14.211	13.764	13.584	13.476
		0.05	20.620	19.129	17.935	16.992	15.670	14.188	13.718	13.531	13.425
		0.1	20.771	19.262	18.045	17.078	15.713	14.160	13.660	13.466	13.361
		0.2	21.091	19.543	18.279	17.261	15.798	14.077	13.505	13.288	13.182
		0.3	21.437	19.848	18.532	17.457	15.881	13.948	13.273	13.017	12.903

The dependences of the fundamental dimensionless frequencies λ_0 of the free axisymmetric ($n = 0$) and non-axisymmetric ($n = 1$) vibrations of the circular plate on selected values of the power-law index and the porosities volume fraction are presented in Figures 3–6 as the two-dimensional (2D) and three-dimensional (3D) graphs for the two various types of distribution of porosity and all considered boundary conditions.

6.2. Perfect FGM Circular Plate

The obtained general solution (48) and the defined boundary conditions (37 ÷ 40) were used to calculate the first three dimensionless frequencies λ of the axisymmetric ($n = 0$) and non-axisymmetric ($n = 1$) vibrations of the perfect ($\psi = 0$) FGM circular plate with various boundary conditions.

The obtained numerical results are presented in Tables 6–9 for selected values of the power-law index g. Numerical results obtained for the clamped and simply supported plates (Tables 6 and 7) were compared with the results presented in the paper [27], where the effect of the coupling in-plane and transverse displacements was omitted.

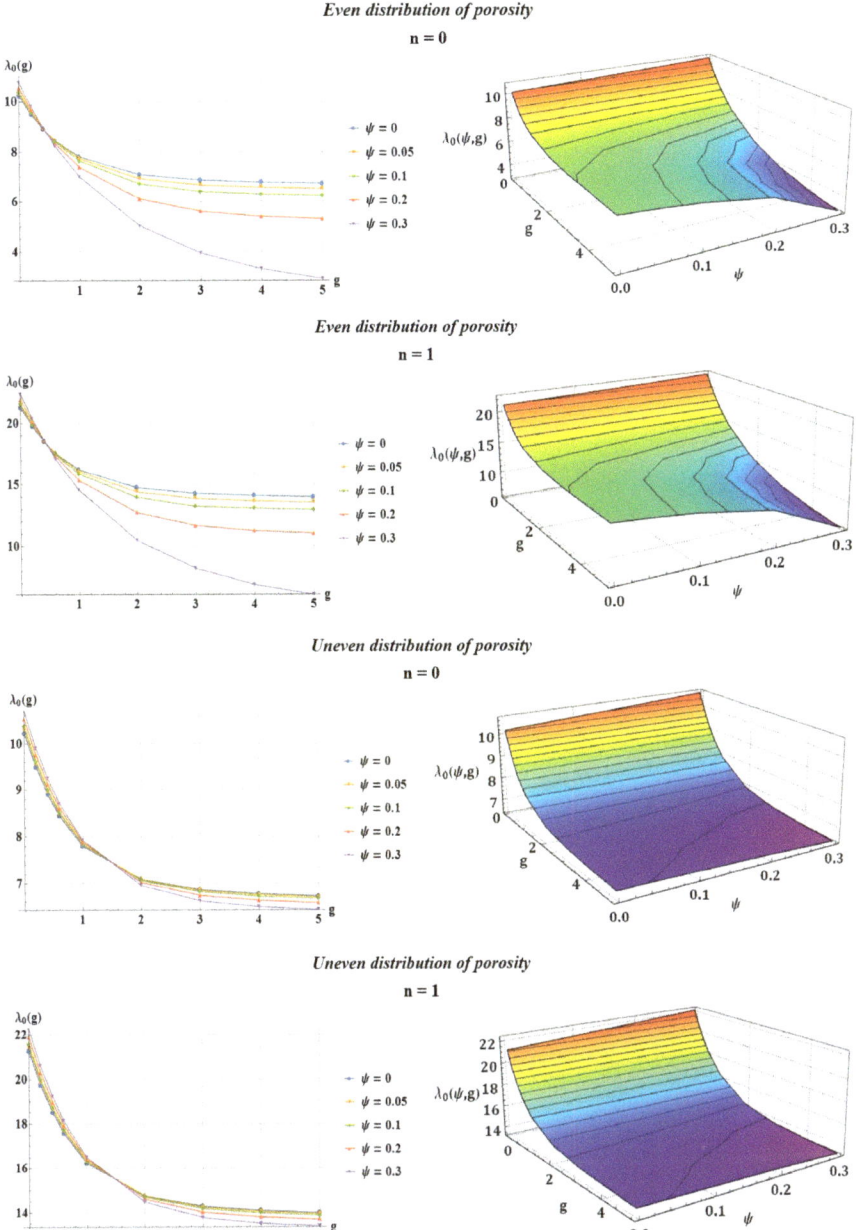

Figure 3. The dependence of the fundamental dimensionless frequencies λ_0 of the free axisymmetric ($n = 0$) and non-axisymmetric ($n = 1$) vibrations on selected values of the power-law index and the porosity volume fraction of the clamped circular plate with evenly and unevenly distributed porosities.

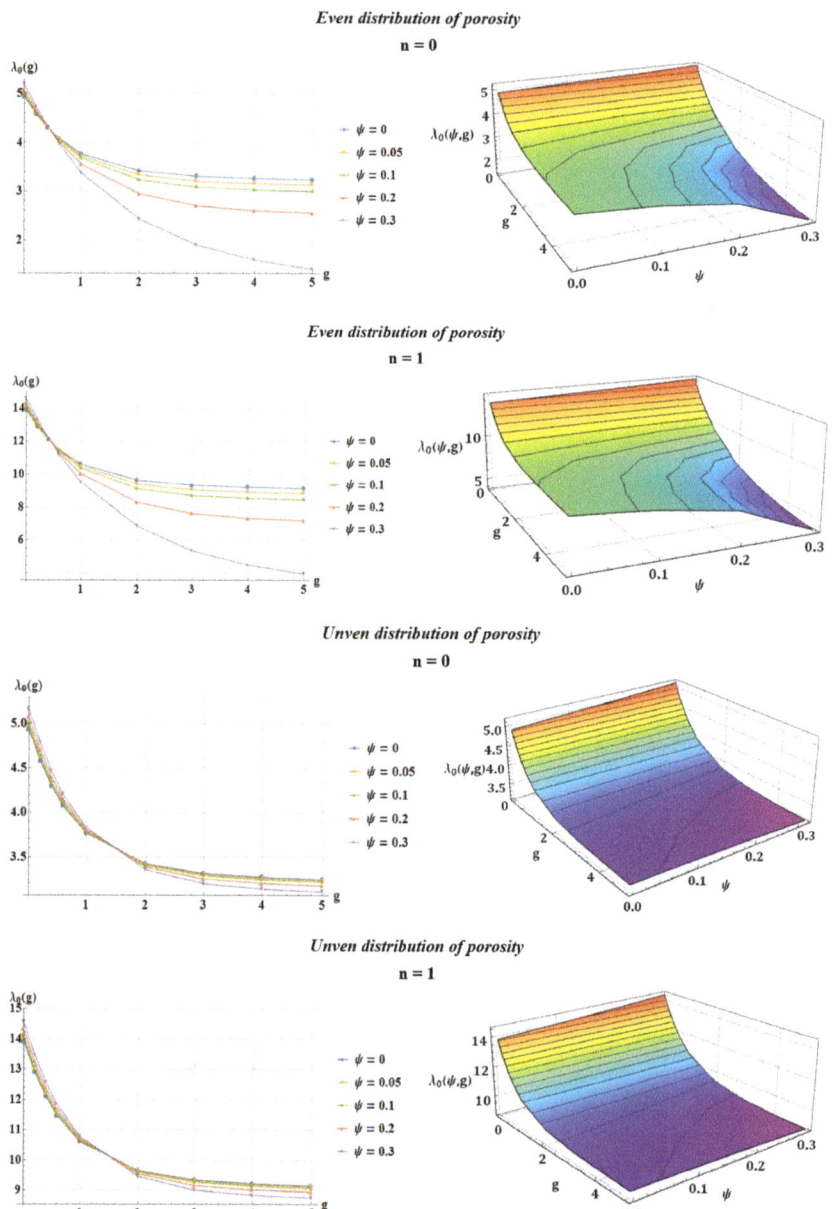

Figure 4. The dependence of the fundamental dimensionless frequencies λ_0 of the free axisymmetric ($n = 0$) and non-axisymmetric ($n = 1$) vibrations on selected values of the power-law index and the porosity volume fraction of the simply supported circular plate with evenly and unevenly distributed porosities.

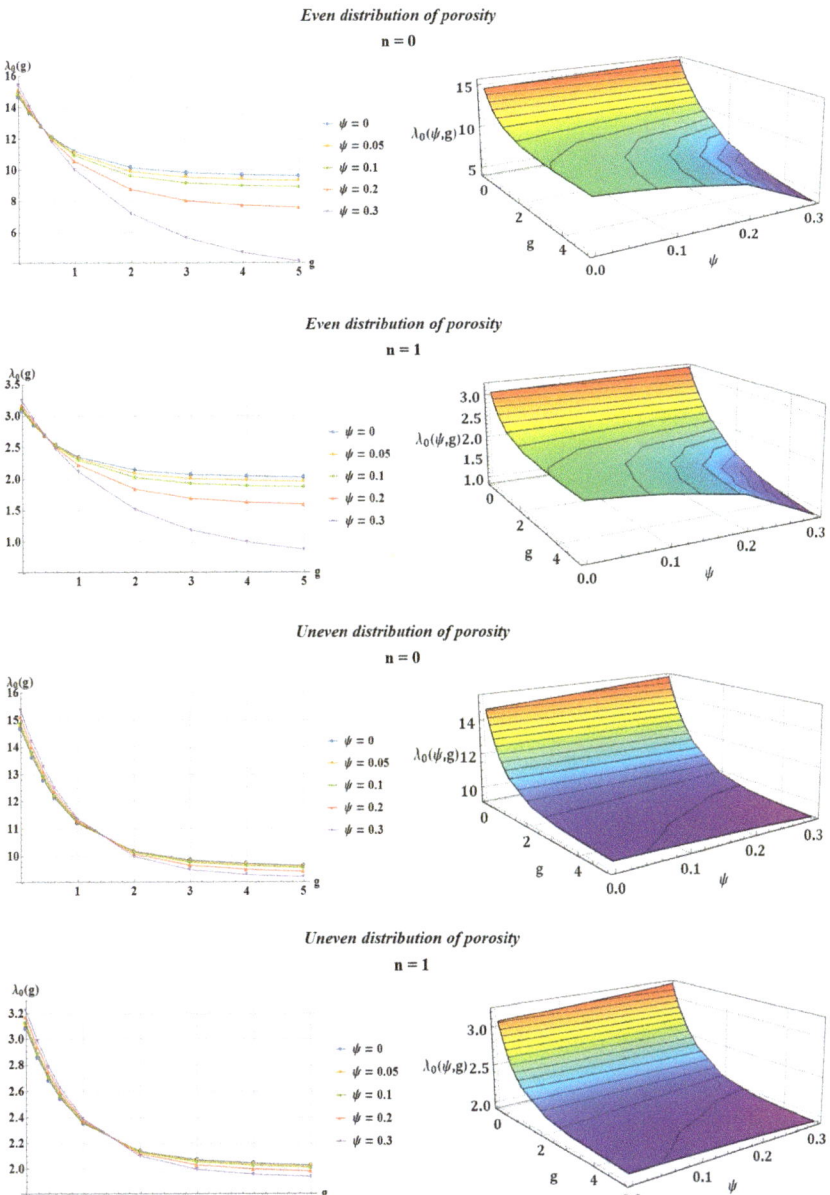

Figure 5. The dependence of the fundamental dimensionless frequencies λ_0 of the free axisymmetric ($n = 0$) and non-axisymmetric ($n = 1$) vibrations on selected values of the power-law index and the porosity volume fraction of the sliding supported circular plate with evenly and unevenly distributed porosities.

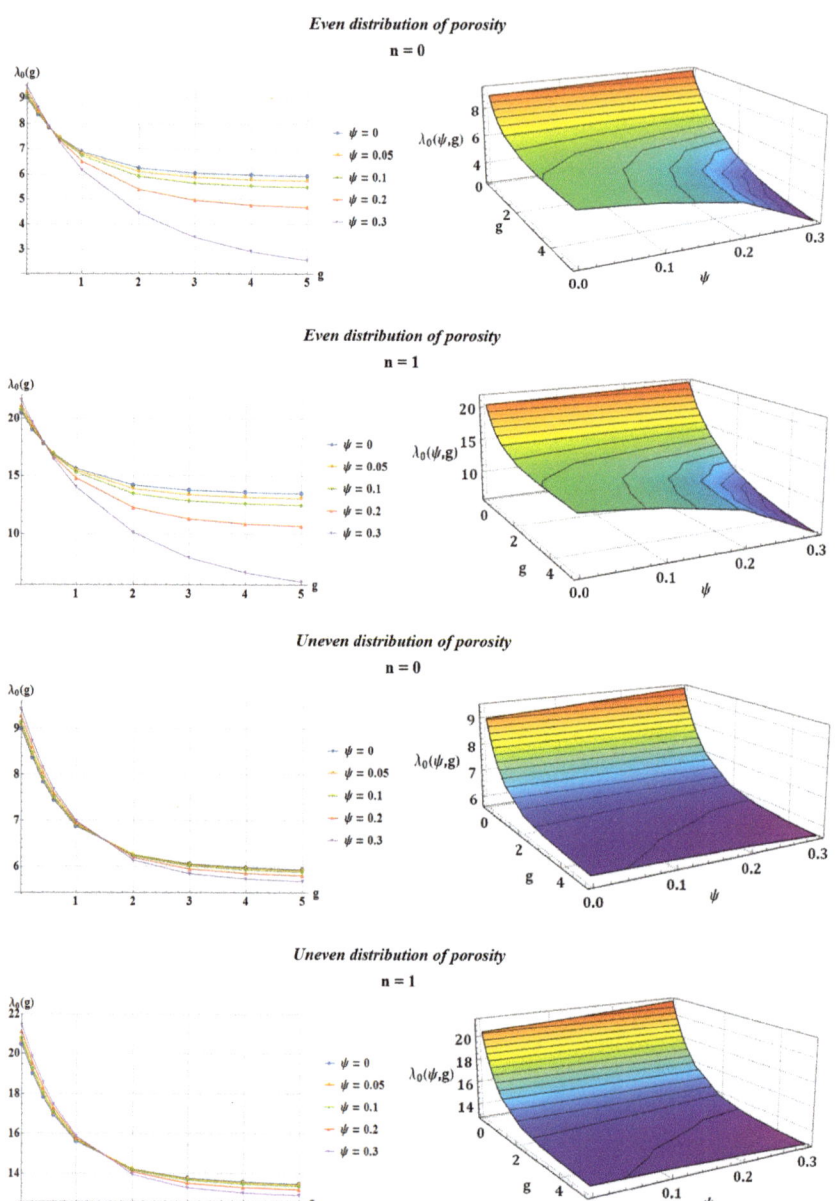

Figure 6. The dependence of the fundamental dimensionless frequencies λ_0 of the free axisymmetric ($n = 0$) and non-axisymmetric ($n = 1$) vibrations on selected values of the power-law index and the porosity volume fraction of the free circular plate with evenly and unevenly distributed porosities.

Table 6. The dimensionless frequencies of the clamped perfect FGM circular plate.

n	λ		\multicolumn{6}{c}{g}					
			1	2	3	4	5	∞
0	λ_0	Present	7.797	7.090	6.867	6.777	6.724	5.199
		[27]	8.498	8.123	7.911	7.733	7.573	-
	λ_1	Present	30.357	27.604	26.737	26.386	26.177	20.243
		[27]	33.086	31.625	30.798	30.107	29.485	-
	λ_2	Present	68.012	61.845	59.902	59.116	58.649	45.352
		[27]	74.127	70.855	69.002	67.453	66.059	-
1	λ_0	Present	16.228	14.756	14.292	14.105	13.993	10.821
	λ_1		46.430	42.219	40.893	40.357	40.038	30.961
	λ_2		91.655	83.344	80.725	79.667	79.037	61.118

Table 7. The dimensionless frequencies of the simply supported perfect FGM circular plate.

n	λ		\multicolumn{6}{c}{g}					
			1	2	3	4	5	∞
0	λ_0	Present	3.767	3.425	3.317	3.274	3.248	2.512
		[27]	4.105	3.924	3.821	3.736	3.658	-
	λ_1	Present	22.685	20.628	19.980	19.717	19.562	15.127
		[27]	24.724	23.633	23.015	22.498	22.033	-
	λ_2	Present	56.602	51.470	49.853	49.199	48.810	37.744
		[27]	61.692	58.968	57.426	56.137	54.977	-
1	λ_0	Present	10.608	9.646	9.343	9.220	9.147	7.074
	λ_1		37.003	33.648	32.591	32.163	31.909	24.675
	λ_2		78.446	71.332	69.091	68.185	67.646	52.310

Table 8. The dimensionless frequencies of the free perfect FGM circular plate.

n	λ	\multicolumn{6}{c}{g}					
		1	2	3	4	5	∞
0	λ_0	6.872	6.248	6.052	5.973	5.926	4.582
	λ_1	29.343	26.682	25.844	25.505	25.303	19.567
	λ_2	66.979	60.905	58.992	58.218	57.757	44.663
1	λ_0	15.628	14.211	13.764	13.584	13.476	10.421
	λ_1	45.653	41.513	40.209	39.682	39.368	30.443
	λ_2	90.799	82.565	79.971	78.922	78.298	60.547

Table 9. The dimensionless frequencies of the perfect FGM circular plate with sliding support.

n	λ	\multicolumn{6}{c}{g}					
		1	2	3	4	5	∞
0	λ_0	11.206	10.190	9.870	9.740	9.663	7.473
	λ_1	37.568	34.161	33.088	32.654	32.396	25.051
	λ_2	79.000	71.836	69.579	68.667	68.124	52.680
1	λ_0	2.352	2.139	2.072	2.045	2.029	1.568
	λ_1	21.676	19.711	19.091	18.841	18.692	14.454
	λ_2	55.612	50.569	48.981	48.338	47.956	37.084

The fundamental dimensionless frequencies of the perfect FGM circular plates with and without the effect of the coupling in-plane and transverse displacements obtained for selected values of the power-law index and diverse boundary conditions are presented in Table 10. Additionally, the differences (errors) between obtained results were calculated according to the equation:

$$\delta(\%) = \left| \frac{\lambda_0^{\mathcal{P}} - \lambda_0^{\mathcal{D}}}{\lambda_0^{\mathcal{P}}} \right| \cdot 100\%, \tag{52}$$

where $\lambda_0^{\mathcal{P}}$ and $\lambda_0^{\mathcal{D}}$ are the fundamental dimensionless frequencies of the perfect FGM circular plate without and with effect of the coupling in-plane and transverse displacements, respectively. Figure 7 presents the dependence of the differences (errors) between obtained results for the power-law index $g \geq 0$.

Table 10. The differences between the fundamental dimensionless frequencies of the perfect FGM circular plates with and without effect of the coupling in-plane and transverse displacements.

BCs	n	λ	g								
			1	2	3	4	5	10	30	60	∞
Clamped	0	$\lambda_0^{\mathcal{P}}$	8.498	8.123	7.911	7.733	7.573	6.977	6.064	5.687	5.199
		$\lambda_0^{\mathcal{D}}$	7.797	7.090	6.867	6.777	6.724	6.512	5.960	5.654	5.199
		$\delta(\%)$	8.2	12.7	13.2	12.3	11.2	6.6	1.7	0.5	0
	1	$\lambda_0^{\mathcal{P}}$	17.687	16.906	16.464	16.094	15.762	14.520	12.621	11.835	10.821
		$\lambda_0^{\mathcal{D}}$	16.228	14.756	14.292	14.105	13.993	13.552	12.404	11.767	10.821
		$\delta(\%)$	8.2	12.7	13.2	12.3	11.2	6.6	1.7	0.5	0
Simply supported	0	$\lambda_0^{\mathcal{P}}$	4.105	3.924	3.821	3.736	3.658	3.370	2.929	2.747	2.512
		$\lambda_0^{\mathcal{D}}$	3.767	3.425	3.317	3.274	3.248	3.145	2.879	2.731	2.512
		$\delta(\%)$	8.2	12.7	13.2	12.3	11.2	6.6	1.7	0.5	0
	1	$\lambda_0^{\mathcal{P}}$	11.562	11.051	10.762	10.521	10.303	9.492	8.250	7.737	7.074
		$\lambda_0^{\mathcal{D}}$	10.608	9.646	9.343	9.220	9.147	8.859	8.108	7.692	7.074
		$\delta(\%)$	8.2	12.7	13.2	12.3	11.2	6.6	1.7	0.5	0
Sliding supported	0	$\lambda_0^{\mathcal{P}}$	12.214	11.675	11.369	11.114	10.885	10.027	8.716	8.173	7.473
		$\lambda_0^{\mathcal{D}}$	11.206	10.190	9.870	9.740	9.663	9.359	8.566	8.126	7.473
		$\delta(\%)$	8.2	12.7	13.2	12.3	11.2	6.6	1.7	0.5	0
	1	$\lambda_0^{\mathcal{P}}$	2.564	2.451	2.387	2.333	2.285	2.105	1.830	1.716	1.569
		$\lambda_0^{\mathcal{D}}$	2.352	2.139	2.072	2.045	2.029	1.964	1.798	1.706	1.569
		$\delta(\%)$	8.2	12.7	13.2	12.3	11.2	6.6	1.7	0.5	0
Free	0	$\lambda_0^{\mathcal{P}}$	7.489	7.159	6.972	6.815	6.674	6.149	5.344	5.012	4.582
		$\lambda_0^{\mathcal{D}}$	6.872	6.248	6.052	5.973	5.926	5.739	5.252	4.983	4.582
		$\delta(\%)$	8.2	12.7	13.2	12.3	11.2	6.6	1.7	0.5	0
	1	$\lambda_0^{\mathcal{P}}$	17.033	16.281	15.855	15.499	15.179	13.984	12.155	11.398	10.421
		$\lambda_0^{\mathcal{D}}$	15.628	14.211	13.764	13.584	13.476	13.051	11.945	11.332	10.421
		$\delta(\%)$	8.2	12.7	13.2	12.3	11.2	6.6	1.7	0.5	0

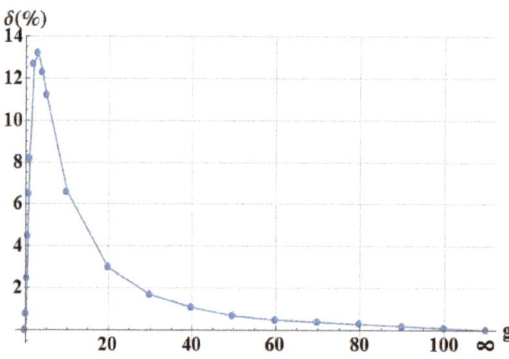

Figure 7. The dependence of the differences (errors) between the fundamental dimensionless frequencies of the perfect FGM circular plate without ($\lambda_0^{\mathcal{P}}$) and with ($\lambda_0^{\mathcal{D}}$) effect of the coupling in-plane and transverse displacements for diverse values of the power-law index.

7. Discussion

7.1. Imperfect FGM Circular Plate

The numerical results for the fundamental dimensionless frequencies of the porous FGM circular plates presented in Tables 2–5 and Figures 3–6 show the following dependences:

- the fundamental eigenfrequency λ_0 of the axisymmetric and non-axisymmetric vibrations of the circular plate decreases with the increasing value of the power-law index g for the two considered distributions of porosities and all considered values of the porosity volume fraction ψ;
- for the evenly distributed porosities, the fundamental eigenfrequency λ_0 of the axisymmetric and non-axisymmetric vibrations of the plate increases with the increasing value of the porosity volume fraction ψ for $g \in [0, 0.4]$ and decreases for $g \in [0.6, 5]$;
- for the unevenly distributed porosities, the fundamental eigenfrequency λ_0 of the axisymmetric and non-axisymmetric vibrations of the plate increases with the increasing value of the porosity volume fraction ψ for $g \in [0, 1]$ and decreases for $g \in [2, 5]$;
- the influence of values of the porosity volume fraction ψ on the values of the fundamental eigenfrequency λ_0 of the axisymmetric and non-axisymmetric vibrations of the plate is smaller for the unevenly distributed porosities than for the evenly distributed porosities;
- for the evenly distributed porosities, the fundamental eigenfrequency λ_0 of the axisymmetric and non-axisymmetric vibrations of plate decreases faster for $\psi = 0.3$ with the increasing values of the power-law index g than for $\psi = \{0, 0.1, 0.2\}$;
- for the unevenly distributed porosities, the fundamental eigenfrequency λ_0 of the axisymmetric and non-axisymmetric vibrations of the plate decreases slowly with the increasing values of the power-law index g for all considered values of the porosity volume fraction ψ.

The observed dependences exist because of the diverse influence of porosity distributions, values of the power-law index and the porosity volume fraction on decreasing (increasing) the ratios of mass to stiffness of the considered circular plates. The all observed dependences are independent of the considered boundary conditions which influence only the values of the dimensionless frequencies of the plate.

7.2. Perfect FGM Circular Plate

It can be observed that the values of dimensionless frequencies of the perfect FGM circular plates obtained by omitting the effect of coupling in-plane and transverse displacements are higher than the values of the dimensionless frequencies of the considered plate with the coupling effect. The differences (errors) between the calculated dimensionless frequencies of free axisymmetric and non-axisymmetric vibration of the perfect FGM circular plate with and without the coupling effect are significant for the power-law index $g \in [0, 20]$, but, for $g \in [20, \infty]$, these differences decrease from 2% to 0%. It can be observed from Table 10 that the differences between the calculated dimensionless frequencies are independent of the modes of vibrations and the boundary conditions of the considered circular plate.

8. Conclusions

This paper presents the influence of two different types of distribution of porosities on the free vibrations of the thin functionally graded circular plate with clamped, simply supported, sliding supported, and free edges. To this aim, the boundary value problem was formulated and a solution was obtained in the exact form. The universal multiparametric characteristic equations were defined using the properties of the multiparametric general solution obtained for the plate with even and uneven distribution of porosities. The effects of the power-law index, the volume fraction index and diverse boundary conditions on the values of the dimensionless frequencies of the free axisymmetric and non-axisymmetric vibrations of the circular plate were comprehensively studied. Additionally, the influences of the power-law index and different boundary conditions on the values of dimensionless frequencies of the FGM circular plate without porosities were also presented.

The presented multiparametric analytical approach can be effectively applying for free vibration of circular and annular plates with other diverse models of an FGM and FGM porous material. The material parameters can be modeled via the exponential or sigmoid functions, as well as Mori–Tanaka functions or other homogenization techniques [39–44]. Diverse applied homogenization techniques only have an influence on the forms of the final replaced plate's stiffnesses and directly on the function $\mathfrak{M}_i$ presented in the obtained general solution in the present paper. It will be the goal of future papers.

The obtained multiparametric general solution will allow for studying the influences of diverse additional complicating effects such as stepped thickness, cracks, additional mounted elements expressed by only additional boundary conditions on the dynamic behavior of the porous functionally graded circular and annular plates. The exact frequencies of vibration presented in non-dimensional form can serve as benchmark values for researchers and engineers to validate their analytical and numerical methods applied in design and analysis of porous functionally graded structural elements.

Author Contributions: Formal analysis, K.K.Ż. and P.J.; Investigation, K.K.Ż. and P.J.; Methodology, K.K.Ż.; Visualization, P.J.; Writing—original draft, K.K.Ż. and P.J.

Funding: This research was funded by the Ministry of Science and Higher Education, Poland, grant number S/WM/4/2017.

Conflicts of Interest: The authors declare no conflict of interest.

References

1. Zhu, J.; Lai, Z.; Yin, Z.; Jeon, J.; Lee, S. Fabrication of ZrO_2-NiCr functionally graded material by powder metallurgy. *Mater. Chem. Phys.* **2001**, *68*, 130–135. [CrossRef]
2. Wattanasakulpong, N.; Prusty, B.G.; Kelly, D.W.; Hoffman, M. Free vibration analysis of layered functionally graded beams with experimental validation. *Mater. Des.* **2012**, *36*, 182–190. [CrossRef]
3. Jabbari, M.; Farzaneh Joubaneh, E.; Khorshidvand, A.R.; Eslami, M.R. Buckling analysis of porous circular plate with piezoelectric actuator layers under uniform radial compression. *Int. J. Mech. Sci.* **2013**, *70*, 50–56. [CrossRef]
4. Khorshidvand, A.R.; Farzaneh Joubaneh, E.; Jabbari, M.; Eslami, M.R. Buckling analysis of a porous circular plate with piezoelectric sensor–actuator layers under uniform radial compression. *Acta Mech.* **2014**, *225*, 179–193. [CrossRef]
5. Mojahedin, A.; Farzaneh Joubaneh, E.; Jabbari, M. Thermal and mechanical stability of a circular porous plate with piezoelectric actuators. *Acta Mech.* **2014**, *225*, 3437–3452. [CrossRef]
6. Behravan Rad, A.; Shariyat, M. Three-dimensional magneto-elastic analysis of asymmetric variable thickness porous FGM circular plates with non-uniform tractions and Kerr elastic foundations. *Compos. Struct.* **2015**, *125*, 558–574. [CrossRef]
7. Barati, M.R.; Sadr, M.H.; Zenkour, A.M. Buckling analysis of higher order graded smart piezoelectric plates with porosities resting on elastic foundation. *Int. J. Mech. Sci.* **2016**, *117*, 309–320. [CrossRef]
8. Mechab, I.; Mechab, B.; Benaissa, S.; Serier, B.; Bachir Bouiadjra, B. Free vibration analysis of FGM nanoplate with porosities resting on Winkler Pasternak elastic foundations based on two-variable refined plate theories. *J. Braz. Soc. Mech. Sci. Eng.* **2016**, *38*, 2193–2211. [CrossRef]
9. Mojahedin, A.; Jabbari, M.; Khorshidvand, A.R.; Eslami, M.R. Buckling analysis of functionally graded circular plates made of saturated porous materials based on higher order shear deformation theory. *Thin-Walled Struct.* **2016**, *99*, 83–90. [CrossRef]
10. Wang, Y.Q.; Zu, J.W. Vibrations behaviors of functionally graded rectangular plates with porosities and moving in thermal environment. *Aerosp. Sci. Technol.* **2017**, *69*, g550–g562. [CrossRef]
11. Gupta, A.; Talha, M. Influence of porosity on the flexural and vibration response of gradient plate using nonpolynomial higher-order shear and normal deformation theory. *Int. J. Mech. Mater. Des.* **2017**, *14*, 277–296. [CrossRef]

12. Wang, Y.Q.; Zu, J.W. Vibration characteristics of moving sigmoid functionally graded plates containing porosities. *Int. J. Mech. Mater. Des.* **2018**, *14*, 473–489. [CrossRef]
13. Ebrahimi, F.; Jafari, A.; Barati, M.R. Free vibration analysis of smart porous plates subjected to various physical fields considering neutral surface position. *Arab. J. Sci. Eng.* **2017**, *42*, 1865–1881. [CrossRef]
14. Feyzi, M.R.; Khorshidvand, A.R. Axisymmetric post-buckling behavior of saturated porous circular plates. *Thin-Walled Struct.* **2017**, *112*, 149–158. [CrossRef]
15. Wang, Y.Q.; Zu, J.W. Large-amplitude vibration of sigmoid functionally graded thin plates with porosities. *Thin-Walled Struct.* **2017**, *119*, 911–924. [CrossRef]
16. Wang, Y.Q.; Wan, Y.H.; Zhang, Y.F. Vibrations of longitudinally travelling functionally graded material plates with porosities. *Eur. J. Mech. A/Solids* **2017**, *66*, 55–68. [CrossRef]
17. Ebrahimi, F.; Jafari, A.; Barati, M.R. Vibration analysis of magneto-electro-elastic heterogeneous porous material plates resting on elastic foundations. *Thin-Walled Struct.* **2017**, *119*, 33–46. [CrossRef]
18. Shahverdi, H.; Barati, M.R. Vibration analysis of porous functionally graded nanoplates. *Int. J. Eng. Sci.* **2017**, *120*, 82–99. [CrossRef]
19. Shojaeefard, M.H.; Googarchin, H.S.; Ghadiri, M.; Mahinzare, M. Micro temperature-dependent FG porous plate: Free vibration and thermal buckling analysis using modified couple stress theory with CPT and FSDT. *Appl. Math. Model.* **2017**, *50*, 633–655. [CrossRef]
20. Barati, M.R.; Shahverdi, H. Nonlinear vibration of nonlocal four-variable graded plates with porosities implementing homotopy perturbation and Hamiltonian methods. *Acta Mech.* **2018**, *229*, 343–362. [CrossRef]
21. Kiran, M.C.; Kattimani, S.C.; Vinyas, M. Porosity influence on structural behaviour of skew functionally graded magneto-electro-elastic plate. *Compos. Struct.* **2018**, *191*, 36–77. [CrossRef]
22. Cong, P.H.; Chien, T.K.; Khoa, N.D.; Duc, N.D. Nonlinear thermomechanical buckling and post-buckling response of porous FGM plates using Reddy's HSDT. *Aerosp. Sci. Technol.* **2018**, *77*, 419–428. [CrossRef]
23. Kiran, M.C.; Kattimani, S.C. Assessment of porosity influence on vibration and static behaviour of functionally graded magneto-electro-elastic plate: A finite element study. *Eur. J. Mech. A/Solids* **2018**, *71*, 258–277. [CrossRef]
24. Arshid, E.; Khorshidvand, A.R. Free vibration analysis of saturated porous FG circular plates integrated with piezoelectric actuators via differential quadrature method. *Thin-Walled Struct.* **2018**, *125*, 220–233. [CrossRef]
25. Shahsavari, D.; Shahsavari, M.; Li, L.; Karami, B. A novel quasi-3D hyperbolic theory for free vibration of FG plates with porosities resting on Winkler/Pasternak/Kerr foundation. *Aerosp. Sci. Technol.* **2018**, *72*, 134–149. [CrossRef]
26. Ebrahimi, F.; Rastgo, A. An analytical study on the free vibration of smart circular thin FGM plate based on classical plate theory. *Thin-Walled Struct.* **2008**, *46*, 1402–1408. [CrossRef]
27. Lal, R.; Ahlawat, N. Axisymmetric vibrations and buckling analysis of functionally graded circular plates via differential transform method. *Eur. J. Mech. A/Solids* **2015**, *52*, 85–94. [CrossRef]
28. Lal, R.; Ahlawat, N. Buckling and vibration of functionally graded non-uniform circular plates resting on Winkler foundation. *Lat. Am. J. Solids Struct.* **2015**, *12*, 2231–2258. [CrossRef]
29. Żur, K.K. Quasi-Green's function approach to free vibration analysis of elastically supported functionally graded circular plates. *Compos. Struct.* **2018**, *183*, 600–610. [CrossRef]
30. Żur, K.K. Free vibration analysis of elastically supported graded annular plates via quasi-Green's function method. *Compos. Part B* **2018**, *144*, 37–55. [CrossRef]
31. Reddy, J.N.; Wang, C.M.; Kitipornchai, S. Axisymmetric bending of functionally graded circular and annular plates. *Eur. J. Mech. A/Solids* **1999**, *18*, 185–199. [CrossRef]
32. Wattanasakulong, N.; Chaikittiratana, A. Flexural vibration of imperfect functionally graded beams based on Timoshenko beam theory: Chebyshev collocation method. *Meccanica* **2015**, *50*, 1331–1342. [CrossRef]
33. Delale, F.; Erdogan, F. The crack problem for a non-homogeneous plane. *Asme J. Appl. Mech.* **1983**, *50*, 609–614. [CrossRef]
34. Reddy, J.N. *Theory and Analysis of Elastic Plates and Shells*; CRC Press: Boca Raton, FL, USA, 2006.
35. Wu, T.Y.; Liu, G.R. Free vibration analysis of circular plates using generalized differential quadrature rule. *Comput. Methods Appl. Mech. Eng.* **2002**, *191*, 5365–5380. [CrossRef]
36. Yalcin, H.S.; Arikoglu, A.; Ozkol, I. Free vibration analysis of circular plates by differential transformation method. *Appl. Math. Comput.* **2009**, *212*, 377–386. [CrossRef]

37. Zhou, Z.H.; Wong, K.W.; Xu, X.S.; Leung, A.Y.T. Natural vibration of circular and annular thin plates by Hamiltonian approach. *J. Sound Vib.* **2011**, *330*, 1005–1017. [CrossRef]
38. Duan, G.; Wang, X.; Jin, C. Free vibration analysis of circular thin plates with stepped thickness by the DSC element method. *Thin-Walled Struct.* **2014**, *85*, 25–33. [CrossRef]
39. Mori, T.; Tanaka, K. Average stress in matrix and average elastic energy of materials with misfitting inclusions. *Acta Metall.* **1973**, *21*, 571–574. [CrossRef]
40. Chung, Y.L.; Chi, S.H. The residual stress of functionally graded materials. *J. Chin. Inst. Civ. Hydraul. Eng.* **2001**, *13*, 1–9.
41. Chi, S.H.; Chung, Y. Mechanical behavior of functionally graded material plates under transverse load—Part I: Analysis. *Int. J. Solids Struct.* **2006**, *43*, 3657–3674. [CrossRef]
42. Ke, L.-L.; Wang, Y.-S.; Yang, J.; Kitipornchai, S. Nonlinear free vibration of size-dependent functionally graded nanobeams. *Int. J. Eng. Sci.* **2012**, *50*, 256–267. [CrossRef]
43. Hornung, U. *Homogenization and Porous Media*; Springer: Berlin, Germany, 1997.
44. Adrianov, I.V.; Awrejcewicz, J.; Danishevskyy, V. *Asymptotical Mechanics of Composites: Modelling Composites without FEM*; Springer: Berlin, Germany, 2018.

© 2019 by the authors. Licensee MDPI, Basel, Switzerland. This article is an open access article distributed under the terms and conditions of the Creative Commons Attribution (CC BY) license (http://creativecommons.org/licenses/by/4.0/).

Article

On the Effect of Thomson and Initial Stress in a Thermo-Porous Elastic Solid under G-N Electromagnetic Theory

Elsayed M. Abd-Elaziz [1], Marin Marin [2,*] and Mohamed I. A. Othman [3]

[1] Zagazig Higher Institute of Engineering & Technology, Ministry of Higher Education, P.O. Box 44519, Zagazig, Egypt; sayed_nr@yahoo.com
[2] Department of Mathematics and Computer Science, Transilvania University of Brasov, Brasov 500036, Romania
[3] Department of Mathematics, Faculty of Science, Zagazig University, P.O. Box 44519, Zagazig, Egypt; m_i_a_othman@yahoo.com
* Correspondence: m.marin@unitbv.ro

Received: 22 January 2019; Accepted: 14 March 2019; Published: 20 March 2019

Abstract: The present work investigated the effect of Thomson and initial stress in a thermo-porous elastic solid under G-N electromagnetic theory. The Thomson coefficient affects the heat condition equation. A constant Thomson coefficient, instead of traditionally a constant Seebeck coefficient, is assumed. The charge density of the induced electric current is taken as a function of time. A normal mode method is proposed to analyze the problem and to obtain numerical solutions. The results that were obtained for all physical sizes are graphically illustrated and we offer a comparison between the type II G-N theory and the G-N theory of type III, both in the present case and in the absence of specific parameters, as initial stress, pores and the Thomson effect. Some particular cases are also discussed in the context of the problem. The results indicate that the effect of initial stress, Thomson coefficient effect, and magnetic field are very pronounced.

Keywords: Thomson effect; initial stress; magneto-thermoelastic; voids; normal mode method; G-N theory

1. Introduction

In the generalized theories, the governing equations involve thermal relaxation times and they are of the hyperbolic type. Green and Naghdi [1–3] considered a new extend theory by including the thermal displacement gradient between the constitutive variables. As we know, the classically coupled thermoelasticity includes the temperature gradient as one of the constitutive variables.

An important feature of this theory is that it does not accommodate the dissipation of thermal energy. In paper by Sharma and Chauhan [4], we find an approach regarding the elastic interactions without considering the energy dissipation due to heat sources and body forces.

An important step in evolution of the classical theory of elasticity was made through the appearance of the theory of poroelasticity, which consider the volume of void, in an elastic body with pores, as a kinematics variable.

This gave the opportunity to investigate some concrete types of biological and geological solids and their useful applications. See, for instance, the applications in the fuel-cell industry [5–10].

We have to point out that the theory of linear elastic bodies with pores allows for the approach of such properties of biological and geological medium that could not be studied in the context of classical theory. It is very important to note that, when the volume of the pores tends to zero, we can see that the poroelastic theory reduces to the theory of classical elasticity.

In the paper, Nunziato and Cowin [11] first established a theory of elastic bodies with pores in the non-linear theory case.

This theory of porous media has gained a great extension over the last period of time, and many authors consider different mathematical models for the mechanical behavior of solids with pores, by combining the poroelasticity theory with other different theories, in other words, combining different effects, [12–16].

The consideration of the dynamic reaction of a thermoelastic body with additional parameters is very helpful in solving many concrete applications. For instance, the initial stresses are considered in a thermoelastic body with pores due to different reasons, such as the gravity variations, the difference of temperature, the process of quenching, etc.

Clearly, the earth is constantly under the influence of high initial stresses. As such, the researchers have allocated great importance to the study the effect of initial stresses regarding the thermal and mechanical state of a solid. For instance, Montanaro in [17] investigated a thermoelastic isotropic body with hydrostatic initial stress.

Of course, the laser pulse has an effect on thermal loading in an elastic body with voids. Othman and Abd-Elaziz studied this effect in the paper [18]. Marin investigated Cesaro means in the thermoelasticity of dipolar bodies [19]. Marin and Oechsner studied the effect of a dipolar structure on the Holder stability in Green-Naghdi thermoelasticity [20].

Other effects, such as the effect of the Earth's electromagnetic field on seismic propagations, the designing of different elements of machine, emissions of electromagnetic radiations from nuclear devices, plasma physics, etc., can be found in [21–27].

In our present study, we approach of the plane strain problem of a half-space body consisting of an electro-magneto-thermoelastic material that possesses voids and is subjected to some initial stress and to the Thomson effect. Our mathematical model is regarding the Green–Naghdi theory of type II and III of thermoelasticity. We assume that the Thomson effect is a constant coefficient and the density of charges that are induced by electric current is a function that depends on time variable.

In order to obtain the expressions for the considered parameters, it used the known normal mode technique. We also have obtained some graphic representations for the repartition of the considered variables.

2. Formulation of the Problem

An isotropic and homogeneous elastic body with pores (voids) is considered, with the temperature T_0, in the reference state, and the half space ($y \geq 0$). The motion referred to a rectangular Cartesian system of coordinates (x, y, z) with origins in the surface ($z = 0$). Additionally, the X-axis is pointing vertically into the body. In the of a two-dimensional problem, we suppose that the evolution of the body will be characterized by the displacement vector u, with components $u = (u, v, 0)$. The functions that are considered in this context are dependent on the time variable t and of the spatial variables x and y.

We consider a magnetic field with components $H = (0, 0, H_3)$, having a constant intensity, which acts parallel to the direction of the Z-axis.

It is known that a magnetic field of the form $H = (0, 0, H_0 + h(x, y, t))$ produces an induced electric field of components $E = (E_1, E_2, 0)$, and an induced magnetic field, as denoted by h, and these satisfy the electromagnetism equations, in the linearized form. We will use the Maxwell's equations [24] in order to characterize the evolution of the electric field and for variation of the magnetic field, as follows:

$$\nabla \times h = J + \dot{D}, \tag{1}$$

$$\nabla \times E = -\dot{B}, \tag{2}$$

$$\nabla \cdot B = 0, \ \nabla \cdot D = \rho_e, \tag{3}$$

$$B = \mu_0 (H + h), \ D = \varepsilon_0 E. \tag{4}$$

The modified Ohm's law for a medium with finite conductivity supplements the above system of coupled equations, namely

$$J = \sigma_0(E + \mu_0 \dot{u} \times H), \tag{5}$$

where μ_0 is the magnetic permeability, B is the magnetic displacement vector, ε_0 is the electric permeability, J is the current density vector, ρ_e is the charge density, D is the electric displacement vector, and E is the induced electric field vector.

For an isotropic and homogeneous thermoelastic body having pores, the constitutive equations receive the following form:

$$\sigma_{ij} = 2\mu\, e_{ij} + (\lambda\, e_{rr} + \lambda_0\, \phi - \beta T)\delta_{ij} - L^*(\delta_{ij} + m^*_{ij}), \tag{6}$$

$$h_i = \alpha \phi_{,i}, \tag{7}$$

$$g = -\lambda_0 e_{rr} - \xi_1 \varphi + mT, \tag{8}$$

$$\rho \eta^* = \beta e_{rr} + a_0 T + m\phi. \tag{9}$$

The strain-displacement relation is

$$e_{ij} = \frac{1}{2}(u_{i,j} + u_{j,i}). \tag{10}$$

The tensor of rotation has the components:

$$m^*_{ij} = \frac{1}{2}(u_{j,i} - u_{i,j}),\ i,j = 1,2,3. \tag{11}$$

In Green-Naghdi (G-N) theories we take into account the Thomson effect, so that the Fourier's law becomes

$$q_i = -[kT_{,i} + k^* \dot{T}_{,i}] + MJ_i, \tag{12}$$

which gives

$$q_{i,i} = -\left[kT_{,ii} + k^* \dot{T}_{,ii}\right] + M J_{i,i}. \tag{13}$$

If we take into account Equations (1) and (3), then from Equation (13), we deduce

$$q_{i,i} = -[kT_{,ii} + k^* \dot{T}_{,ii}] + M\dot{\rho}_e. \tag{14}$$

where T is the temperature above the reference temperature T_0 is chosen so that $|(T - T_0)/T_0| < 1$, λ, μ are the counterparts of Lame' constants, t is the time, σ_{ij} are the components of the stress tensor, $h_{,i}$ is the equilibrated stress vector, ψ is the equilibrated inertia, g is the intrinsic equilibrated body force, $\alpha, \lambda_0, \xi_1, \omega_0, m$ are constants of material that are due to the presence of the pores, $\beta = (3\lambda + 2\mu)\alpha_t$, such that α_t is the coefficient of thermal expansion, δ_{ij} is the Kronecker delta, ρ is the mass density, C_E is the specific heat at the constant strain, k is the thermal conductivity, η^* is entropy per unit mass, k^* is a constant, and q_{ij} are the components of the first heat flux moment vector, we write the equation of continuity for the charges in the body in the form

$$\dot{\rho}_e + \nabla \cdot (\rho_e v_i) = 0, \tag{15}$$

where the velocity of the charges has the components v_i.

Let us now consider that the charge density is a function that does not depend on spatial variables, but only on time variable. Thus, Equation (15) will reduce to

$$\dot{\rho}_e + \rho_e \nabla \cdot (v_i) = 0. \tag{16}$$

We will assume that the charges have the speed of components v_i, which are proportional to the components of the velocity for particles $\dot{u}_i$, so that we can write

$$v_i = p_0 \dot{u}_i, \tag{17}$$

which gives

$$\nabla \cdot v_i = p_0 \nabla \cdot \dot{u}_i = p_0 \dot{e}, \tag{18}$$

where p_0 is a positive constant (non-dimensional).

If we take into account Equation (18), from Equation (16), we are led to

$$\dot{\rho}_e = -\rho_e p_0 \dot{e}, \tag{19}$$

which gives

$$\int \frac{d\rho_e}{\rho} = -p_0 \int de. \tag{20}$$

Hence, we obtain

$$\rho_e = \rho_e^0 \exp(-p_0 e) \simeq \rho_e^0 (1 - p_0 e), \tag{21}$$

where ρ_e^0 is the charge density when the strain vanishes.

Then, we obtain

$$\dot{\rho}_e = -\rho_e^0 (1 - p_0 e) p_0 \dot{e}. \tag{22}$$

While taking into account the Equation (22), from Equation (14) we deduce that the Fourier's law, in its generalized form, receives the form:

$$q_{i,i} = -[kT_{,ii} + k^* \dot{T}_{,ii}] + M \rho_e^0 (1 - p_0 e) p_0 \dot{e}. \tag{23}$$

In the case of null heat supply, the balance energy becomes

$$\rho T_0 \dot{\eta}^* = - q_{i,i}. \tag{24}$$

Taking into account Equations (9) and (23), from Equation (24), we deduce that the equation of heat conduction can be written in the form

$$kT_{,ii} + k^* \dot{T}_{,ii} - m T_0 \dot{\phi} = \rho \, C_e \ddot{T} + \beta \, T_0 \ddot{u}_{i,i} + M \rho_e^0 (1 - p_0 e) p_0 \dot{e}. \tag{25}$$

This equation can be substitute by an approximate form

$$kT_{,ii} + k^* \dot{T}_{,ii} - m T_0 \dot{\phi} = \rho \, C_e \ddot{T} + \beta \, T_0 \ddot{u}_{i,i} + M \rho_e^0 p_0 \dot{e}. \tag{26}$$

As a consequence, we can obtain the stress components in a simplified form. Accordingly, from Equations (6), (10), and (11), we are led to

$$\sigma_{xx} = \lambda \left[\frac{\partial u}{\partial x} + \frac{\partial v}{\partial y}\right] + 2\mu \frac{\partial u}{\partial x} + \lambda_0 \phi - \beta T - L^*, \tag{27}$$

$$\sigma_{yy} = \lambda \left[\frac{\partial u}{\partial x} + \frac{\partial v}{\partial y}\right] + 2\mu \frac{\partial v}{\partial y} + \lambda_0 \phi - \beta T - L^*, \tag{28}$$

$$\sigma_{zz} = \lambda \left[\frac{\partial u}{\partial x} + \frac{\partial v}{\partial y}\right] + \lambda_0 \phi - \beta T - L^*, \tag{29}$$

$$\sigma_{xy} = \left(\mu + \frac{L^*}{2}\right) \frac{\partial u}{\partial y} + \left(\mu - \frac{L^*}{2}\right) \frac{\partial v}{\partial x}, \tag{30}$$

$$\sigma_{yx} = (\mu + \frac{L^*}{2})\frac{\partial v}{\partial x} + (\mu - \frac{L^*}{2})\frac{\partial u}{\partial y}, \sigma_{xz} = \sigma_{yz} = 0. \quad (31)$$

The equations of motion, taking into account the Lorentz force

$$\sigma_{ji,j} + F_i = \rho u_{i,tt}. \quad (32)$$

The Lorentz force is given by

$$F_i = \mu_0 (\mathbf{J} \times \mathbf{H})_i. \quad (33)$$

The current density vector $\mathbf{J}$ is parallel to electric intensity vector $\mathbf{E}$, thus $\mathbf{J} = (J_1, J_2, 0)$. The Ohm's law (5) after linearization gives

$$J_1 = \sigma_0 (E_1 + \mu_0 H_0 \dot{v}), \quad J_2 = \sigma_0 (E_2 - \mu_0 H_0 \dot{u}). \quad (34)$$

Equations (1), (4), and (34) give

$$\frac{\partial h}{\partial y} = \sigma_0 (E_1 + \mu_0 H_0 \frac{\partial v}{\partial t}) + \varepsilon_0 \frac{\partial E_1}{\partial t}, \quad (35)$$

$$\frac{\partial h}{\partial x} = -\sigma_0 (E_1 - \mu_0 H_0 \frac{\partial u}{\partial t}) - \varepsilon_0 \frac{\partial E_2}{\partial t}. \quad (36)$$

From Equations (2) and (5), we get

$$\frac{\partial E_1}{\partial y} - \frac{\partial E_2}{\partial x} = \mu_0 \frac{\partial h}{\partial t}. \quad (37)$$

From Equations (33) and (34), we obtain

$$F_1 = \sigma_0 \mu_0 H_0 (E_2 - \mu_0 H_0 \frac{\partial u}{\partial t}), F_2 = -\sigma_0 \mu_0 H_0 (E_1 + \mu_0 H_0 \frac{\partial v}{\partial t}), F_3 = 0. \quad (38)$$

From Equations (27)–(32) and (38), we get

$$(\mu - \frac{L^*}{2})\nabla^2 u + (\lambda + \mu + \frac{L^*}{2})\frac{\partial e}{\partial x} + b\frac{\partial \phi}{\partial x} - \beta \frac{\partial T}{\partial x} + \sigma_0 \mu_0 H_0 (E_2 - \mu_0 H_0 \frac{\partial u}{\partial t}) = \rho \frac{\partial^2 u}{\partial t^2} \quad (39)$$

$$(\mu - \frac{L^*}{2})\nabla^2 v + (\lambda + \mu + \frac{L^*}{2})\frac{\partial e}{\partial y} + b\frac{\partial \phi}{\partial y} - \beta \frac{\partial T}{\partial y} - \sigma_0 \mu_0 H_0 (E_1 + \mu_0 H_0 \frac{\partial v}{\partial t}) = \rho \frac{\partial^2 v}{\partial t^2}, \quad (40)$$

in which we used the notation $e = \frac{\partial u}{\partial x} + \frac{\partial v}{\partial y}$.

For the equation of the equilibrated forces, we obtain

$$\rho \psi \phi_{,tt} = h_{i,i} + g. \quad (41)$$

Also, while taking into account Equations (7), (8), and (41), we are led to

$$\alpha \phi_{,jj} - \lambda_0 u_{i,i} - \xi_1 \phi - \omega_0 \phi_{,t} + m T = \rho \psi \ddot{\phi}. \quad (42)$$

Let us define the non-dimensional sizes

$$(x'_i, u'_i) = \frac{w_1^*}{c_1}(x_i, u_i), (t', t'_0) = w_1^*(t, t_0), \phi' = \frac{\psi w_1^{*2}}{c_1^2}\phi, \sigma'_{ij} = \frac{\sigma_{ij}}{\mu}, p'_1 = \frac{p_1}{\mu}, L^{*'} = \frac{L^*}{\mu},$$

$$\theta' = \frac{T - T_0}{T_0}, h' = \frac{w_1^*}{\sigma_0 H_0 \mu_0 c_1^2} h, E'_i = \frac{w_1^*}{\sigma_0 H_0^2 \mu_0 c_1^2} E_i, \quad (43)$$

where $w_1^* = \frac{\rho c_e c_1^2}{K}$, $c_1^2 = \frac{\lambda + 2\mu}{\rho}$.

For dimensionless sizes that are defined in Equation (43), we can write the above basic equations in the following from

$$a_1 \nabla^2 u + a_2 \frac{\partial e}{\partial x} + a_3 \frac{\partial \phi}{\partial x} - a_4 \frac{\partial \theta}{\partial x} + a_5(a_6 E_2 - \frac{\partial u}{\partial t}) = a_7 \frac{\partial^2 u}{\partial t^2}, \tag{44}$$

$$a_1 \nabla^2 v + a_2 \frac{\partial e}{\partial y} + a_3 \frac{\partial \phi}{\partial y} - a_4 \frac{\partial \theta}{\partial y} - a_5(a_6 E_1 + \frac{\partial v}{\partial t}) = a_7 \frac{\partial^2 v}{\partial t^2}, \tag{45}$$

$$(\nabla^2 - a_8 - a_9 \frac{\partial}{\partial t} - a_{10} \frac{\partial^2}{\partial t^2})\phi - a_{11} e + a_{12} \theta = 0, \tag{46}$$

$$[(k + k^* w_1^* \frac{\partial}{\partial t})\nabla^2 - a_{13} \frac{\partial^2}{\partial t^2}]\theta + a_{14} \frac{\partial \phi}{\partial t} - \beta c_1^* \frac{\partial^2 e}{\partial t^2} - M_0 \frac{\partial e}{\partial t} = 0, \tag{47}$$

$$\frac{\partial h}{\partial y} = a_{15} E_1 + a_{16} \frac{\partial E_1}{\partial t} + \frac{\partial v}{\partial t}, \tag{48}$$

$$\frac{\partial h}{\partial x} = -a_{15} E_2 - a_{16} \frac{\partial E_2}{\partial t} + \frac{\partial u}{\partial t}, \tag{49}$$

$$\frac{\partial E_1}{\partial y} - \frac{\partial E_2}{\partial x} = \frac{\partial h}{\partial t}, \tag{50}$$

by dropping the dashed, for convenience. Here, $M_0 = \frac{M \rho_e^0 p_0 c_1^2}{w_1^* T_0}$, is the Peltier coefficient at T_0 and $a_1 = \mu - \frac{L^* \mu}{2}$, $a_2 = \lambda + \mu + \frac{L^* \mu}{2}$, $a_3 = \frac{bc_1^2}{\psi w_1^{*2}}$, $a_4 = \beta T_0$, $a_5 = \frac{\mu_0^2 H_0^2 \sigma_0 c_1^2}{w_1^*}$, $a_6 = \frac{\mu_0 \sigma_0 c_1^2}{w_1^*}$, $a_7 = \rho c_1^2$, $a_8 = \frac{\xi_1 c_1^2}{\alpha w_1^{*2}}$, $a_9 = \frac{w_0 c_1^2}{\alpha w_1^*}$, $a_{12} = \frac{m \psi T_0}{\alpha}$, $a_{13} = \rho C_e c_1^2$, $a_{14} = \frac{mc_1^4}{\psi w_1^{*3}}$, $a_{15} = \frac{\sigma_0 \mu_0 c_1^2}{w_1^*}$, $a_{16} = \mu_0 \varepsilon_0 c_1^2$, $a_{10} = \frac{\rho \psi c_1^2}{\alpha}$, $a_{11} = \frac{\lambda_0 \psi}{\alpha}$.

From Equations (44) and (45), we obtain

$$[(a_1 + a_2)\nabla^2 - a_7 \frac{\partial}{\partial t} - a_5 \frac{\partial^2}{\partial t^2}] e + a_3 \nabla^2 \phi - a_4 \nabla^2 \theta - a_6 a_5 \frac{\partial h}{\partial t} = 0. \tag{51}$$

From Equations (48) and (49), we obtain

$$(\nabla^2 - a_{15} \frac{\partial}{\partial t} - a_{16} \frac{\partial^2}{\partial t^2})h - \frac{\partial e}{\partial t} = 0. \tag{52}$$

3. The Solution of the Problem

3.1. Decomposition by Normal Mode Analysis

Using the normal mode analysis, we can decompose the solution of the above physical parameters in the following form

$$[e, \theta, \phi, h, \sigma_{ij}](x, y, t) = [e^*, \theta^*, \phi^*, h^*, \sigma_{ij}^*](y) \exp[i(a x - \omega t)], \tag{53}$$

in which $e^*, \phi^*, \theta^*, h^*, \sigma_{ij}^*$ are the amplitudes of the respective fields, $i = \sqrt{-1}$, ω is the frequency, and a is the wave number.

By taking into account Equation (53) in Equations (46), (47), (51), and (52), we are led to

$$(r_1 D^2 - r_2) e^* + a_3(D^2 - a^2)\phi^* - a_4(D^2 - a^2)\theta^* + r_3 h^* = 0, \tag{54}$$

$$(D^2 - r_4) \phi^* - a_{11} e^* + a_{12} \theta^* = 0, \tag{55}$$

$$(r_5 D^2 - r_6) \theta^* + r_7 \phi^* + r_8 e^* = 0, \tag{56}$$

$$(D^2 - r_9) h^* + i\omega e^* = 0, \tag{57}$$

where $r_1 = a_1 + a_2$, $r_2 = ra^2 - i\omega a_5 - a_7\omega^2$, $r_3 = i\omega a_6 a_5$, $r_4 = a^2 + a_8 - i\omega a_9 - a_{10}\omega^2$, $r_5 = k - i\omega k^* w^*$, $r_6 = r_5 a^2 - a_{13}\omega^2$, $r_7 = i\omega a_{14}$, $r_8 = \beta\omega^2 c_1^2 - i\omega M_0$, and $r_9 = a^2 - i\omega a_{15} - a_{10}\omega^2$.

Eliminating (55), we get the following ordinary Equation (52) between $e^*(y)$ $\phi^*(y)$, $\theta^*(y)$, the differential Equation satisfied by $h^*(y)$:

$$\left(D^8 - N_1 D^6 + N_2 D^4 - N_3 D^2 + N_4\right)\{h^*(y), \phi^*(y), e^*(y), \theta^*(y)\} = 0. \tag{58}$$

We can write the Equation (58) in a decomposed form, as follows

$$(D^2 - k_1^2)(D^2 - k_2^2)(D^2 - k_3^2)(D^2 - k_4^2)\{h^*(y), \phi^*(y), e^*(y), \theta^*(y)\} = 0, \tag{59}$$

where, k_n^2 ($n = 1, 2, 3, 4$) are roots of the characteristic equation of Equation (58) and $s_1 = r_5 r_9 + r_4 r_5 + r_6$, $s_2 = r_6 r_9 + r_4 r_5 r_9 - a_{12} r_7 + r_6 r_4$, $s_3 = r_4 r_6 r_9 + a_{12} r_7 r_9$, $s_4 = r_5 r_6 - a_{12} r_8 + r_6$, $s_5 = a_{11} r_6 r_9 - a_{12} r_8 r_9$, $s_6 = r_8 r_9 + r_4 r_8 - a_{11} r_3$, $s_7 = r_8 r_4 r_9 - a_{11} r_7 r_9$, $s_8 = r_6 + r_4 r_5$, $s_9 = r_4 r_6 - a_{12} r_7$, $s_{10} = \frac{1}{r_1 r_5}$, $N_1 = s_{10}(r_1 s_1 + r_2 r_5 - a_3 a_{11} r_5 + a_4 r_8)$, $N_2 = s_{10}(r_1 s_2 + r_2 s_1 - a_3 a_{11} r_5 a^2 - a_3 s_4 + a_4 s_6 + a_4 r_8 a^2 - i\omega r_3 r_5)$, $N_3 = s_{10}(r_1 s_3 + r_2 s_2 - a_3 s_5 - a_3 s_4 a^2 + a_4 s_7 + a_4 s_6 a^2 - i\omega r_3 r_8)$, and $N_4 = s_{10}(r_2 s_3 - a_3 s_5 a^2 + a_4 s_7 a^2 - i\omega r_3 r_9)$.

The general solutions of the Equation (59), bound at $y \to \infty$,

$$e(x, y, t) = \sum_{n=1}^{4} R_n \exp[-k_n y + i(ax - \omega t)], \tag{60}$$

$$h(x, y, t) = \sum_{n=1}^{4} H_{1n} R_n \exp[-k_n y + i(ax - \omega t)], \tag{61}$$

$$\theta(x, y, t) = \sum_{n=1}^{4} H_{2n} R_n \exp[-k_n y + i(ax - \omega t)], \tag{62}$$

$$\phi(x, y, t) = \sum_{n=1}^{4} H_{3n} R_n \exp[-k_n y + i(ax - \omega t)], \tag{63}$$

where

$$H_{1n} = \frac{-i\omega}{k_n^2 - s_9}, \quad H_{2n} = \frac{-r_8(k_n^2 - r_4) - r_7 a_{11}}{(k_n^2 - r_4)(r_5 k_n^2 - r_6) - r_7 a_{12}}, \quad H_{3n} = \frac{a_{11} - a_{12} H_{2n}}{k_n^2 - r_4}. \tag{64}$$

3.2. Boundary Conditions

In the following, we will consider some boundary conditions on the surface of Equation $y = 0$, which will help us to determine the above constants R_1, R_2, R_3 and R_4.

3.2.1. The mechanical boundary condition

The mechanical boundary condition that the bounding plane to the surface $y = 0$ has zero strain, so we have

$$e(x, 0, t) = 0$$

3.2.2. The Boundary Restriction of Heat

We assume that the boundary surface of the body is subject to a thermal shock described by the function

$$\theta(x, 0, t) = \theta_0 \exp(i(ax - \omega t)), \tag{65}$$

where θ_0 is constant.

3.2.3. Voids Conditions

$$\frac{\partial \phi}{\partial y} = 0. \tag{66}$$

3.2.4. The Boundary Restriction for Electromagnetic Field

On the surface of the half-space $h(x, 0, t) = h^*$, we consider that the electromagnetic field intensity is a continuous function. Here, the intensity of the magnetic field in free space is h^*.

We now assume there is no magnetic or electric field in the free space, that is,

$$h(x, 0, t) = h^* = 0. \tag{67}$$

In order to obtain the constants R_1, R_2, R_3 and R_4, we will use the dimensionless size $\theta'_0 = \frac{\theta_0}{T_0}$ and the expressions of the variables into the boundary restrictions imposed above. Additionally, we will use the normal mode analysis in order to obtain the system of equations

$$\sum_{n=1}^{4} R_n = 0, \tag{68}$$

$$\sum_{n=1}^{4} H_{2n} R_n = \theta_0, \tag{69}$$

$$\sum_{n=1}^{4} k_n H_{3n} R_n = 0, \tag{70}$$

$$\sum_{n=1}^{4} H_{1n} R_n = 0. \tag{71}$$

after suppressing the primes.

After applying the inverse of matrix method for the above equations, we get the values of the constants R_n ($n = 1, 2, 3, 4$), hence; we obtain the expressions of strain, magnetic intensity, temperature distribution and the change in the volume fraction field for the generalized thermoelastic medium with voids.

4. Special Cases

4.1. Pores Neglect

First, we will neglect the presence of the voids, that is, we have ($\alpha = \lambda_0 = \xi_1 = \omega_0 = m = \psi = 0$). While putting ($\alpha = \lambda_0 = \xi_1 = \omega_0 = m = \psi = 0$) in Equations (54)–(57), we get:

$$(r_1 D^2 - r_2) e^* - a_4 (D^2 - a^2) \theta^* + r_3 h^* = 0, \tag{72}$$

$$(r_5 D^2 - r_6) \theta^* + r_8 e^* = 0, \tag{73}$$

$$(D^2 - r_9) h^* + i \omega e^* = 0. \tag{74}$$

Eliminating e^*, θ^*, and h^* among Equations (72)–(74), we obtain the following sixth order differential Equation, which is satisfied by $e^*(y)$, $\theta^*(y)$ and $h^*(y)$

$$[D^6 - B_1 D^4 + B_2 D^2 - B_3] \{e^*(y), \theta^*(y), h^*(y)\} = 0, \tag{75}$$

where $s_{10} = \frac{1}{r_1 r_5}$, $B_2 = s_{10}[r_1 r_6 r_9 + r_2 r_5 r_9 + r_2 r_6 - r_8 a_4(r_9 + a^2) + i\omega r_3 r_5]$, $B_1 = s_{10}(r_1 r_9 r_5 + r_1 r_6 + r_2 r_5 - r_8 a_4)$, $B_3 = s_{10}(r_2 r_5 r_9 - r_8 a_4 r_9 a^2 + i\omega r_3 r_6)$.

The solutions of Equation (75) are

$$(e^*, h^*, \theta^*)(y) = \sum_{n=1}^{3} (1, H_{4n}, H_{5n}) R_n^* \exp(-\alpha_n^* y), \tag{76}$$

where α_n^{*2} ($n = 1, 2, 3$) are the roots of the characteristic equation of Equation (75) and $H_{4n} = \frac{-i\omega}{\alpha_n^{*2} - r_9}$, $H_{5n} = \frac{-r_8}{r_5 \alpha_n^{*2} - r_6}$.

The expressions for the strain, the induced magnetic field, and the temperature field in the generalized initially stressed the electro-magneto-thermoelastic half-space solid with voids are:

$$(e^*, h^*, \theta^*)(y) = \sum_{n=1}^{3} (1, H_{4n}, H_{5n}) R_n^* \exp(-\alpha_n^* y + i(a x - \omega t)). \tag{77}$$

We wish to determine the above coefficients R_n^*, ($n = 1, 2, 3$).

To this end, we will keep in mind the boundary conditions in Equations (64), (65), and (67), and we will use the method of inverse of the matrix, as following:

$$\begin{pmatrix} R_1^* \\ R_2^* \\ R_3^* \end{pmatrix} = \begin{pmatrix} 1 & 1 & 1 \\ H_{51} & H_{52} & H_{53} \\ H_{41} & H_{42} & H_{43} \end{pmatrix}^{-1} \begin{pmatrix} 0 \\ \theta_0 \\ 0 \end{pmatrix}. \tag{78}$$

4.2. Neglecting the Initial Stress

By taking $(L^* = 0)$ in the governing equations, the corresponding expressions of the physical variables can be obtained without initial stress.

5. Numerical Results and Discussion

For numerical computations, following Dhaliwal and Singh [28], magnesium material was chosen for the purposes of numerical evaluations. All of the units of parameters that were used in the calculation are given in SI units. The constants were taken as $\lambda = 2.17 \times 10^{10}$ N/m², $\mu = 3.278 \times 10^{10}$ N/m², $k = 1.7 \times 10^2$ W/m·deg, $\rho = 1.74 \times 10^3$ Kg/m³, $\beta = 2.68 \times 10^6$ N/m²·deg, $C_e = 1.04 \times 10^3$ J/Kg·deg, $\omega_1^* = 3.58 \times 10^{11}$ /s, $\alpha_t = 1.78 \times 10^{-5}$ N/m² and $T_0 = 298$ K.

The voids parameters are $\psi = 1.753 \times 10^{-15}$ m², $\alpha = 3.688 \times 10^{-5}$ N, $\xi_1 = 1.475 \times 10^{10}$ N/m², $\lambda_0 = 1.13849 \times 10^{10}$ N/m², $m = 2 \times 10^6$ N/m²·deg and $\omega_0 = 0.0787 \times 10^{-3}$ N/m²s.

The Magnetic field parameters are $\varepsilon_0 = 10^{-9}/(36\pi)$ F/M, $\mu_0 = 4\pi \times 10^{-7}$ H/M, $H_0 = 10^5$ A/M and $\sigma_0 = 9.36 \times 10^5$ Col²/Cal·cm·sec.

The comparisons were carried out for $a = 0.7$ m, $\theta_0 = 0.1$ k, $\omega = \chi_0 + i\chi_1$, $\chi_0 = 2$ rad/s, $\chi_1 = 0.09$ rad/s, $x = 0.2$ m, and $0 \le y \le 4$ m.

Since, we have $\exp(\omega t) = [\cos(\chi_1 t) + i \sin(\chi_1 t)] \exp(\chi_0 t)$, and for small values of time we can take $\omega = \chi_0$, which is a real constant.

The above comparisons have been made in the context of two (G-N) theories of type II and III, in three situations:

(i) whether we have an initial stress or not $[L^* = 0$ and 10^5 at $M_0 = 0.5$ and $H_0 = 10^5]$;
(ii) whether we have a Thomson effect or not $[M_0 = 0$ and 0.5 at $H_0 = 10^5$ and $L^* = 10^5]$;
(iii) whether we have some void parameters or not $[M_0 = 0.5, H_0 = 10^5$ and $L^* = 10^5]$.

Case i: In the Figures 1–4, we made the calculations for $t = 0.2$, at $x = 0.2$. The values of the deformation e, the values of the temperature θ, the electromagnetic field h, and the values of the voids function ϕ are graphically represented, for different values of y in some graphs of two-dimensional space. In these figures, we use the solid lines for the results in the case without initial stress for the (G-N) theory of type II. For the results in the case of the (G-N) theory of type II with initial stress, we have used the large dashes line. In the case without initial stress for the (G-N) theory of type III, we have used the small dashes. Finally, for the results in the case with initial stress for the (G-N) theory of type III, we have used the small dashes line with dot.

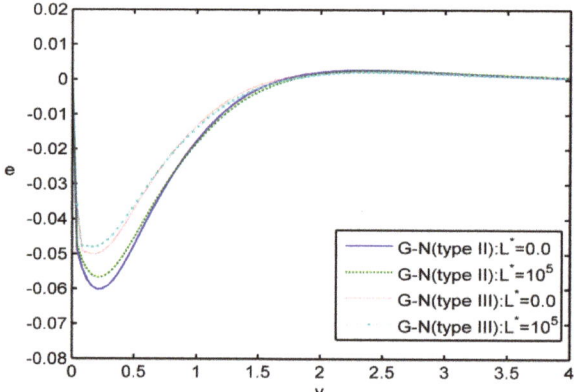

Figure 1. The strain e distribution at $M_0 = 0.5$ and $H_0 = 10^5$.

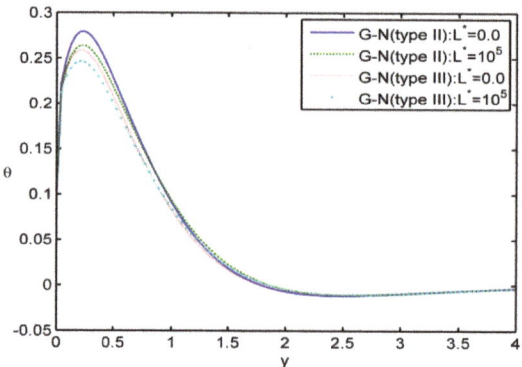

Figure 2. The temperature θ distribution at $M_0 = 0.5$ and $H_0 = 10^5$.

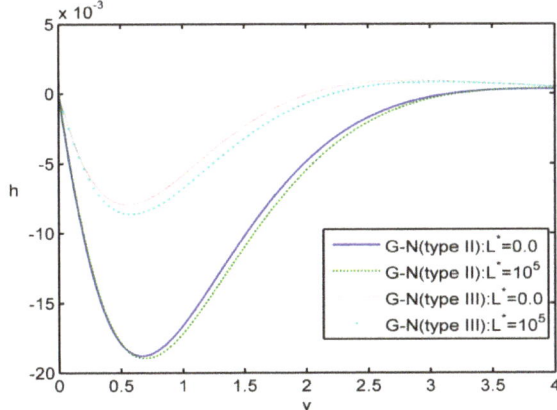

Figure 3. The induced magnetic field h distribution at $M_0 = 0.5$, and $H_0 = 10^5$.

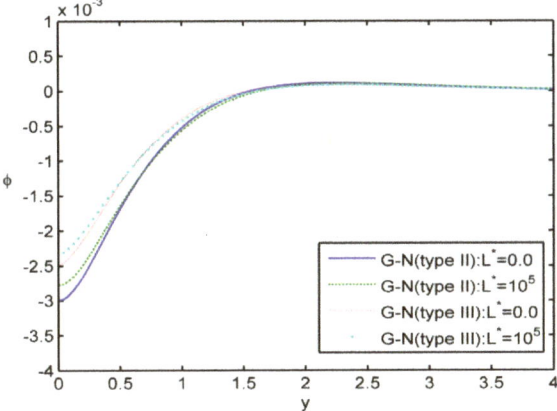

Figure 4. The change in the volume fraction field ϕ distribution at $M_0 = 0.5$ and $H_0 = 10^5$.

Figure 1 depicts the variation of the strain e versus y. The magnitude of the strain is found to be large for the G-N theory of type III. It can be seen that the initial stress shows an increasing effect on the magnitude of strain. In Figure 2, the parameter for the initial stress is decreasing as an effect. Additionally, when the initial stress parameter L^* is increasing, the value of the temperature θ is decreasing. Figure 3 shows the variation of the induced magnetic field h versus y. The value of strain is found to be large for the theory of G-N of type III. It can be seen that the presence of the initial stress shows a decreasing effect on the magnitude of the induced magnetic field. Figure 4 expresses the distribution of the change in the volume fraction field ϕ versus y. It was observed that the initial stress has a great effect on the distribution of ϕ.

Case ii: In the Figures 5–8, calculations were made for $M_0 = 0.0, 0.5$ at $H_0 = 10^5$, and $t = 0.2$. The strain e, the temperature θ, the electromagnetic field h and the voids function ϕ are graphically represented in some of the graphs for different values of y. Here, the solid lines is for results in the G-N theory of type II at $M_0 = 0.0$, which gives the classical Fourier's law of heat conduction, the large dashes line is for results in the type II G-N theory at $M_0 = 0.5$, which gives the generalized Fourier's law of heat conduction. We also use a small dashes line for the results for the type III G-N theory in the case $M_0 = 0.0$, while the line with small dashes and the dot is for results for the type III G-N theory for $M_0 = 0.5$. Figure 5 is for the effect of M_0 in the case that it exists and we can see that the value of the deformation e is increasing when M_0 increases with the corresponding difference. Figure 6 is for the value of the temperature θ, which is decreasing for the parameter M_0, which is increasing. In Figure 7, the effect of parameter M_0 exists and the value of the induced magnetic field h increases when the parameter M_0 increases. In Figure 8, the value of the voids function ϕ is increasing for the case that the parameter M_0 is increasing.

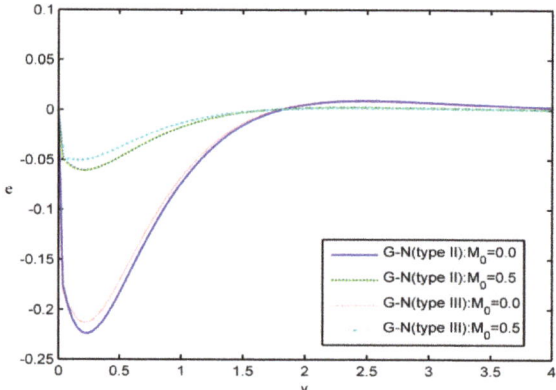

Figure 5. The strain e distribution at $H_0 = 10^5$ and $L^* = 10^5$.

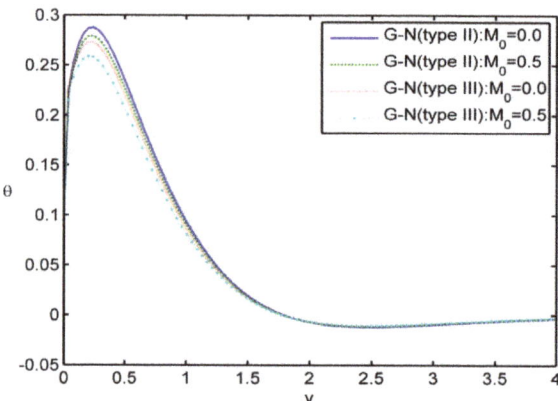

Figure 6. The temperature θ distribution at $H_0 = 10^5$ and $L^* = 10^5$.

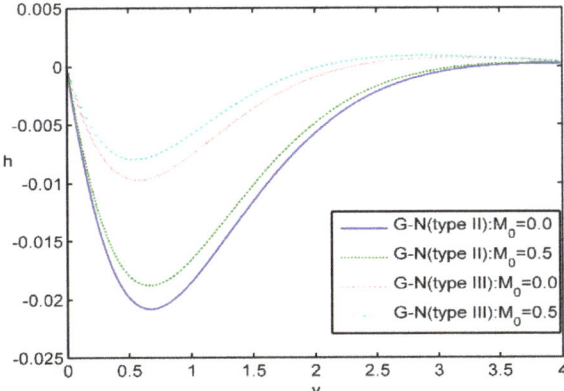

Figure 7. The induced magnetic field h distribution at $H_0 = 10^5$ and $L^* = 10^5$.

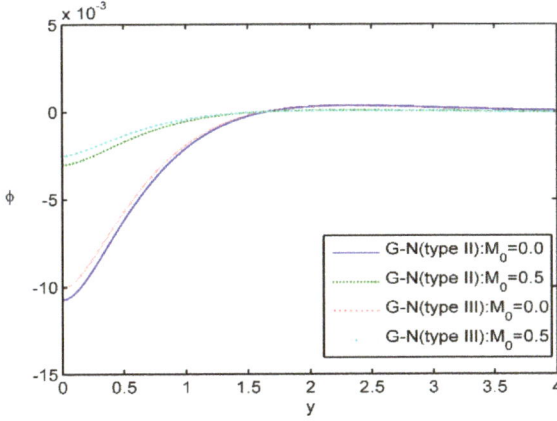

Figure 8. The change in the volume fraction field ϕ distribution at $H_0 = 10^5$ and $L^* = 10^5$.

Case iii: Figures 9–11 present the evolution of the physical sizes with regards to the distance y in 2D during $M_0 = 0.5$, $H_0 = 10^5$, $L^* = 10^5$, and with and without void parameter, in the context of G-N theory of type II and type III G-N theories. In this case of comparison, the solid lines is for results of pore effect 1.9 in the type II G-N theory, the small dashes line is also the voids effect, but in the case of the type III G-N theory. We use a large dashes line for results in the G-N theory of type II by neglecting the effect of pores, while the small dashes line with dot is used for results in the type III G-N theory, neglecting the effect of pores. In Figure 9, we find the repartition of the strain e, and we have a comparison between the values of the strain in the case of the presence of the pores to those in the case of neglecting the voids, in the range $0 \leq y \leq 1.7$; while, the values are the same for two cases at $y \geq 1.7$. Figure 10 illustrates the repartition of the temperature θ, and a comparison between the temperature in the case of presence of pores to those in the case of neglecting the voids, for y in the range $0 < y < 1.9$. In the case $y > 1.9$, the values are the same for two cases. Figure 11 depicts the repartition of the magnetic field h, an the values of the magnetic field h in the case of the presence of pores are compared to those in the case of neglecting the voids, for y in the range $0 \leq y \leq 3$; in the case $y \geq 3$. The values are the same for the two cases.

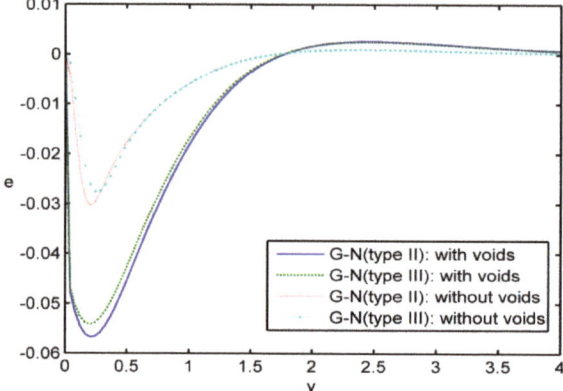

Figure 9. The strain e distribution at $H_0 = 10^5$, $M_0 = 0.5$ and $L^* = 10^5$.

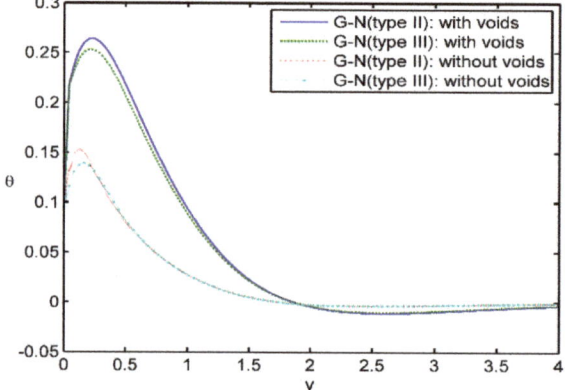

Figure 10. The temperature θ distribution at $H_0 = 10^5$, $M_0 = 0.5$, and $L^* = 10^5$.

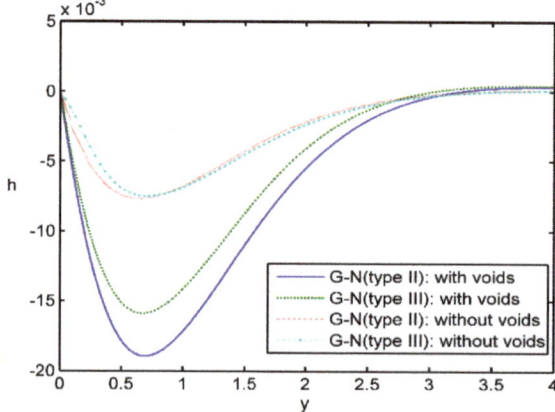

Figure 11. The induced magnetic field h distribution at $H_0 = 10^5$, $M_0 = 0.5$, and $L^* = 10^5$.

Figures 12–15 are giving 3D surface curves for the physical quantities i.e., the strain e, the temperature θ, the magnetic field h, and the voids fuction ϕ for the thermoelastic theory of electromagnetic bodies with pores, by taking into account the Thomson effect and the effect of the initial stress. The importance of these figures is that they give the dependence of the above physical sizes regarding the vertical component of distance.

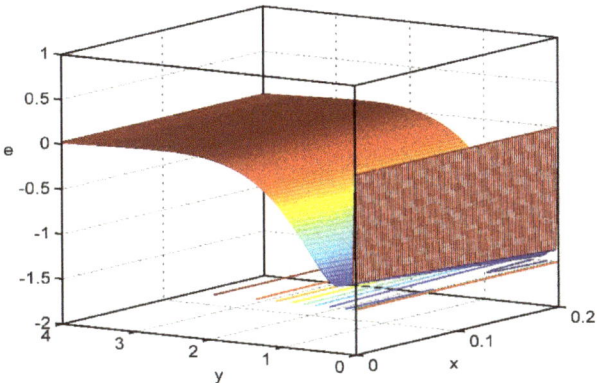

Figure 12. Three-dimensional (3D) curve distribution of the strain e versus the distances at: $M_0 = 0.5$, $H_0 = 10^5$, and $L^* = 10^5$.

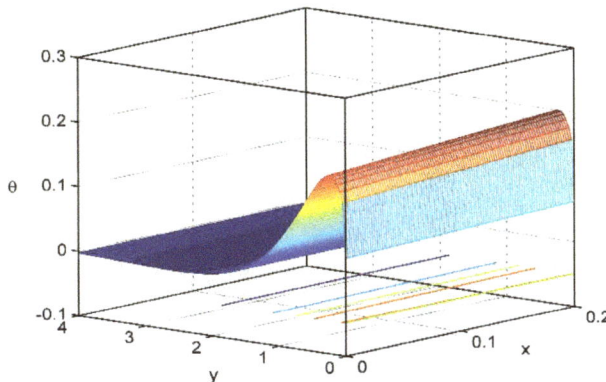

Figure 13. 3D Curve distribution of the temperature θ versus the distances, at: $M_0 = 0.5$, $H_0 = 10^5$ and $L^* = 10^5$.

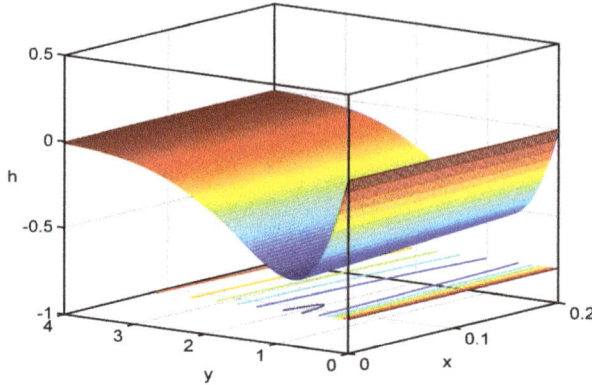

Figure 14. 3D Curve distribution of the induced magnetic field h versus the distances, at: $M_0 = 0.5$, $H_0 = 10^5$ and $L^* = 10^5$.

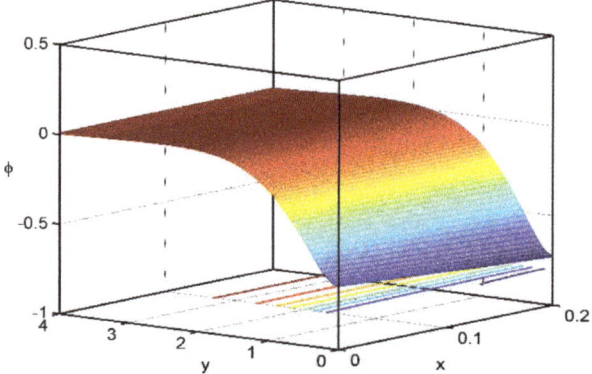

Figure 15. 3D Curve distribution of the change in the volume fraction field ϕ versus the distances, at: $M_0 = 0.5$, $H_0 = 10^5$ and $L^* = 10^5$.

6. Conclusions

The results concluded from the above analysis can be summarized, as follows:

(1) We have derived the field equations of homogeneous, isotropic, electro-magneto-thermo- porous elastic half-plane with the Thomson effect and initial stress.
(2) The analytical solutions that are based upon normal mode analysis for the thermoelastic problem in solids have been developed and utilized.
(3) The presence of initial stress, void parameters, and Thomson effect play significant roles in all of the physical quantities.
(4) The value of all physical quantities converges to zero with the increase in distance y and all of the functions are continuous.
(5) The deformation of a body depends on the nature of the applied forces and Thomson effect, as well as the type of boundary conditions.

Author Contributions: All three authors conceived the framework and structured the whole paper, checked the results of the paper and completed the revision of the article. The authors have equally contributed to the elaboration of this manuscript. All authors have read and approved the final manuscript.

Funding: This research received no external funding.

Acknowledgments: The authors would like to thank the anonymous referee for the comments that helped us improve this article.

Conflicts of Interest: The authors declare no conflict of interest.

References

1. Green, A.E.; Naghdi, P.M. A re-examination of the basic postulates of thermo-mechanics. *Proc. R. Soc. Lond. A* **1991**, *432*, 171–194. [CrossRef]
2. Green, A.E.; Naghdi, P.M. On undamped heat wave in elastic solids. *J. Therm. Stresses* **1992**, *15*, 253–264. [CrossRef]
3. Green, A.E.; Naghdi, P.M. Thermoelasticity without energy dissipation. *J. Elast.* **1993**, *31*, 189–209. [CrossRef]
4. Sharma, J.N.; Chouhan, R.N. On the problems of body forces and heat sources in thermoelasticity without energy dissipation, Indian. *J. Pure Appl. Math.* **1999**, *30*, 595–610.
5. Liang, M.; Liu, Y.; Xiao, B.; Yang, S.; Wang, Z.; Han, H. An analytical model for the transverse permeability of gas diffusion layer with electrical double layer effects in proton exchange membrane fuel cells. *Int. J. Hydrogen Energy* **2018**, *43*, 17880–17888. [CrossRef]
6. Xiao, B.; Zhang, X.; Wang, W.; Long, G.; Chen, H.; Kang, H.; Ren, W. Afractal model for water flow through unsaturated porous rocks. *Fractals* **2018**, *26*, 1840015. [CrossRef]
7. Hassan, M.; Marin, M.; Alsharif, A.; Ellahi, R. Convective heat transfer flow of nanofluid in a porous medium over wavy surface. *Phys. Lett. A Gen. At. Solid State Phys.* **2018**, *382*, 2749–2753. [CrossRef]
8. Long, G.; Liu, S.; Xu, G.; Won, S.W. A perforation-erosion model for hydraulic-fracturing applications. *SPE Prod. Oper.* **2018**, *33*, 770–783. [CrossRef]
9. Xiao, B.; Chen, H.; Xiao, S.; Cai, J. Research on relative permeability of nanofibers with capillary pressure effect by means of fractal-monte Carlo technique. *J. Nanosci. Nanotechnol.* **2017**, *17*, 6811–6817. [CrossRef]
10. Xiao, B.; Wang, W.; Fan, J.; Chen, H.; Hu, X.; Zhao, D.; Zhang, X.; Ren, W. Optimization of the fractal-like architecture of porous fibrous materials related to permeability, diffusivity and thermal conductivity. *Fractals* **2017**, *25*, 1750030. [CrossRef]
11. Nunziato, J.W.; Cowin, S.C. A non-linear theory of elastic materials with voids. *Arch. Rat. Mech. Anal.* **1979**, *72*, 175–201. [CrossRef]
12. Cowin, S.C.; Nunziato, J.W. Linear theory of elastic materials with voids. *J. Elast.* **1983**, *13*, 125–147. [CrossRef]
13. Puri, P.; Cowin, S.C. Plane waves in linear elastic materials with voids. *J. Elast.* **1985**, *15*, 167–183. [CrossRef]
14. Iesan, D. A theory of thermoelastic materials with voids. *Acta Mech.* **1986**, *60*, 67–89. [CrossRef]
15. Cicco, S.D.; Diaco, M. A theory of thermoelastic material with voids without energy dissipation. *J. Therm. Stress.* **2002**, *25*, 493–503. [CrossRef]
16. Othman, M.I.A.; Edeeb, E.R.M. The effect of rotation on thermoelastic medium with voids and temperature dependent under three theories. *J. Eng. Mech.* **2018**, *144*, 04018003. [CrossRef]
17. Montanaro, A. On singular surfaces in isotropic linear thermoelasticity with initial stress. *J. Acoust. Soc. Am.* **1999**, *106*, 1586–1588. [CrossRef]
18. Othman, M.I.A.; Abd-Elaziz, E.M. The effect of thermal loading due to laser pulse on generalized thermoelastic medium with voids in dual phase lag model. *J. Therm. Stress.* **2015**, *38*, 1068–1082. [CrossRef]
19. Marin, M. Cesaro means in thermoelasticity of dipolar bodies. *Acta Mechanica* **1997**, *122*, 155–168. [CrossRef]
20. Marin, M.; Oechsner, A. The effect of a dipolar structure on the Holder stability in Green-Naghdi thermoelasticity. *Contin Mech Thermodyn* **2017**, *29*, 1365–1374. [CrossRef]
21. Nowinski, J. *Theory of Thermoelasticity with Applications*; Sijthoff and Noordhoff International Publishers: Alphen Aan Den Rijn, The Netherlands, 1978.
22. Chadwick, P. *Progress in Solid Mechanics*; Hill, R., Sneddon, I.N., Eds.; North Holland Publishing Company: North Holland, Amsterdam, 1960.
23. Hussain, F.; Ellahi, R.; Zeeshan, A. Mathematical models of electromagnetohydrodynamic multiphase flows synthesis with nanosized hafnium particles. *Appl. Sci.* **2018**, *8*, 275. [CrossRef]
24. Bhatti, M.M.; Zeeshan, A.; Ellahi, R.; Ijaz, N. Heat and mass transfer of two-phase flow with Electric double layer effects induced due to peristaltic propulsion in the presence of transverse magnetic field. *J. Mol. Liq.* **2017**, *230*, 237–246. [CrossRef]

25. Othman, M.I.A.; Abd-Elaziz, E.M. Plane waves in a magneto-thermoelastic solids with voids and microtemperatures due to hall current and rotation. *Results Phys.* **2017**, *7*, 4253–4263. [CrossRef]
26. Marin, M.; Craciun, E.M.; Pop, N. Consideration on mixed initial-boundary value problems for micro-polarporous bodies. *Dyn. Syst. Appl.* **2016**, *25*, 175–196.
27. Marin, M.; Craciun, E.M. Uniqueness results for a boundary value problem in dipolar thermoelasticity to model composite materials. *Compos. Part B Eng.* **2017**, *126*, 27–37. [CrossRef]
28. Dhaliwal, R.S.; Singh, A. *Dynamic Coupled Thermoelasticity*; Hindustan Publ. Corp.: New Delhi, India, 1980.

© 2019 by the authors. Licensee MDPI, Basel, Switzerland. This article is an open access article distributed under the terms and conditions of the Creative Commons Attribution (CC BY) license (http://creativecommons.org/licenses/by/4.0/).

Article

New Exact Solutions of the Generalized Benjamin–Bona–Mahony Equation

Behzad Ghanbari [1,*], Dumitru Baleanu [2,3] and Maysaa Al Qurashi [4]

1. Department of Engineering Science, Kermanshah University of Technology, Kermanshah 6713954658, Iran
2. Department of Mathematics, Faculty of Arts and Sciences, Cankaya University, Ankara 06530, Turkey; dumitru@cankaya.edu.tr
3. Institute of Space Sciences, P.O. Box, MG-23, R 76900 Magurele-Bucharest, Romania
4. Department of Mathematics, College of Science, King Saud University, Riyadh 11495, Saudi Arabia; maysaa@ksu.edu.sa
* Correspondence: b.ghanbary@yahoo.com

Received: 5 December 2018; Accepted: 25 December 2018; Published: 27 December 2018

Abstract: The recently introduced technique, namely the generalized exponential rational function method, is applied to acquire some new exact optical solitons for the generalized Benjamin–Bona–Mahony (GBBM) equation. Appropriately, we obtain many families of solutions for the considered equation. To better understand of the physical features of solutions, some physical interpretations of solutions are also included. We examined the symmetries of obtained solitary waves solutions through figures. It is concluded that our approach is very efficient and powerful for integrating different nonlinear pdes. All symbolic computations are performed in Maple package.

Keywords: exact solutions; the generalized Benjamin–Bona–Mahony equation; generalized exponential rational function method; solitary wave solutions; symbolic computation

1. Introduction

The Benjamin–Bona–Mahony (BBM) equation has been studied by Benjamin, Bona, and Mahony in 1972 as the improved KdV equation for the description of long surface gravity waves having a small amplitude. They have also investigated the stability and uniqueness of solutions to the BBM equation [1]. The description of the drift of waves in plasma physics, the propagation of wave in semi-conductors and optical devices [2], and the behavior of Rossby waves in rotating fluids [3] are some other phenomena that are modeled by this equation.

Let us consider the dimensionless form of the (1 + 1) the generalized Benjamin–Bona–Mahony (GBBM) equation as follows [4]:

$$u_t + \alpha u_x + (\beta u^\vartheta + \gamma u^{2\vartheta})u_x - \delta u_{xxt} = 0, \qquad (1)$$

with the unknown function u and the constants of α, δ, n, β and ϑ.

It is also known that the multi-soliton solutions for the Equation (1) only exist with the conditions $\gamma = 0$ and $\vartheta = 1$, i.e., whenever we have

$$u_t + \alpha u_x + (\beta u)u_x - \delta u_{xxt} = 0. \qquad (2)$$

The main application of the BBM equation is related to model the hydromagnetic waves in cold plasma, the acoustic waves in anharmonic crystals and the acoustic-gravity waves in compressible fluids [5,6].

Recently, the investigation of exact solutions of nonlinear PDEs has begun to attract mathematicians and physicists' attention because of the onset of soliton [7–9]. Therefore, several efficient techniques for handling NPDEs have been developed. Among them, we can list the traditional methods: the Hirotas bilinear method [10] and the Darboux transformation method [11]. There are also some recent direct and algebraic methods: the variational iteration method [12], the exp-function method [13], various extended tanh-function methods [14] and Lie symmetry analysis [15–17].

The relatively new technique called the generalized exponential rational function method (or GERFM in short) was firstly suggested by Ghanbari et al. in Ref. [18] to solve the resonance nonlinear Schrödinger equation as

$$i\psi_t + \alpha \psi_{xx} + \beta \mathcal{F}\left(|\psi|^2\right)\psi + \gamma \left(\frac{|\psi|_{xx}}{|\psi|}\right)\psi = 0, i = \sqrt{-1}. \quad (3)$$

Another application of the method has been carried out in Ref. [19] where the authors have implemented the method to solve the Fokas–Lenells equation in the presence of the perturbation terms, as follows:

$$i\frac{\partial \psi}{\partial t} + \frac{\partial^2 \psi}{\partial x^2} + \alpha |\psi|^2 \psi + i\left[\gamma_1 \frac{\partial \psi^3}{\partial x^3} + \gamma_2 \frac{\partial \psi}{\partial x}|\psi|^2 + \gamma_3 \frac{\partial |\psi|^2}{\partial x}\psi\right] = 0. \quad (4)$$

The method also has been successfully implemented to retrieve traveling wave solutions to the nonlinear Schrödinger's equation in the presence of Hamiltonian perturbations [20] as

$$i\psi_t + a\psi_{xx} + \left(b_1|\psi| + b_2|\psi|^2\right)\psi = i\{\alpha \psi_x + \lambda \left(|\psi|^2 \psi\right)_x + \theta \left(|\psi|^2\right)_x \psi\}. \quad (5)$$

In all cases, the authors have declared that the method introduces some new solutions that have not been reported in previous works. In addition, it deduces that the method can be applied to study many other nonlinear PDEs in many branches of physics, biology, engineering. This research aims to integrate the GBBM Equation (1) using the GERFM. For this reason, our paper is organized as below: Section 2 deals with the presentation of the method. Section 3 is devoted to the application of GERFM to the GBBM equation. Eventually, the conclusion of the present research is outlined in the last section.

2. The Main Steps of GERFM

In this subsection, we review the routine description of GERFM.

1. Let us take into account the nonlinear PDE in the form:

$$\mathcal{L}(\psi, \psi_x, \psi_t, \psi_{xx}, \ldots) = 0. \quad (6)$$

Using the transformations $\psi = \psi(\eta)$ and $\eta = \sigma x - lt$, Equation (6) is reduced to following NODE as:

$$\mathcal{L}(\psi, \psi', \psi'', \ldots) = 0, \quad (7)$$

where the values of σ and l will be found later.

2. Now, the structure of the wave solution of Equation (7) is assumed to be

$$\psi(\eta) = p_0 + \sum_{k=1}^{M} p_k \Theta(\eta)^k + \sum_{k=1}^{M} \frac{q_k}{\Theta(\eta)^k}, \quad (8)$$

where

$$\Theta(\eta) = \frac{\iota_1 e^{\kappa_1 \eta} + \iota_2 e^{\kappa_2 \eta}}{\iota_3 e^{\kappa_3 \eta} + \iota_4 e^{\kappa_4 \eta}}. \quad (9)$$

The values of constants $\iota_i, \kappa_i (1 \leq i \leq 4)$, p_0, p_k and $q_k (1 \leq k \leq M)$ are determined, in such a way that solution (8) always persuade Equation (7). By considering the homogenous balance principle, the value of M is determined.

3. Substituting Equation (8) into Equation (7), an algebraic equation $P(S_1, S_2, S_3, S_4) = 0$ in terms of $S_i = e^{q_i \eta}$ for $i = 1, \ldots, 4$ is constructed. Then, making each coefficient for the powers of P to zero, we acquire a series of nonlinear equations in terms of $p_i, q_i (1 \leq i \leq 4)$, and σ, l, p_0, p_k and $q_k (1 \leq k \leq M)$ is generated.
4. By solving the above system of equations using any computer package like Maple (18, Waterloo Maple, Canada), the values of $\iota_i, \kappa_i (1 \leq i \leq 4)$, p_0, p_k, and $q_k (1 \leq k \leq M)$ are determined, replacing these values in Equation (8) provides us the exact solutions of the nonlinear PDE (6).

3. Utilization of GERFM for the GBBM Equation

Let us consider the following dependent variable transformation

$$u(x,t) = \psi(\eta), \quad \eta = kx - \theta t, \tag{10}$$

where k and θ are constants need to be calculated. Under the transformation of Equation (10), Equation (1) can be reduced to the following NODE:

$$(\alpha k - \theta) \psi_\eta + k \left(\beta \psi^\vartheta + \gamma \psi^{2\vartheta} \right) \psi_\eta + \delta k^2 \theta \psi_{\eta\eta\eta} = 0. \tag{11}$$

We may now integrate Equation (11) to have

$$(\alpha k - \theta) \psi + k \left(\frac{\beta}{\vartheta + 1} \psi^{\vartheta+1} + \frac{\gamma}{2\vartheta + 1} \psi^{2\vartheta+1} \right) + \delta k^2 \theta \psi_{\eta\eta} = 0. \tag{12}$$

Using the transformation $\psi = \psi^\vartheta$ in (12), we attain

$$\psi \psi_{\eta\eta} + \sigma_1 \psi^2 + \sigma_2 \psi^3 + \sigma_3 \psi^4 + \sigma_4 \left(\psi_\eta \right)^2 = 0, \tag{13}$$

where

$$\sigma_1 = \frac{(\alpha k - \theta) \vartheta}{\delta k^2 \theta},$$
$$\sigma_2 = \frac{\beta \vartheta}{\delta k (\vartheta + 1) \theta},$$
$$\sigma_3 = \frac{\gamma \vartheta}{\delta k (2\vartheta + 1) \theta},$$
$$\sigma_4 = \frac{1 - \vartheta}{\vartheta}.$$

In this section, GERFM will be used to determine solitary wave solutions of (1). To this end, if we apply the balancing principle for the terms of ψ^4 and $(\psi_\eta)^2$ in (13), (i.e., $4M = 2(M+1)$), we get $M = 1$. This implies that Equation (1) has the solution given by

$$\psi(\eta) = p_0 + p_1 \Theta(\eta) + \frac{q_1}{\Theta(\eta)}. \tag{14}$$

We now exert the GERFM to derive the following categories of solutions for Equation (1):

Family 1: We attain $\iota = [1,1,1,-1]$ and $\kappa = [1,-1,1,-1]$, so we will obtain

$$\Theta(\eta) = \frac{-3 - 2e^\eta}{1 + e^\eta}. \tag{15}$$

Case 1:

$$\theta = -\frac{\sqrt{2\vartheta + 1} \sqrt{(\gamma \vartheta + \gamma)(\vartheta + 2)^2 \alpha - (2\vartheta + 1) \beta^2 \vartheta \beta}}{\sqrt{-\delta}(\gamma \vartheta + \gamma)(\vartheta + 2)^2},$$

$$k = -\frac{\sqrt{2\vartheta + 1}\vartheta\beta}{\sqrt{-\delta}\sqrt{(\gamma\vartheta + \gamma)(\vartheta + 2)^2 \alpha - (2\vartheta + 1)\beta^2}},$$

$$p_0 = \frac{(-6\vartheta - 3)\beta}{\gamma(\vartheta + 2)}, p_1 = 0, q_1 = \frac{(-12n - 6)\beta}{\gamma(\vartheta + 2)}.$$

These resulting values direct us to have

$$\psi(\eta) = \frac{(-6\vartheta - 3)\beta}{\gamma(\vartheta + 2)(3 + 2e^\eta)}.$$

Consequently, we can get the following exact wave solution

$$u_1(x,t) = \left(\frac{(-6\vartheta - 3)\beta}{\gamma(\vartheta + 2)(3 + 2e^\eta)}\right)^{\frac{1}{\vartheta}}, \tag{16}$$

where

$$\eta = -\frac{\sqrt{2\vartheta + 1}\vartheta\beta}{\sqrt{-\delta}\sqrt{(\gamma\vartheta + \gamma)(\vartheta + 2)^2 \alpha - (2\vartheta + 1)\beta^2}}x +$$

$$\frac{\sqrt{2\vartheta + 1}\sqrt{(\gamma\vartheta + \gamma)(\vartheta + 2)^2 \alpha - (2\vartheta + 1)\beta^2}\vartheta\beta}{\sqrt{-\delta}(\gamma\vartheta + \gamma)(\vartheta + 2)^2}t.$$

Case 2:

$$\theta = -\frac{\sqrt{2\vartheta + 1}\sqrt{25(\gamma\vartheta + \gamma)(\vartheta + 2)^2 \alpha - (2\vartheta + 1)\beta^2}\vartheta\beta}{25\sqrt{-\delta}(\gamma\vartheta + \gamma)(\vartheta + 2)^2},$$

$$k = -\frac{\sqrt{2\vartheta + 1}\vartheta\beta}{\sqrt{-\delta}\sqrt{25(\gamma\vartheta + \gamma)(\vartheta + 2)^2 \alpha - (2\vartheta + 1)\beta^2}},$$

$$p_0 = \frac{(-2n - 1)\beta}{\gamma(\vartheta + 2)}, p_1 = -\frac{\beta(2\vartheta + 1)}{5\gamma(\vartheta + 2)}, q_1 = -\frac{6\beta(2\vartheta + 1)}{5\gamma(\vartheta + 2)}.$$

led

$$\psi(\eta) = \frac{(-2\vartheta - 1)\beta e^\eta}{(\vartheta + 2)(1 + e^\eta)\gamma(15 + 10e^\eta)}.$$

Hence, we get the following solitary wave solution for GBBM as

$$u_2(x,t) = \left(\frac{(-2\vartheta - 1)\beta e^\eta}{(\vartheta + 2)(1 + e^\eta)\gamma(15 + 10e^\eta)}\right)^{\frac{1}{\vartheta}}, \tag{17}$$

where

$$\eta = -\frac{\sqrt{2\vartheta + 1}\vartheta\beta}{\sqrt{-\delta}\sqrt{25(\gamma\vartheta + \gamma)(\vartheta + 2)^2 \alpha - (2\vartheta + 1)\beta^2}}x +$$

$$\frac{\sqrt{2\vartheta + 1}\sqrt{25(\gamma\vartheta + \gamma)(\vartheta + 2)^2 \alpha - (2\vartheta + 1)\beta^2}\vartheta\beta}{25\sqrt{-\delta}(\gamma\vartheta + \gamma)(\vartheta + 2)^2}t.$$

Family 2: We attain $\iota = [1,1,1,-1]$ and $\kappa = [1,-1,1,-1]$, so we will obtain

$$\Theta(\eta) = -\frac{\sin(\eta)}{\cos(\eta)}. \tag{18}$$

Case 1:
$$\theta = \frac{\sqrt{2\vartheta+1}\sqrt{(\gamma\vartheta+\gamma)(\vartheta+2)^2\alpha - (2\vartheta+1)\beta^2\vartheta\beta}}{2\sqrt{\delta}(\gamma\vartheta+\gamma)(\vartheta+2)^2},$$

$$k = \frac{\sqrt{2\vartheta+1}\vartheta\beta}{2\sqrt{\delta}\sqrt{(\gamma\vartheta+\gamma)(\vartheta+2)^2\alpha - (2\vartheta+1)\beta^2}},$$

$$p_0 = -\frac{\beta(2\vartheta+1)}{2\gamma(\vartheta+2)}, p_1 = 0, q_1 = \frac{i(2\vartheta+1)\beta}{2\gamma(\vartheta+2)}.$$

These resulting values help us to have

$$\psi(\eta) = -\frac{k\sinh(\eta)}{\sqrt{2\gamma^2 k^2 - \gamma^2 q^2 + 2q\gamma - 1}\sqrt{\vartheta}\cosh(\eta)}.$$

Consequently, the following exact wave solution is determined

$$u_3(x,t) = \left(-\frac{k\sinh(\eta)}{\sqrt{2\gamma^2 k^2 - \gamma^2 q^2 + 2q\gamma - 1}\sqrt{\vartheta}\cosh(\eta)}\right)^{\frac{1}{\vartheta}}, \quad (19)$$

where

$$\eta = \frac{\sqrt{2\vartheta+1}\vartheta\beta}{2\sqrt{\delta}\sqrt{(\gamma\vartheta+\gamma)(\vartheta+2)^2\alpha - (2\vartheta+1)\beta^2}}x -$$

$$\frac{\sqrt{2\vartheta+1}\sqrt{(\gamma\vartheta+\gamma)(\vartheta+2)^2\alpha - (2\vartheta+1)\beta^2\vartheta\beta}}{2\sqrt{\delta}(\gamma\vartheta+\gamma)(\vartheta+2)^2}t.$$

Family 3: We attain $\iota = [1,1,1,-1]$ and $\kappa = [1,-1,1,-1]$, so we will obtain

$$\Theta(\eta) = \frac{\sin(\eta) + \cos(\eta)}{\cos(\eta)}. \quad (20)$$

Case 1:
$$\theta = -\frac{\sqrt{2\vartheta+1}\sqrt{(\gamma\vartheta+\gamma)(\vartheta+2)^2\alpha + (2\vartheta+1)\beta^2\vartheta\beta}}{\sqrt{-\delta}(\gamma\vartheta+\gamma)(\vartheta+2)^2},$$

$$k = -\frac{\sqrt{2\vartheta+1}\vartheta\beta}{\sqrt{-\delta}\sqrt{(\gamma\vartheta+\gamma)(\vartheta+2)^2\alpha + (2\vartheta+1)\beta^2}},$$

$$p_0 = \frac{(-2\vartheta-1)\beta}{\gamma(\vartheta+2)}, p_1 = \frac{\beta(2\vartheta+1)}{2\gamma(\vartheta+2)}, q_1 = \frac{\beta(2\vartheta+1)}{\gamma(\vartheta+2)}.$$

These resulting values led us to obtain

$$\psi(\eta) = \frac{\beta(2\vartheta+1)}{2(\vartheta+2)\gamma\cos(\eta)(\sin(\eta)+\cos(\eta))}.$$

Hence, one arrives to the following exact wave solution:

$$u_4(x,t) = \left(\frac{\beta(2\vartheta+1)}{2(\vartheta+2)\gamma\cos(\eta)(\sin(\eta)+\cos(\eta))}\right)^{\frac{1}{\vartheta}}, \quad (21)$$

where
$$\eta = -\frac{\sqrt{2\vartheta+1}\vartheta\beta}{\sqrt{-\delta}\sqrt{(\gamma\vartheta+\gamma)(\vartheta+2)^2\alpha+(2\vartheta+1)\beta^2}}x+$$

$$\frac{\sqrt{2\vartheta+1}\sqrt{(\gamma\vartheta+\gamma)(\vartheta+2)^2\alpha+(2\vartheta+1)\beta^2}\vartheta\beta}{\sqrt{-\delta}(\gamma\vartheta+\gamma)(\vartheta+2)^2}t.$$

Family 4: We attain $\iota = [1,1,1,-1]$ and $\kappa = [1,-1,1,-1]$, so we will obtain

$$\Theta(\eta) = \frac{-\sin(\eta)+\cos(\eta)}{\sin(\eta)}. \tag{22}$$

Case 1:

$$\theta = -\frac{\sqrt{2\vartheta+1}\sqrt{(\gamma\vartheta+\gamma)(\vartheta+2)^2\alpha+(2\vartheta+1)\beta^2}\vartheta\beta}{2\sqrt{-\delta}(\gamma\vartheta+\gamma)(\vartheta+2)^2},$$

$$k = -\frac{\sqrt{2\vartheta+1}\vartheta\beta}{2\sqrt{-\delta}\sqrt{(\gamma\vartheta+\gamma)(\vartheta+2)^2\alpha+(2\vartheta+1)\beta^2}},$$

$$p_0 = \frac{(-2\vartheta-1)\beta}{\gamma(\vartheta+2)}, p_1 = -\frac{\beta(2\vartheta+1)}{2\gamma(\vartheta+2)}, q_1 = \frac{(-2\vartheta-1)\beta}{\gamma(\vartheta+2)}.$$

These solutions direct us to get

$$\psi(\eta) = \frac{\beta(2\vartheta+1)}{2\gamma(\vartheta+2)\sin(\eta)(\sin(\eta)-\cos(\eta))}.$$

As a result, we can get the following exact wave solution:

$$u_5(x,t) = \left(\frac{\beta(2\vartheta+1)}{2\gamma(\vartheta+2)\sin(\eta)(\sin(\eta)-\cos(\eta))}\right)^{\frac{1}{\vartheta}}, \tag{23}$$

where
$$\eta = -\frac{\sqrt{2\vartheta+1}\vartheta\beta}{2\sqrt{-\delta}\sqrt{(\gamma\vartheta+\gamma)(\vartheta+2)^2\alpha+(2\vartheta+1)\beta^2}}x+$$

$$\frac{\sqrt{2\vartheta+1}\sqrt{(\gamma\vartheta+\gamma)(\vartheta+2)^2\alpha+(2\vartheta+1)\beta^2}\vartheta\beta}{2\sqrt{-\delta}(\gamma\vartheta+\gamma)(\vartheta+2)^2}t.$$

Family 5: We attain $\iota = [1,1,1,-1]$ and $\kappa = [1,-1,1,-1]$, so we will obtain

$$\Theta(\eta) = -\frac{1}{1+e^\eta}. \tag{24}$$

Case 1:

$$\theta = -\frac{\sqrt{2\vartheta+1}\sqrt{(\gamma\vartheta+\gamma)(\vartheta+2)^2\alpha-(2\vartheta+1)\beta^2}\vartheta\beta}{\sqrt{-\delta}(\gamma\vartheta+\gamma)(\vartheta+2)^2},$$

$$k = -\frac{\sqrt{2\vartheta+1}\vartheta\beta}{\sqrt{-\delta}\sqrt{(\gamma\vartheta+\gamma)(\vartheta+2)^2\alpha-(2\vartheta+1)\beta^2}},$$

$$p_0 = 0, p_1 = \frac{\beta(2\vartheta+1)}{\gamma(\vartheta+2)}, q_1 = 0.$$

Then, we arrived to
$$\psi(\eta) = \frac{(-2\vartheta - 1)\beta}{\gamma(\vartheta+2)(1+e^\eta)}.$$

Therefore, the following exact wave solution for the equation is achieved

$$u_6(x,t) = \left(\frac{(-2\vartheta-1)\beta}{\gamma(\vartheta+2)(1+e^\eta)}\right)^{\frac{1}{\vartheta}}, \qquad (25)$$

where
$$\eta = -\frac{\sqrt{2\vartheta+1}\vartheta\beta}{\sqrt{-\delta}\sqrt{(\gamma\vartheta+\gamma)(\vartheta+2)^2\alpha - (2\vartheta+1)\beta^2}}x +$$

$$\frac{\sqrt{2\vartheta+1}\sqrt{(\gamma\vartheta+\gamma)(\vartheta+2)^2\alpha - (2\vartheta+1)\beta^2}\vartheta\beta}{\sqrt{-\delta}(\gamma\vartheta+\gamma)(\vartheta+2)^2}t.$$

Family 6: We attain $\iota = [1,1,1,-1]$ and $\kappa = [1,-1,1,-1]$, so we will obtain

$$\Theta(\eta) = \frac{3e^\eta + 2}{1 + e^\eta}. \qquad (26)$$

Case 1:
$$\theta = -\frac{\sqrt{2\vartheta+1}\sqrt{(\gamma\vartheta+\gamma)(\vartheta+2)^2\alpha - (2\vartheta+1)\beta^2}\vartheta\beta}{\sqrt{-\delta}(\gamma\vartheta+\gamma)(\vartheta+2)^2},$$

$$k = -\frac{\sqrt{2\vartheta+1}\vartheta\beta}{\sqrt{-\delta}\sqrt{(\gamma\vartheta+\gamma)(\vartheta+2)^2\alpha - (2\vartheta+1)\beta^2}},$$

$$p_0 = \frac{(-4n-2)\beta}{\gamma(\vartheta+2)}, p_1 = 0, q_1 = \frac{(-4n-2)\beta}{\gamma(\vartheta+2)}.$$

These values let us to consider

$$\psi(\eta) = \frac{(-4n-2)\beta}{\gamma(\vartheta+2)(e^\eta+2)}.$$

Thus, we obtain

$$u_7(x,t) = \left(\frac{(-4n-2)\beta}{\gamma(\vartheta+2)(e^\eta+2)}\right)^{\frac{1}{\vartheta}}, \qquad (27)$$

where
$$\eta = -\frac{\sqrt{2\vartheta+1}\vartheta\beta}{\sqrt{-\delta}\sqrt{(\gamma\vartheta+\gamma)(\vartheta+2)^2\alpha - (2\vartheta+1)\beta^2}}x +$$

$$\frac{\sqrt{2\vartheta+1}\sqrt{(\gamma\vartheta+\gamma)(\vartheta+2)^2\alpha - (2\vartheta+1)\beta^2}\vartheta\beta}{\sqrt{-\delta}(\gamma\vartheta+\gamma)(\vartheta+2)^2}t.$$

Family 7: We attain $\iota = [1,1,1,-1]$ and $\kappa = [1,-1,1,-1]$, so we will obtain

$$\Theta(\eta) = \frac{-e^\eta - 2}{1 + e^\eta}. \qquad (28)$$

Case 1:
$$\theta = -\frac{\sqrt{2\vartheta+1}\sqrt{(\gamma\vartheta+\gamma)(\vartheta+2)^2\alpha - (2\vartheta+1)\beta^2}\vartheta\beta}{\sqrt{-\delta}(\gamma\vartheta+\gamma)(\vartheta+2)^2},$$

$$k = -\frac{\sqrt{2\vartheta+1}\vartheta\beta}{\sqrt{-\delta}\sqrt{(\gamma\vartheta+\gamma)(\vartheta+2)^2\alpha-(2\vartheta+1)\beta^2}},$$

$$p_0 = \frac{(-6\vartheta-3)\beta}{\gamma(\vartheta+2)}, p_1 = 0, q_1 = \frac{(12n+6)\beta}{\gamma(\vartheta+2)}.$$

These resulting values help us to consider

$$\psi(\eta) = \frac{(-6\vartheta-3)\beta e^\eta}{\gamma(\vartheta+2)(3e^\eta+2)}.$$

Accordingly, we can get the following exact wave solution

$$u_8(x,t) = \left(\frac{(-6\vartheta-3)\beta e^\eta}{\gamma(\vartheta+2)(3e^\eta+2)}\right)^{\frac{1}{\vartheta}}, \tag{29}$$

where

$$\eta = -\frac{\sqrt{2\vartheta+1}\vartheta\beta}{\sqrt{-\delta}\sqrt{(\gamma\vartheta+\gamma)(\vartheta+2)^2\alpha-(2\vartheta+1)\beta^2}}x+$$

$$\frac{\sqrt{2\vartheta+1}\sqrt{(\gamma\vartheta+\gamma)(\vartheta+2)^2\alpha-(2\vartheta+1)\beta^2}\vartheta\beta}{\sqrt{-\delta}(\gamma\vartheta+\gamma)(\vartheta+2)^2}t.$$

Family 8: We attain $\iota = [1,1,1,-1]$ and $\kappa = [1,-1,1,-1]$, so we will obtain

$$\Theta(\eta) = \frac{2e^\eta+1}{1+e^\eta}. \tag{30}$$

Case 1:

$$\theta = -\frac{\sqrt{2\vartheta+1}\sqrt{9(\gamma\vartheta+\gamma)(\vartheta+2)^2\alpha-(2\vartheta+1)\beta^2}\vartheta\beta}{9\sqrt{-\delta}(\gamma\vartheta+\gamma)(\vartheta+2)^2},$$

$$k = -\frac{\sqrt{2\vartheta+1}\vartheta\beta}{\sqrt{-\delta}\sqrt{9(\gamma\vartheta+\gamma)(\vartheta+2)^2\alpha-(2\vartheta+1)\beta^2}},$$

$$p_0 = \frac{(-2\vartheta-1)\beta}{\gamma(\vartheta+2)}, p_1 = \frac{\beta(2\vartheta+1)}{3\gamma(\vartheta+2)}, q_1 = \frac{2\beta(2\vartheta+1)}{3\gamma(\vartheta+2)}.$$

From these results, one has

$$\psi(\eta) = \frac{(-2\vartheta-1)\beta e^\eta}{3\gamma(\vartheta+2)(1+e^\eta)(2e^\eta+1)}.$$

Thus, we can get the following exact wave solution

$$u_9(x,t) = \left(\frac{(-2\vartheta-1)\beta e^\eta}{3\gamma(\vartheta+2)(1+e^\eta)(2e^\eta+1)}\right)^{\frac{1}{\vartheta}}, \tag{31}$$

where

$$\eta = -\frac{\sqrt{2\vartheta+1}\vartheta\beta}{\sqrt{-\delta}\sqrt{9(\gamma\vartheta+\gamma)(\vartheta+2)^2\alpha-(2\vartheta+1)\beta^2}}x+$$

$$\frac{\sqrt{2\vartheta+1}\sqrt{9(\gamma\vartheta+\gamma)(\vartheta+2)^2\alpha-(2\vartheta+1)\beta^2}\vartheta\beta}{9\sqrt{-\delta}(\gamma\vartheta+\gamma)(\vartheta+2)^2}t.$$

Family 9: We attain $\iota = [1, 1, 1, -1]$ and $\kappa = [1, -1, 1, -1]$, so we will obtain

$$\Theta(\eta) = \frac{\cos(\eta) - 2\sin(\eta)}{\sin(\eta)}. \tag{32}$$

Case 1:

$$\theta = -\frac{\sqrt{2\vartheta + 1}\sqrt{4(\gamma\vartheta + \gamma)(\vartheta + 2)^2 \alpha + (2\vartheta + 1)\beta^2 \vartheta \beta}}{8\sqrt{-\delta}(\gamma\vartheta + \gamma)(\vartheta + 2)^2},$$

$$k = -\frac{\sqrt{2\vartheta + 1}\vartheta \beta}{2\sqrt{-\delta}\sqrt{4(\gamma\vartheta + \gamma)(\vartheta + 2)^2 \alpha + (2\vartheta + 1)\beta^2}},$$

$$p_0 = \frac{(-2\vartheta - 1)\beta}{\gamma(\vartheta + 2)}, p_1 = -\frac{\beta(2\vartheta + 1)}{4\gamma(\vartheta + 2)}, q_1 = -\frac{5\beta(2\vartheta + 1)}{4\gamma(\vartheta + 2)}.$$

These results suggest us to have

$$\psi(\eta) = \frac{(2\vartheta + 1)\beta}{4(\vartheta + 2)\gamma \sin(\eta)(\cos(\eta) + 2\sin(\eta))}.$$

At this point, the following exact wave solution is formulated

$$u_{10}(x, t) = \left(\frac{(2\vartheta + 1)\beta}{4(\vartheta + 2)\gamma \sin(\eta)(\cos(\eta) + 2\sin(\eta))}\right)^{\frac{1}{\vartheta}}, \tag{33}$$

where

$$\eta = \frac{\sqrt{2\vartheta + 1}\vartheta \beta}{2\sqrt{-\delta}\sqrt{4(\gamma\vartheta + \gamma)(\vartheta + 2)^2 \alpha + (2\vartheta + 1)\beta^2}} x -$$

$$\frac{\sqrt{2\vartheta + 1}\sqrt{4(\gamma\vartheta + \gamma)(\vartheta + 2)^2 \alpha + (2\vartheta + 1)\beta^2 \vartheta \beta}}{8\sqrt{-\delta}(\gamma\vartheta + \gamma)(\vartheta + 2)^2} t.$$

To analyze the dynamic behavior of the obtained solutions, some three-dimensional figures have been depicted in some special cases. The moduli of s $u_3(x,t)$, $u_4(x,t)$, $u_6(x,t)$, $u_8(x,t)$, $u_9(x,t)$ and $u_{10}(x,t)$ are depicted in Figures 1–6, respectively. The analytical results and profiles obtained in this contribution provide us a different physical interpretation for the considered equation. As we observe, the absolute value of solutions displayed in Figure 1 is a bright solitary wave, in Figure 2 is a periodic wave, in Figure 3 is a kink solitary wave, in Figure 4 is a dark wave, in Figure 5 is a periodic wave soliton, and finally in Figure 6 is a singular periodic wave.

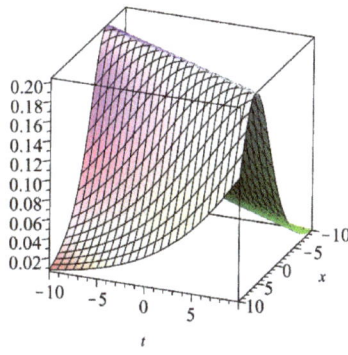

Figure 1. Perspective view of the modulus of $u_3(x,t)$ with $\alpha = 0.5, \beta = 2, \delta = 1, \gamma = -0.5$ and $\vartheta = 1.5$.

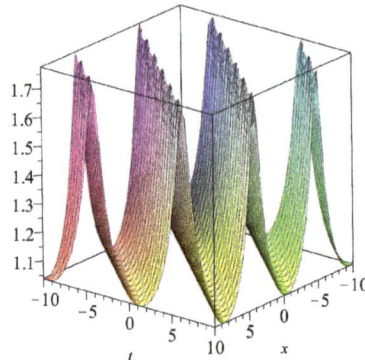

Figure 2. Perspective view of the modulus of $u_4(x,t)$ with $\alpha = 1, \beta = 1, \delta = 1, \gamma = 0.5$, and $\vartheta = 3$.

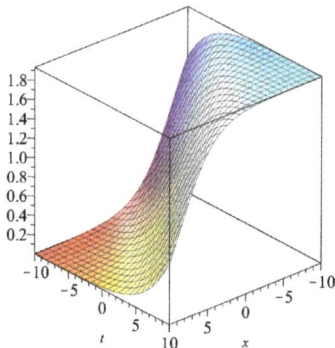

Figure 3. Perspective view of the modulus of $u_6(x,t)$ with $\alpha = 0.5, \beta = 1, \delta = 2.0, \gamma = -0.5$, and $\vartheta = 1.1$.

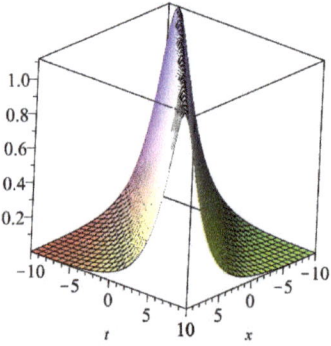

Figure 4. Perspective view of the modulus of $u_8(x,t)$ with $\alpha = 0.5, \beta = 1, \delta = 1, \gamma = -1.5$, and $\vartheta = 1.5$.

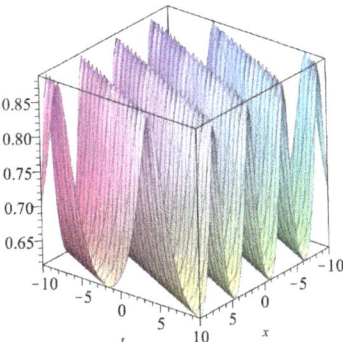

Figure 5. Perspective view of the modulus of $u_9(x,t)$ with $\alpha = 0.5, \beta = 1, \delta = 0.5, \gamma = 2$, and $\vartheta = 3$.

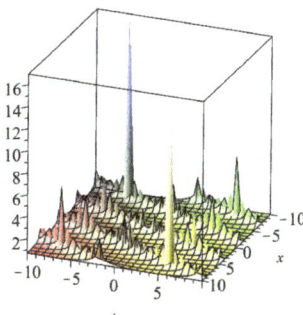

Figure 6. Perspective view of the modulus of $u_{10}(x,t)$ with $\alpha = 1, \beta = 1, \delta = 1, \gamma = -0.5$, and $\vartheta = 3$.

4. Conclusions

The study in this paper was devoted to the derivation of new exact solitary wave solutions of the generalized BBM equation through the GERFM. The correctness of the whole solutions $u_1(x,t)$-$u_{10}(x,t)$ has been verified with a symbolic Maple package, and it is found that all are satisfied with their corresponding original equations. The obtained solutions could be classified as periodic solutions and soliton solutions. Some graphical representations reveal the fact that the wave profile u behaves as bright and kink, multi-soliton solutions. These new obtained solutions could help for a deeper understanding of systems described by the BBM equation. All obtained solutions in the present work are new, and have not been previously reported in the literature. This is the main advantage of the GERFM over existing methods for solving GBBM equations, and indicates that GERFM is an efficient and easy to use tool that can help physicists and mathematicians handle and explore various sets of nonlinear PDEs.

Author Contributions: Investigation, B.G.; Project administration, D.B.; Software, B.G.; Validation, M.A.Q.

Funding: This research received no external funding.

Conflicts of Interest: The authors declare no conflict of interest.

References

1. Benjamin, T.B.; Bona, J.L.; Mahony, J.J. Model equations for long waves in nonlinear dispersive systems. *Philos. Trans. R. Soc. Lond. Ser. A* **1972**, *272*, 47. [CrossRef]
2. Yan, C. Regularized long wave equation and inverse scattering transform. *J. Math. Phys.* **1993**, *24*, 2618–2630. [CrossRef]

3. Meiss, J.; Horton, W. Solitary drift waves in the presence of magnetic shear. *Phys. Fluids* **1982**, *25*, 1838. [CrossRef]
4. Belobo, D.B.; Das, T. Solitary and Jacobi elliptic wave solutions of the generalized Benjamin–Bona–Mahony equation. *Commun. Nonlinear Sci. Numer. Simulat.* **2017**, *48*, 270–277. [CrossRef]
5. Muatjetjeja, B.; Khalique, C.M. Benjamin–Bona–Mahony Equation with Variable Coefficients: Conservation Laws. *Symmetry* **2014**, *6*, 1026–1036. [CrossRef]
6. Alsayyed, O.; Jaradat, H.M.; Jaradat, M.M.M.; Mustafa, Z.; Shata, F. Multi-soliton solutions of the BBM equation arisen in shallow water. *J. Nonlinear Sci. Appl.* **2016**, *9*, 1807–1814. [CrossRef]
7. Lam, L. *Introduction to Nonlinear Physics*; Springer: New York, NY, USA, 2003.
8. Wazwaz, A.M. *Partial Differential Equations and Solitary Waves Theory*; Higher Education Press, Springer: Berlin, Germany, 2009.
9. Fitz Hugh, R. Mathematical models of threshold phenomena in the nerve membrane. *Bull. Math. Biophys.* **1955**, *17*, 257–278. [CrossRef]
10. Hirota, R. Exact Solution of the Korteweg-de Vries Equation for Multiple Collisions of Solitons. *Phys. Rev. Lett.* **1971**, *27*, 1192. [CrossRef]
11. Suzo, A.A. Intertwining technique for the matrix Schrodinger equation. *Phys. Lett. A* **2005**, *335*, 88–102. [CrossRef]
12. He, J.H. The variational iteration method for eight-order initial-boundary value problems. *Phys. Scr.* **2007**, *76*, 680–682. [CrossRef]
13. Biazar, J.; Ayati, Z. Application of Exp-function method to EW-Burgers equation. *Numer. Methods Partial Differ. Equ.* **2010**, *26*, 1476–1482. [CrossRef]
14. Fan, E. Extended tanh-function method and its applications to nonlinear equations. *Phys. Lett. A* **2000**, *277*, 212–218. [CrossRef]
15. Didovych, M. A (1+2)-Dimensional Simplified Keller–Segel Model: Lie Symmetry and Exact Solutions. *Symmetry* **2015**, *7*, 1463–1474. [CrossRef]
16. Cherniha, R.; King, J.R. Lie and Conditional Symmetries of a Class of Nonlinear (1 + 2)-Dimensional Boundary Value Problems. *Symmetry* **2015**, *7*, 1410–1435. [CrossRef]
17. Cherniha, R.; Didovych, M. A (1 + 2)-Dimensional Simplified Keller–Segel Model: Lie Symmetry and Exact Solutions. II. *Symmetry* **2017**, *9*, 13. [CrossRef]
18. Ghanbari, B.; Inc, M. A new generalized exponential rational function method to find exact special solutions for the resonance nonlinear Schrödinger equation. *Eur. Phys. J. Plus* **2018**, *133*, 142. [CrossRef]
19. Osman, M.S.; Ghanbari, B. New optical solitary wave solutions of Fokas-Lenells equation in presence of perturbation terms by a novel approach. *Optik* **2018**, *175*, 328–333. [CrossRef]
20. Ghanbari, B.; Raza, N. An analytical method for soliton solutions of perturbed Schrödinger's equation with quadratic-cubic nonlinearity. *Mod. Phys. Lett. B* **2019**, *33*. [CrossRef]

 © 2018 by the authors. Licensee MDPI, Basel, Switzerland. This article is an open access article distributed under the terms and conditions of the Creative Commons Attribution (CC BY) license (http://creativecommons.org/licenses/by/4.0/).

Article

Hysteretically Symmetrical Evolution of Elastomers-Based Vibration Isolators within α-Fractional Nonlinear Computational Dynamics

Silviu Nastac [1,*], Carmen Debeleac [1] and Sorin Vlase [2]

[1] Engineering and Agronomy Faculty in Braila, Research Center for Mechanics of Machines and Technological Equipments, "Dunarea de Jos" University of Galati, 810017 Braila, Romania
[2] Department of Mechanical Engineering, Faculty of Mechanical Engineering, Transilvania University of Brașov, B-dul Eroilor 29, 500036 Brașov, Romania
* Correspondence: snastac@ugal.ro; Tel.: +40-374-652-572

Received: 8 June 2019; Accepted: 12 July 2019; Published: 15 July 2019

Abstract: This study deals with computational analysis of vibration isolators' behavior, using the fractional-order differential equations (FDE). Numerical investigations regarding the influences of α-fractional derivatives have been mainly focused on the dissipative component within the differential constitutive equation of rheological model. Two classical models were considered, Voigt-Kelvin and Van der Pol, in order to develop analyses both on linear and nonlinear formulations. The aim of this research is to evaluate the operational capability, provided by the α-fractional derivatives within the viscous component of certain rheological model, to enable an accurate response regarding the realistic behavior of elastomeric-based vibration isolators. The hysteretic response followed, which has to be able to assure the symmetry of dynamic evolution under external loads, and at the same time, properly providing dissipative and conservative characteristics in respect of the results of experimental investigations. Computational analysis was performed for different values of α-fractional order, also taking into account the integer value, in order to facilitate the comparison between the responses. The results have shown the serviceable capability of the α-fractional damping component to emulate, both a real dissipative behavior, and a virtual conservative characteristic, into a unitary way, only by tuning the α-order. At the same time, the fractional derivative models are able to preserve the symmetry of hysteretic behavior, comparatively, e.g., with rational-power nonlinear models. Thereby, the proposed models are accurately able to simulate specific behavioral aspects of rubber-like elastomers-based vibration isolators, to the experiments.

Keywords: α-fractional calculus; vibration isolation; fractional-order differential equation; rubber-like elastomers; Riemann–Liouville/Caputo/Grünwald-Letnikov fractional derivative

1. Introduction

Fractional calculus (FC) studies the differentiation and integration of an arbitrary real or complex order of the differential operator. The starting point of FC's use was a discussion between Leibniz and L'Hospital (1695) concerning the calculation and significance of the derivative of the power function, which also became the name of FC. FC has been largely applied to pure mathematics and its application in physics or engineering has remained extremely limited until, by association with the theory of chaos or fractals, it has increased the interest in this branch of analysis. The reputation enjoyed by FC is still exotic, and for many researchers, it is still unclear what its practical use is. Fractional derivatives take into account the previous evolution of the variable, unlike the standard differential operator, which is limited to taking into account local time history. This feature allows for a better description of system dynamics [1].

Taking the previous aspects into account, the next paragraphs contain a briefly survey of some relevant works within the areas of both computational aspects, and some practical applications on vibration isolation analysis, of FC. Thus, in article [1], there is presented a dynamic statistical analysis of a system composed of a large number of masses. The conclusion of this article shows that, while the individual dynamics of each material point has an integer-order behavior, global dynamics reveal both the existence of the integer dynamics and the fractional dynamics. Babiarz and Legowski, in the paper [2], developed a fractional model of the human arm, and is an approach representing an attempt to consider the properties of damping in the simplest form.

Inside the editorial paper [3], Zhou et al. had proposed a selection of papers presenting many applications of fractional dynamics in different research domains. Three areas are distinguishable: Dynamical analysis (leading to fractional differential equations and systems), fractional-order control theory, and applications considering models with fractional calculus.

The work [4] presents some ways to use Matlab software with the associated numerical methods to simulate and compute the fractional derivatives. These methods are based on series technique (Fourier and Taylor in our case). In addition, some numerical examples were presented. The fractional calculus offers a number of axioms and methods to extend the derivative notions from integer n to arbitrary order α, under favorable conditions [5]. In paper [5], the author has intended to solve a linear homogeneous differential equation with constant coefficients by means of N-fractional calculus method, and the results show that singular differential equations can be solved by means of the N-fractional calculus method.

Koh and Kelly, in Reference [6], are studying a mechanical oscillator consisting of a concentrated mass and a Kelvin-type element with fractional behavior, modeling an elastomeric bearing in groundwork isolation systems. For a dynamic analysis of a fractional oscillator developed efficient numerical multi-step schemes in the time domain, in good agreement with the Laplace and Fourier solutions. The model with fractional derivative agrees, to a satisfactory extent, with the experimental results.

Freundlich, in paper [7], is studying the vibration of a simply supported beam with a viscoelastic behavior of fractional order. The beam, considering the Bernoulli-Euler hypotheses, was excited by the supports movement. In this model, the Riemann–Liouville fractional derivative (RLFD) of order $0 < \alpha \leq 1$ was used. Firstly, the steady-state vibrations were analyzed and therefore the RLFD with lower terminal at -∞ was considered, in order to simplify solution of the FDE and to enable direct computation of the amplitude-frequency characteristic. Final results showed that the appropriate selection of both damping coefficients, and respectively, of fractional derivative order, offers more accurate fitting of the dynamic characteristic comparatively with the model using the integer order derivative.

The fractional equations of the mass-spring-damper system were considered by Gómez-Aguilar et al. in Reference [8]. The results had shown viscoelastic behaviors of the mechanical components (producing, at different scales, temporal fractality). This proves the material heterogeneities in the mechanical components.

Use of the fractional order models in bioengineering applications is presented in paper [9], where a method to obtain a numerical solution, considering the power series expansion was provided. Fractional dynamic of such system leads to different equations and the applications are solved with Matlab/Simulink software. The study offers several illustrative cases that can be used not only in bioengineering, but so too in other disciplines, where it is possible to apply the fractional calculus. In addition, with the previous mentioned paper [9], within chapter [10], Petráš had developed methods to obtain the fractional derivative, fractional integral solution of a fractional differential equations using Matlab. Illustrative examples, together the correspondent Matlab code sources, were also presented.

The researches of Rossikhin et al. dealt with dynamic behavior of mechanical oscillators (linear and nonlinear)—in [11], respectively, with force-driven vibrations of nonlinear oscillators—in Reference [12], where the constitutive equations use fractional derivatives. The authors had concluded that using fractional derivatives is possible to do an approximate analysis of the vibratory regimes of the oscillator.

The nonlinear response of a plate endowed loaded with a random force using fractional derivative is presented in Reference [13] by Malara and Spanos. A statistical linearization is used to determine an approximate solution of the governing equations of the plate vibration. Eigenvalues analysis offers a time-dependent representation of the response. In this way, it is possible to obtain the set of nonlinear fractional differential equation. The linearization is considered in a mean square sense.

Inside study [14], the stochastic response of a structural systems has been studied, with a single-degree-of-freedom. The excitation was stationary and non-stationary. The damping is considered with fractional behavior. The final equation of motion becomes a set of coupled differential linear equations, involving more degrees of freedom. The number of the additional degrees of freedom depends on the discretization of the fractional derivative operator. Using these methods, the stochastic analysis becomes straightforward and simple. It is mentioned that, for the most current engineering interest, it has been proven that the second-order statistics response can be obtained in a closed form, in a simple manner.

In paper [15], Cajić et al. makes an analysis of a nanobeam free damped transverse vibration. The method used is the non-local theory of Eringen and fractional derivative model for viscoelasticity. The classic Euler–Bernoulli beam theory and constitutive equation using fractional order derivatives that offer the form of the motion equation of a nanobeam. Different values of fractional order parameter are used in order to obtain the time dependent behavior.

The actual literature also provides a suite of theoretical approaches related to fractional differential equations. Hereby, Ma [16] proposed the Adomian decomposition method and a computational technique for the study of the impulsive fractional differential equations. Continuing this research in Reference [17], to obtain an exact solution for nonlinear fractional problems, the authors combined the double Laplace transforms with the Adomian decomposition method. The new method is applied to solve regular and singular conformable fractional coupled Burgers' equations, and illustrated the effectiveness of the proposed method with some examples.

Finally, in this section, and in addition to previously mentioned works, comprehensive information regarding both theoretical and computational aspects of FC can be found within the texts [18–22]. These works also contain some applications of FC on various important areas of engineering and physics.

Linear and nonlinear models regarding the elastomers-based vibration insulation devices can be found in papers [23–25], where they had earnestly presented the influence of dissipative characteristic on isolator dynamic behavior. In addition, the research trials, as well as the efforts of the author, within the area of vibration isolation using the passive systems, was presented by the papers [26–30], which contain both computational and experimental results, according with proposed linear and nonlinear approaches.

Usually, numerical investigations have a sensible aspect related to the required computational resources, optimal balancing between hardware performances, and computing time. Most of the researchers made additional searches into computing time optimization, beyond that of the mainly numerical procedures development [31]. Regarding this aspect and within the present study, the authors noted that the computational cost required by evaluation of α-fractional derivative increase the total simulation time to a certain measure, directly depending on both FC formulation and solving the numerical approach had adopted within the computational model. Computational cost optimization was not framed by the aim of this study, but nevertheless, some related-to investigations done by the authors show a relatively small increasing of total simulation time for FC-based algorithm comparatively with the linear case. Taking into account an average 6000 points single dimension grid for each simulation, it has to be noted that the FC had required 1.12...1.23 s, in the same time with 0.88...0.93 s by the classical algorithm (hardware and numerical FC formulation were kept identical).

This paper was structured based on the next sections as follows: Theoretical basics of α-fractional derivative calculus and classical models related on FDE's within vibratory systems analysis, computational assessments regarding fractional-order dissipation, discussions, and final concluding remarks.

2. Materials and Methods

The basic approach of fractional calculus consists by the Riemann–Liouville fractional integral (RLFI) [4,9,10]

$$_aJ_t^\alpha f(t) = \frac{1}{\Gamma(\alpha)} \int_a^t (t-\tau)^{\alpha-1} f(\tau) d\tau, \quad (1)$$

with $\alpha \in (0,1]$ and $t > a$, where a is the initial value (usually $a = 0$). Associated derivative is able to result through the Lagrange rule for differential operators. Evaluating the nth order derivative over the $(n - \alpha)$ order integral, yields the α order derivative (related to RLFD) as follows

$$_aD_t^\alpha f(t) = D_t^n {}_aJ_t^{n-\alpha} f(t) = \frac{1}{\Gamma(n-\alpha)} \frac{d^n}{dt^n} \left[\int_a^t (t-\tau)^{n-\alpha-1} f(\tau) d\tau \right], \quad (2)$$

with $\alpha \in (n-1, n]$, $\alpha > 0$, notation D denoting d/dt and $\Gamma(\bullet)$ means the Gamma function, defined as follows

$$\Gamma(t) = \int_0^\infty e^{-z} z^{t-1} dz. \quad (3)$$

Taking into account the initial value a set to 0, that denotes the most common case, the RLFD, Equation (2), becomes [2]

$$_0D_t^\alpha f(t) = \frac{1}{\Gamma(1-\alpha)} \frac{d}{dt} \int_0^t (t-\tau)^{-\alpha} f(\tau) d\tau, \quad (4)$$

where in it was assumed the restriction of derivative order $\alpha \in (0,1]$.

Following the RLFI previous expressions, a commonly compact definition, used for fractional derivatives, is [6]

$$D^\alpha f \equiv {}_aD_t^\alpha [f(t)] \equiv \frac{d^\alpha f(t)}{[d(t-a)]^\alpha} \equiv \begin{cases} \frac{1}{\Gamma(n-\alpha)} \frac{d^n}{dt^n} \left[\int_a^t (t-\tau)^{n-\alpha-1} f(\tau) d\tau \right], & \begin{array}{l} n > \alpha \geq 0 \\ n \text{ integer} \end{array} \\ \frac{1}{\Gamma(-\alpha)} \int_a^t (t-\tau)^{-\alpha-1} f(\tau) d\tau, & \alpha < 0 \\ \frac{d^n}{dt^n} f(t), & a = 0, \alpha = n, n \text{ integer} \end{cases} \quad (5)$$

Caputo, in 1990's, supposed another way to define the fractional derivatives, actually known as Caputo Fractional Derivative (CFD) [4,8]

$$_aD_t^\alpha f(t) = \frac{1}{\Gamma(1-\beta)} \int_a^t (t-\tau)^{-\beta} f^{(m+1)}(\tau) d\tau, \quad (6)$$

where $\alpha = \beta + m$, $\beta \in (0,1]$ and m denotes the nearest integer less than α. The Caputo definition does not require the fractional order initial conditions to define, and thus, it is often preferred in solving differential equations.

A.K. Grünwald and A.V. Letnikov propose a new generalization of integer derivative by approaching the integer difference of the fractional case [4]. Supposing the binomial coefficients as an iterative formulation [10]

$$\binom{x}{y} = c_y^{(x)} = \left(1 - \frac{1+x}{y}\right) c_{y-1}^{(x)}, \quad c_0^{(x)} = 1, \quad y = 0,1,2,\ldots,k, \quad (7)$$

and considering the real valued binomial coefficients based on Gamma function [10,16] because that allows a generalization to non-integer arguments

$$\binom{x}{y} = \frac{\Gamma(x)}{\Gamma(x-y)\,\Gamma(y)}, \quad x, y \in \mathcal{R}, \tag{8}$$

the Grünwald-Letnikov Fractional Derivative (GLFD) is [4,9,10]

$$_aD_t^\alpha f(t) = \lim_{\substack{h \to 0 \\ k = (t-a)/h}} h^{-\alpha} \sum_{j=0}^{[k]} (-1)^j \binom{\alpha}{j} f(t - jh), \tag{9}$$

with the notation [•] meaning the integer part of the argument. Obviously, $y(t) = 0$ for $t < 0$ within transient problems, thus $_{-\infty}D_t^\alpha y(t)$ becomes identical to $_0D_t^\alpha y(t)$, [6]. For analyses within the area of vibration isolation, into the context of the viscoelasticity theory, the fractional order α, known as memory parameter, was assumed to have a value varying from zero to one.

The fractional model, used within computational analyses, followed the conventional Voigt-Kelvin rheological schematization. The constitutive equation, embedding fractional order formulation into dissipative term, were assumed to have the following expression [30]

$$\tau(t) = G_o(1 + c\,D^\alpha)\gamma(t), \tag{10}$$

with damping constant $c > 0$, G_o denoting the shear modulus, and τ, γ meaning the stress and the shear strain, respectively. It has to be noted that the authors of the paper [6], using the less-square fit procedure for a set of experimental data, showed that the fractional Voigt-Kelvin model fits the data much better than the hysteretic, classical Kelvin and standard linear solid models.

Three practical applications of FDE for mechanical vibration analyses [30], embedding fractional order terms, are briefly presented as follows. The first case is the fractional order Van der Pol oscillator, having both inertial and dissipative α-fractional terms

$$D^{1+\alpha}x + \delta(x^2 - 1)D^\alpha x + b\,x = f(t), \tag{11}$$

or supposing only the fractional damping

$$\ddot{x} + \delta(x^2 - 1)D^\alpha x + b\,x = f(t). \tag{12}$$

Second case relates to the equations of fractional order Duffing-like systems, with a dissipative term

$$D^{1+\alpha}x + \delta D^\alpha x + b\,x + a\,x^3 = f(t), \tag{13}$$

or without damping

$$D^\alpha x + b\,x + a\,x^3 = f(t). \tag{14}$$

The last exemplification case presents an interesting system with small fractional and delayed damping, modeled by the following expression

$$\ddot{x}(t) + \delta\left[\beta_1 \dot{x}(t) + \beta_2 D^\alpha x(t) + \beta_3 \dot{x}(t-1)\right] + b\,x(t) + a\,x^3(t) = f(t). \tag{15}$$

In Equations (11)–(15), having mass normalized formulations, δ denotes damping coefficient, $a > 0$ is the control parameter of nonlinearity, b relates to squared pulsation of equivalent linear system, and $f(t)$ is the excitation term. The parameters β_i are the control coefficients related to each damping component. It has to be noted that extensive examples of FDE can be found in the works [11,12] and within theirs bibliography.

One challenge, especially for practical simulations, consists of settling the correct dimensionality of the differential equations. In order to supply this, an additional parameter µ that must be included within the fractional temporal operator was proposed in the work [8]

$$\frac{d}{dt} \to \frac{1}{\mu^{1-\alpha}} \frac{d^\alpha}{dt^\alpha} \equiv {}_0D_t^\alpha, \quad \frac{d^2}{dt^2} \to \frac{1}{\mu^{2(1-\alpha)}} \frac{d^{2\alpha}}{dt^{2\alpha}} \equiv {}_0D_t^{2\alpha} \tag{16}$$

where parameter µ has a dimension of seconds. According to Reference [8], parameter µ relates to the temporal components within the system (those components implying changes to the time constant of system). Clearly, taking µ = 1 into Equation (16) yields ordinary integer temporal derivative operators.

According to Reference [8], µ can be estimated by following expressions

$$\alpha = \sqrt{k/m}\,\mu, \tag{17}$$

for a mass—spring ensemble,

$$\alpha = \frac{k}{c}\mu, \tag{18}$$

for a damper—spring ensemble, and

$$\alpha = \sqrt{\frac{k}{m} - \frac{c^2}{4m^2}}\,\mu, \tag{19}$$

for a complete mass—spring—damper ensemble, with k, m denoting the rigidity and mass, respectively, and c meaning the damping coefficient.

In the view of this papers goal and taking into account the previous assertions regarding α-fractional derivatives, a α-fractional derivative SDoF system based on Voigt-Kelvin linear schematization, can be modeled with the expression [30]

$$\frac{1}{\mu^{2(1-\alpha)}} {}_0D_t^{2\alpha}x(t) + \frac{\delta}{\mu^{1-\alpha}} {}_0D_t^\alpha x(t) + bx(t) = f(t), \tag{20}$$

where the parameters have aforementioned significations.

Following the aim of this study, to characterize the hysteretic behavior of vibration isolator based on α-fractional derivative approaches, the authors consider a differential formulation related to Equation (20), but ignore the inertial term. This hypothesis were justified by the assertions in paper [11], and was also mentioned and detailed by the analyses in paper [30], where an extended investigation regarding the implications of α-fractional derivative within the classic linear was developed, Duffing and Van der Pol oscillators, respectively. Different values for the main parameters were adopted and the responses, in terms of time, spectral, and hysteretic characteristics were discussed.

Into this paper, the symmetrical trends in evolution of a vibration isolator subjected to harmonic excitation were studied. The capability of this α-fractional derivative approach to simulate both dissipative and conservative character of an isolator through a singular term was investigated, involving only the damping, but α-fractional-typed. The analyses were focused on evaluation of the equivalence between α-fractional-derivative and pseudo-linear systems, but underlined the maintaining of symmetry of hysteretic behavior. This last aspect is very important because of the practical observations (including instrumental investigations), which highlight the symmetry in hysteresis loops independently by dynamic excitation type or material characteristics. In this view, the α-fractional approach comes with the advantage of a unitary expression that intrinsically embeds damping and stiffness parameters, but keeping the realistic behavior (in term of hysteresis symmetry).

3. Results

As was previously mentioned, a simple dissipative oscillator was adopted, according to classical Voigt-Kelvin schematization, Equation (20), with α-fractional derivative in damping term $\left(c \cdot {}_0D_t^\alpha x(t)\right)$ and null stiffness. The inertial term was ignored, the excitation was supposed to be kinematical, in terms of harmonic displacement $x(t)$ with 1 Hz frequency, and the damping coefficient was unitary valued ($c = 1$). Computational investigations were conducted for order $\alpha \in \{0.1, 0.3, 0.5, 0.7, 0.9\}$, and in addition, the ordinary case ($\alpha = 1$) was considered. Fractional derivatives were evaluated using a GLFD-based algorithm.

Timed evolution and spectral composition of the response function were, respectively, presented in Figures 1 and 2. The results related to the regular derivative case ($\alpha = 1$) were differently marked onto the graphs in order to facilitate comparative analysis. For spectral distribution, both magnitude and phase components because some changes in frequency evolution were considered and have more visibility into the phase diagram than the magnitude.

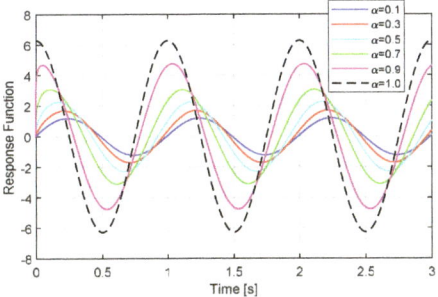

Figure 1. Timed response of α-order fractional damping model with respect to 1 Hz harmonic displacement excitation. Fractional orders were mentioned on graph legend.

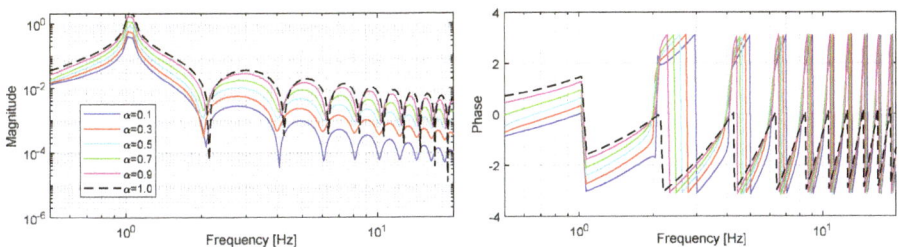

Figure 2. Spectral composition of α-order fractional damping model with respect to 1 Hz harmonic displacement excitation, in terms of magnitude and phase, respectively.

Diagram in Figure 3 presents the response function in respect of the excitation displacement. The presence of the hysteretic loop can be observed, which acquires changes as a function of fractional order. In order to evaluate the linear-like evolution of the system, a Poincare map, with a snapshot increment identical to the time period of excitation signal, was overlapped on diagrams in Figure 3. Null scattering of dots related to Poincare maps highlight the correlated evolution of the system with the excitation frequency. It has to be mentioned that, for the investigations presented into this paper, the evaluation of Poincare maps is justified by the fact that the response function is identical with the system velocity, because the model only incorporates the unitary gained dissipative term (in fact, the response function in respect to the displacement is similar with the phase plane diagram).

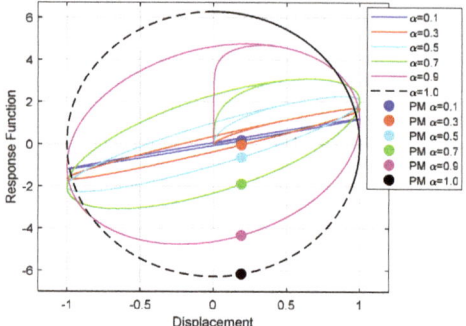

Figure 3. The response of α-fractional damping model with respect to 1 Hz harmonic displacement excitation. Poincare maps, with snapshots at 1s, were doted on graphs (named PM on graph legend).

The changes in slope of each hysteresis in Figure 3 indicate that, in the same time with increasing of α order, the system provide both dissipative and conservative characteristics, which quantitatively depend on the effective α value. This fact leads to the necessity of the pseudo-linear equivalent system evaluation, in terms of stiffness and damping of Voigt-Kelvin schematization. Taking into account the proportionality between the dissipated energy and the hysteresis loop area, it results in the equivalent-damping coefficient. In order to evaluate the stiffness, the hysteresis axis between the extreme horizontal coordinates was considered. For the loops in Figure 3, the correspondent axes were depicted in Figure 4. Analyzing the slopes of axes in Figure 4 clearly result in that α-fractional derivatives provide an additionally elastic component within system behavior. In the same time, the symmetry of the hysteresis loops through both the equality of the extremely values, and the common point of all axes on origin (0, 0) is evident.

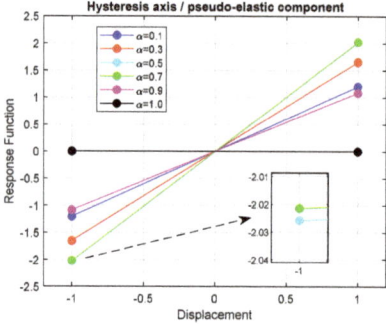

Figure 4. Hysteresis axes for loops in Figure 3, denoting the pseudo-elastic component, virtually emulated by the α-fractional damping model.

An additional analysis was conducted for largest number of α order values. The results, presented in Figure 5, in terms of timed evolution, hysteresis loops (included Poincare maps dots-based representations) and hysteresis axis, highlight the previously mentioned observations.

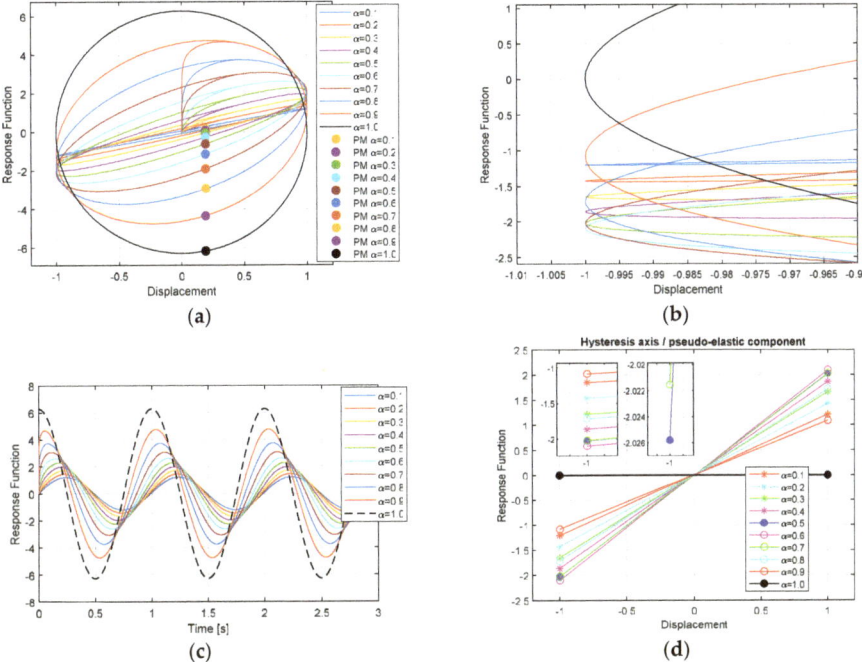

Figure 5. The results of additional analysis, supposing large number of α-fractional orders, in terms of hysteresis, including Poincare maps (**a**), with detail on the left hand side (**b**), timed evolution (**c**), and correspondent hysteresis axes (**d**).

At this point, it is evident that the α-fractional-based dissipative model is able to provide a behavior equivalent to a linear Voigt-Kelvin model, and vice-versa (see Figure 6). In fact, the fractional dissipative model is able to emulate a linear visco-elastic model, with direct linkage between α order and the pair of viscous and elastic parameters, respectively.

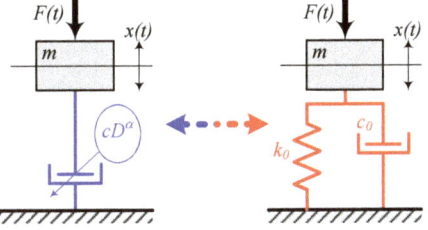

Figure 6. Schematic diagram related to the equivalence between α-fractional dissipative model and emulated linear Voigt-Kelvin model.

A computational application was developed for identification and evaluation of the proper values of the pseudo-linear system parameters. Different values for α-order between zero and one were considered, and both fractional and equivalent systems responses have been compared. The results indicate that the estimated relative error tends to zero, obviously depending by the computation precision. The diagrams in Figure 7 present two cases related to α-order valued to 0.4 and 0.7, respectively. It can be observed that both plots perfectly overlapped.

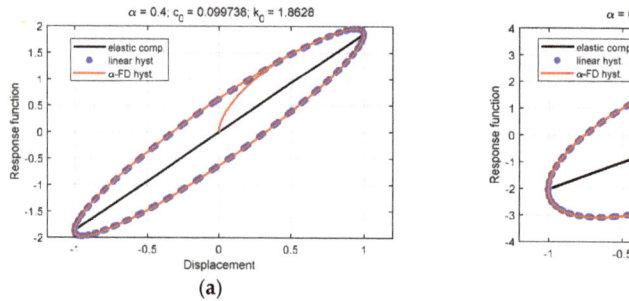

Figure 7. Comparative hysteresis of initial α-fractional dissipative and emulated linear Voigt-Kelvin models respectively, for α-order valued to 0.4 (**a**) and 0.7 (**b**). The identified values of virtual linear system were mentioned on each graph.

Taking into account aforementioned observation, it has to identify the correlation between α-order and the parameters of equivalent linear system. This analysis was performed by considering the equivalent stiffness and damping, respectively, and the results were depicted in Figure 8. In addition, the losses coefficient, as the ratio between dissipative and conservative parameters, was evaluated. The evolution of losses coefficient in respect to α-order was presented in Figure 9, where semi-log representation was adopted taking into account the relative large domain of function variability. It is evident that the losses coefficient $g \to +\infty$ for $\alpha = 1$ and $g = 0$ for $\alpha = 0$.

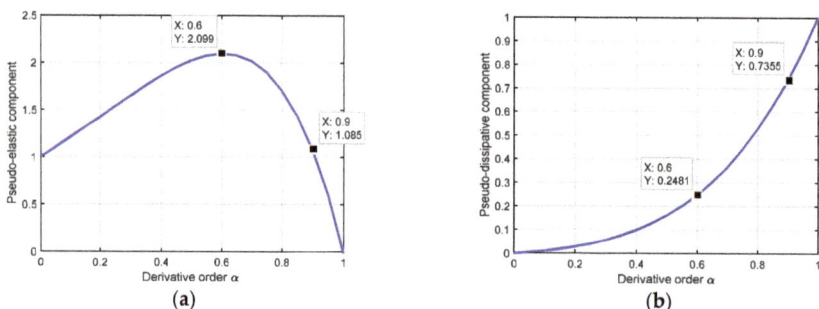

Figure 8. The evolution of pseudo-elastic component (**a**) and pseudo-dissipative component (**b**), respectively, for the emulated linear Voigt-Kelvin model, in respect to α-order of initial fractional derivative model.

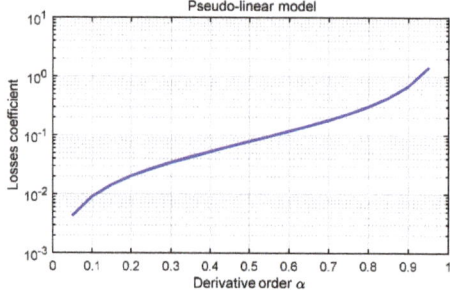

Figure 9. The evolution of losses coefficient for the emulated linear Voigt-Kelvin model, in respect to α-order of initial fractional derivative model.

An extended analysis was developed by taking into account the classical Van der Pol (VdP) expression for dissipative component, obviously supposing the α-fractional order derivative. The effects of fractional derivative in model behavior was followed, related to the observations and acquired for the previous simplest pseudo-linear model. This analysis was adopted based on the previous work evaluations [30], where it was considered for both Duffing and VdP non-linear models. The dissipative term within Duffing-type model enables only linear dependence in respect to the displacement derivative

$$\frac{c_0}{\mu^{1-\alpha}} {}_0D_t^\alpha x(t) + k_0 \left[x(t) + k_3 x^3(t) \right] = f(t), \tag{21}$$

thus that the dynamic behavior cannot be differently influenced by the fractional derivative than the Voigt-Kelvin-based model. Hereby, in the frame of this study and taking into account the nonlinear damping expression, becomes interesting to analyze a VdP-type model, embedding fractional order derivative, related to its capability that emulate a classical VdP model.

The following expression of VdP nonlinear model including α-fractional derivative was proposed for computational investigations [30]

$$-\frac{c_0}{\mu^{1-\alpha}} \left[1 - x^2(t) \right] {}_0D_t^\alpha x(t) + k_0 x(t) = f(t). \tag{22}$$

where $f(t)$ denotes the response of the system, $x(t)$ is the instantaneously displacement, and c_0, k_0 denote damping and stiffness coefficients, respectively.

In respect of the hypothesis within previous case analysis, the parameters c_0, k_0 were identically valued as follows: $c_0 = 1$, $k_0 = 0$. Taking into account the nonlinear term in parenthesis within the left side hand of Equation (22), the analysis has to be conducted related to displacement magnitude. In this study, random values for magnitude were taken. The results presented correspond to the 3.3, 1.0, and 0.3 valued displacement magnitudes. The monitored parameters, for each case, were similarly adopted than the previous linear case: Timed evolution, spectral composition, hysteretic loops with Poincare maps overlapped, and hysteresis axis according to the pseudo-elastic component provided by the model. The results for the three cases were presented in Figures 10–12, respectively (details were mentioned within the figure captions).

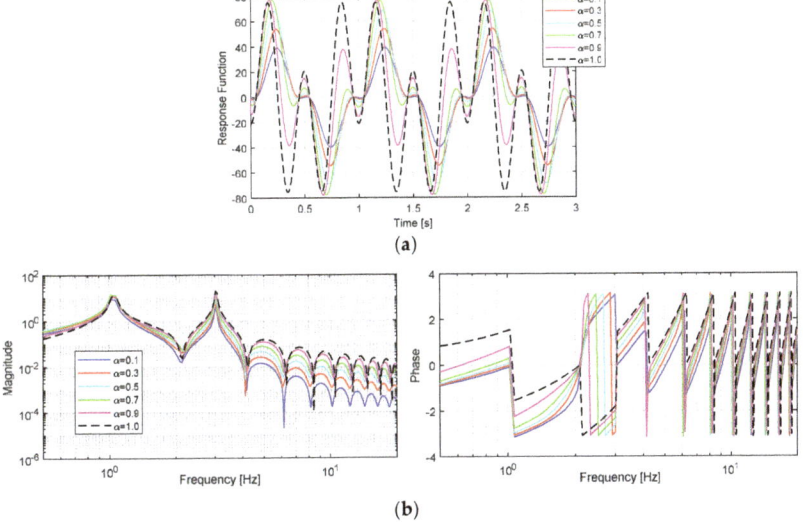

Figure 10. *Cont.*

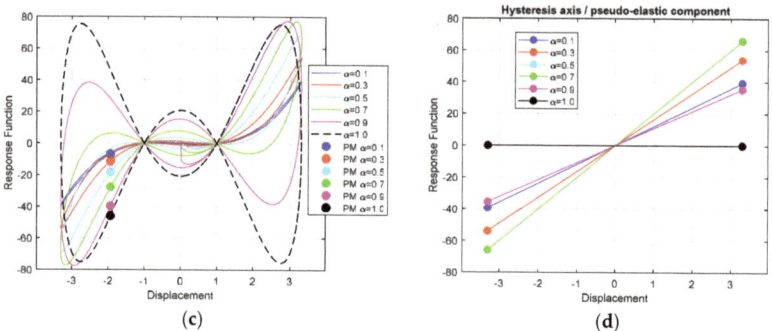

Figure 10. The response of α-order fractional damping VdP-based model, with respect to 1 Hz and 3.3 magnitude harmonic displacement excitation, in terms of timed evolution (**a**), magnitude and phase spectral composition (**b**), hysteresis loops with Poincare map (**c**), and pseudo-elastic component (**d**).

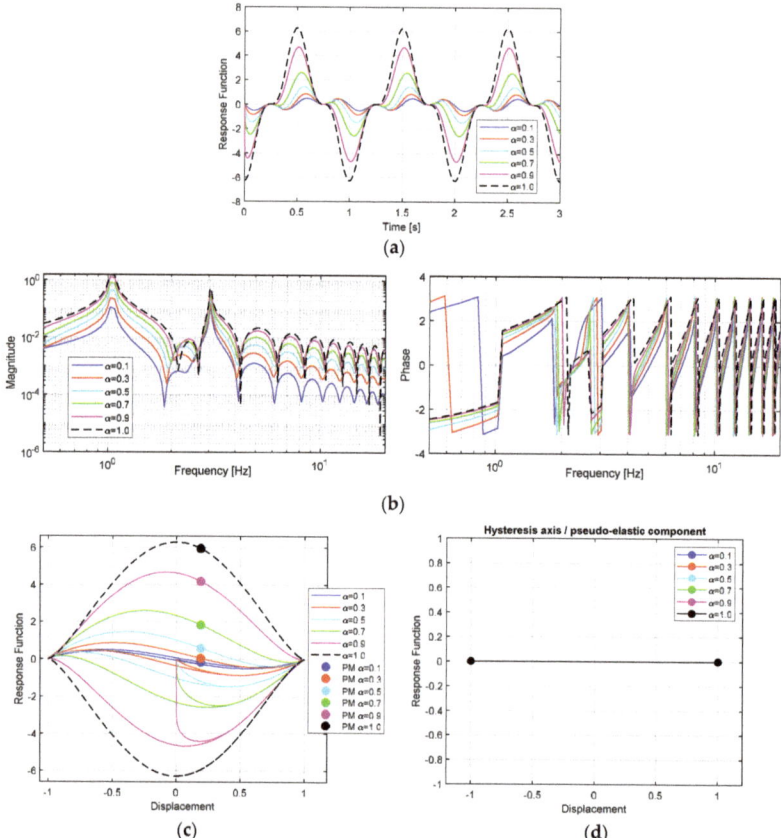

Figure 11. The response of α-order fractional damping VdP-based model, with respect to 1 Hz and unitary magnitude harmonic displacement excitation, in terms of timed evolution (**a**), magnitude and phase spectral composition (**b**), hysteresis loops with Poincare map (**c**), and pseudo-elastic component (**d**).

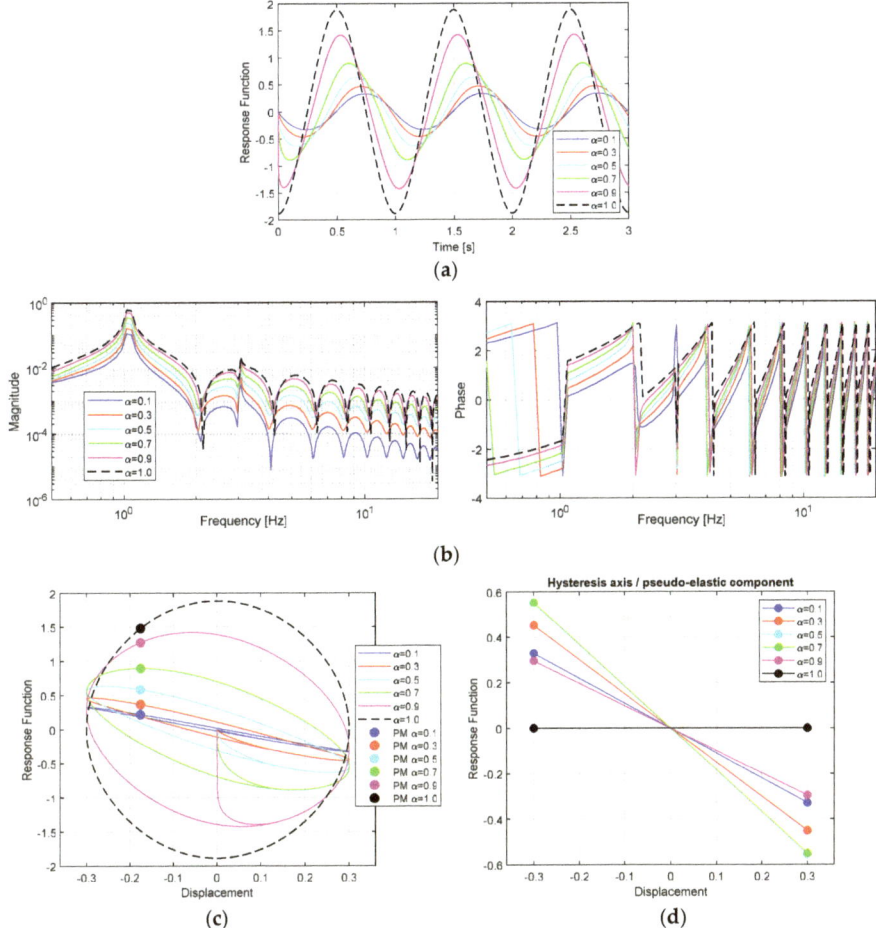

Figure 12. The response of α-order fractional damping VdP-based model, with respect to 1 Hz and 0.3 magnitude harmonic displacement excitation, in terms of timed evolution (**a**), magnitude and phase spectral composition (**b**), hysteresis loops with Poincare map (**c**), and pseudo-elastic component (**d**).

A comparative analysis between this group of results highlight the fact that the VdP model embedding the fractional derivative is also able to emulate a pseudo-conservative behavior, depending on the α-order. This aspect is dependent to the excitation magnitude and appears for any of its values except the unitary ones—see Figure 11c,d.

As was performed for the previous linear case, the authors developed a computational application in order to identify the proper formulation of the equivalent VdP visco-elastic model and to evaluate the parameters values of equivalent models. Identification procedure was imposed by the nonlinear formulation of the dissipative term in VdP model and by the fact that this expression has propagated to the conservative term for the equivalent model. In this view, the authors supposed two forms of the equivalent model. The first case was in respect of the classical VdP expression as follows

$$-c_{01}\left[1 - x^2(t)\right]\dot{x}(t) + k_{01}\,x(t), \tag{23}$$

in the same time that the second case supposes a specific nonlinear terms for equivalent rigidity as follows

$$-c_{02}\left[1-x^2(t)\right]\dot{x}(t)+k_{02}\left[1-x^2(t)\right]x(t). \qquad (24)$$

The analyses were performed for different values of the α-fractional order and displacement magnitudes, respectively. For presentation, the 1.8, 1.0, and 0.8 valued displacement magnitudes were adopted, maintaining $\alpha = 0.7$. The results were depicted as pair diagrams of the first and second equivalent model, for each displacement magnitude, in Figures 13–15, respectively.

Comparative analysis of this last group of results (depicted in Figures 10–12) show that identification of the right equivalent model formulation is easily done, taking into account the differences between the loops evolutions in diagrams. Even if the first equivalent model (emulated VdP-based model with linear rigidity) is able to provide the same energy dissipation for a proper damping coefficient and supplies certain pseudo-stiffness, the hysteretic evolution is clearly different. This fact can produce certain difficulties on the tuning process between theoretical behavior and the experimental investigations.

In the same time, the second approach (emulated VdP-based model with propagated nonlinear rigidity term) provides the best approximation with the α-fractional derivative model. Obviously, similar to the case of the linear model, the estimated relative error tends to zero, only and directly depending by the computation precision.

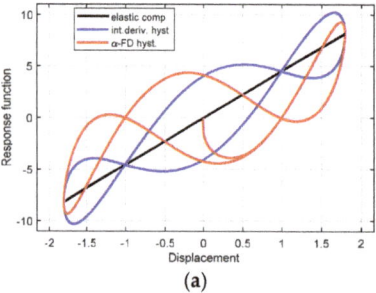
(a)

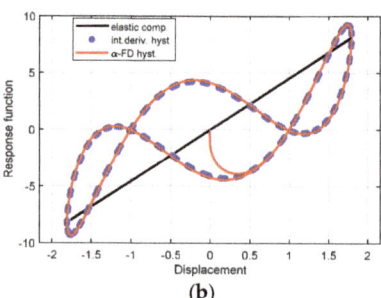
(b)

Figure 13. Responses hysteresis provided by the initial α-fractional dissipative and the emulated VdP-based models, respectively. The simulations were performed for displacement magnitude gained to 1.8, and in respect to emulated VdP model according to Equation (23) (**a**), and respectively, to Equation (24) (**b**). The α-order was valued to 0.7.

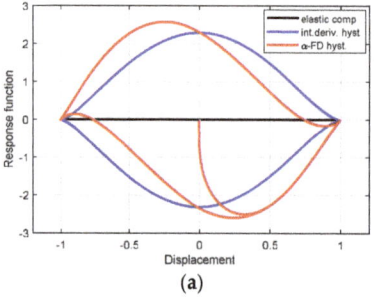
(a)

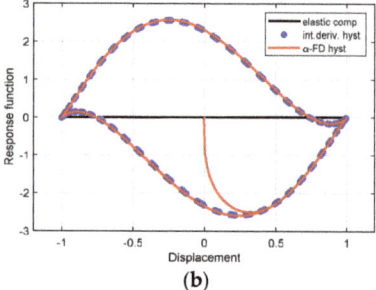
(b)

Figure 14. Responses hysteresis provided by the initial α-fractional dissipative and the emulated VdP-based models, respectively. The simulations were performed for displacement magnitude unitary gained, and in respect to emulated VdP model according to Equation (23) (**a**), and respectively, to Equation (24) (**b**). The α-order was valued to 0.7.

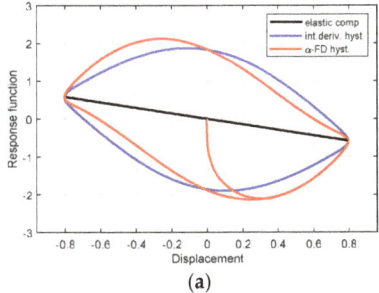 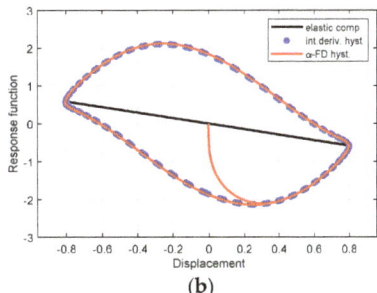

Figure 15. Responses hysteresis provided by the initial α-fractional dissipative and the emulated VdP-based models, respectively. The simulations were performed for displacement magnitude gained to 0.8, and in respect to emulated VdP model according to Equation (23) (**a**), and respectively, to Equation (24) (**b**). The α-order was valued to 0.7.

4. Discussion

Each case of analysis was briefly discussed on the same time with its presentation. Besides this, the main findings within this study has to be detailed, in a general view, taking into account the intrinsic availability of α-fractional derivatives that virtually modify the behavioral characteristics of linear/nonlinear vibration isolators models. The results in Figure 8, in terms of equivalent damping and stiffness, respectively, are relevant into this way. From the diagram of damping evolution in respect of the α-order results, the fractional derivative model is able to provide a full scale of dissipation, from weak to strong, related to the full range of α. In the same time, from the diagram of pseudo-elastic component results that also provided a relative very large scale for rigidity, from stiff, by very stiff (more of doubled initial value), to zero stiffness. These two observations are very useful, e.g., for investigations of elastomers-based vibration isolators, which provide strong nonlinear evolutions under the external dynamic loads and involve many parameters within computational (experimentally augmented) models.

Taking into account the comparative analysis of diagrams in Figure 8, it has to be mentioned that the evolutions of the two components are strongly correlated. In this sense, for small α orders, the FDE-based model enables only weak-dissipation with stiff to very stiff behavior. For medium to 0.9 values of α, the FDE-based model still enables a stiff behavior with damping transition from medium to strong. Finally, for 0.9...1.0 valued α, the model provides only a soft to zero stiffness, but very damping, behavior. These last observations have restricted the area of practical using for the FDE-based dynamic models. Nevertheless, these theoretical models are able to provide enough of a range of dynamic characteristics, in respect of a single parameter, and was thus suitable for inclusion into the computational approaches of vibration isolator dynamics.

In addition, the availability of the α-fractional derivatives has to be underlined, embedded to the differential constitutive equations, which do not change the hysteretic evolution of a dynamic system with symmetrical evolution in respect to its static state equilibrium point. This is a very useful observation, because the investigations done by the authors highlights that the insertion of fractional-order derivatives into the classical differential formulations of dynamic systems actually provides the preservation of the symmetry of the hysteretic behavior.

For the simple, rubber-like, composites without any augmentations related to differentiation of their reversible evolution under dynamic loads, the previously mentioned fact focuses on the interest to these computational approaches, because of theirs accurate capability to simulate the realistic behavior, experimentally acquired, in the same time with keeping the nonlinear characteristics identified to the microstructural and phenomenological analyses.

For the sake of comparison in the view of previous paragraph remarks, the diagrams in Figure 16 show the evolution of a rational-power simulated system, with quite identical material parameters to those used within this study. It was adopted in the force to displacement graphs, considering both the conservative and dissipative force components, in order to highlight the unsymmetrical hysteresis of such computational models. The conservative component of the model was supposed to have linear expression to displacement, with a constant value of rigidity.

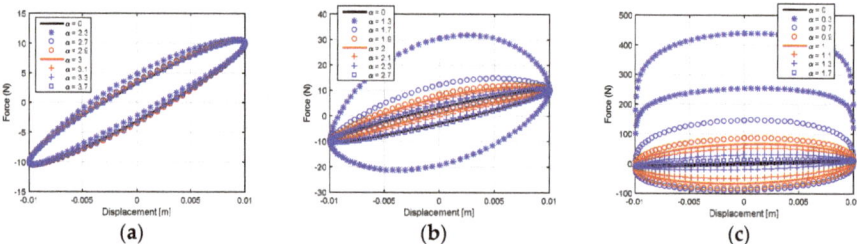

Figure 16. Hysteretic response of rational-power model, for $\alpha \in$ [2.3; 3.7] (**a**), [1.3; 2.7] (**b**) and [0.3; 1.7] (**c**). It was assumed a single time-period of 1 Hz excitation signal, and ratio between rigidity to damping, for linear components, valued to 20 s^{-1}. Within these diagrams, the symbol α must be considered as the rational power of additional dissipative component (multiplied by 20 from linear damping coefficient) into the computational model, with specific values mentioned on graph legends.

5. Conclusions

The computational investigations within this study highlight a very interesting aspect of the α-fractional derivatives embedded in classical linear or nonlinear oscillators. This aspect consists by the capability of emulating a pseudo-conservative characteristic for a computational implementation of the only dissipative component. In the view of practical using these models for the evaluation of vibration isolators dynamic behavior, this capability is very useful because it provides a single tuning parameter (in term of the α-order) instead of two (for linear case) or more parameters (for nonlinear case), for classical rheological dynamic models. Moreover, these FDE-based approaches are able to maintain the hysteresis symmetry, specific to the real systems, within its dynamic behavior under external loads.

Two cases of classical rheological models were analyzed in this work, fractional-derivative augmented. The future investigations will be focused on enlarging the area of fractional calculus applicability within other basic rheological models, and the potential useful results will be discussed into forthcoming works.

Author Contributions: The all authors contributed equally and all authors read the manuscript and approved the final submission.

Funding: This research received no external funding.

Acknowledgments: The authors feel grateful to the editor and thank the referees for the valuable comments that helped us to improve the manuscript.

Conflicts of Interest: The authors declare no conflict of interest.

References

1. Tenreiro Machado, J.A.; Galhano, A. Fractional Dynamics: A Statistical Perspective. *J. Comput. Nonlinear Dyn.* **2008**, *3*, 021201. [CrossRef]
2. Babiarz, A.; Legowski, A. Human arm fractional dynamics. In *Trends in Advanced Intelligent Control, Optimization and Automation, Advances in Intelligent Systems and Computing 577*; Mitkowski, W., Ed.; Springer International Publishing AG: Cham, Switzerland, 2017.
3. Zhou, Y.; Ionescu, C.; Tenreiro Machado, J.A. Fractional dynamics and its applications. *Nonlinear Dyn.* **2015**, *80*, 1661–1664. [CrossRef]

4. Chi, C.; Gao, F. Simulating Fractional Derivatives using Matlab. *J. Softw.* **2013**, *8*, 572–578. [CrossRef]
5. Ozturk, O. A Study on the Damped Free Vibration with Fractional Calculus. *IJAMEC* **2016**, *4*, 156–159.
6. Koh, C.G.; Kelly, J.M. Application of fractional derivatives to seismic analysis of base-isolated models. *Earthq. Eng. Struct. Dyn.* **1990**, *19*, 229–241. [CrossRef]
7. Freundlich, J. Vibrations of a simply supported beam with a fractional viscoelastic material model–supports movement excitation. *Shock Vib.* **2013**, *20*, 1103–1112. [CrossRef]
8. Gómez-Aguilar, J.F.; Yépez-Martínez, H.; Calderón-Ramón, C.; Cruz-Orduña, I.; Escobar-Jiménez, R.F.; Olivares-Peregrino, V.H. Modeling of a Mass-Spring-Damper System by Fractional Derivatives with and without a Singular Kernel. *Entropy* **2015**, *17*, 6289–6303. [CrossRef]
9. Petráš, I. An Effective Numerical Method and Its Utilization to Solution of Fractional Models Used in Bioengineering Applications. *Adv. Differ. Equ.* **2011**, *2011*, 652789. [CrossRef]
10. Petráš, I. Fractional Derivatives, Fractional Integrals, and Fractional Differential Equations in Matlab. In *Engineering Education and Research Using Matlab*; Assi, A., Ed.; InTech: Rijeka, Croatia, 2011.
11. Rossikhin, Y.A.; Shitikova, M.V. New approach for the analysis of damped vibrations of fractional oscillators. *Shock Vib.* **2009**, *16*, 365–387. [CrossRef]
12. Rossikhin, Y.A.; Shitikova, M.V.; Shcheglova, T. Forced Vibrations of a Nonlinear Oscillator with Weak Fractional Damping. *J. Mech. Mater. Struct.* **2009**, *4*, 1619–1636. [CrossRef]
13. Malara, G.; Spanos, P.D. Nonlinear random vibrations of plates endowed with fractional derivative elements. *Probab. Eng. Mech.* **2017**. [CrossRef]
14. Di Paola, M.; Failla, G.; Pirrotta, A. Stationary and non-stationary stochastic response of linear fractional viscoelastic systems. *Probab. Eng. Mech.* **2012**, *28*, 85–90. [CrossRef]
15. Cajić, M.; Karličić, D.; Lazarević, M. Nonlocal Vibration of a Fractional Order Viscoelastic Nanobeam with Attached Nanoparticle. *Theor. Appl. Mech.* **2015**, *42*, 167–190. [CrossRef]
16. Ma, C. A Novel Computational Technique for Impulsive Fractional Differential Equations. *Symmetry* **2019**, *11*, 216. [CrossRef]
17. Eltayeb, H.; Bachar, I.; Kılıçman, A. On Conformable Double Laplace Transform and One Dimensional Fractional Coupled Burgers' Equation. *Symmetry* **2019**, *11*, 417. [CrossRef]
18. Hilfer, R. Foundations of Fractional Dynamics: A Short Account. In *Fractional Dynamics: Recent Advances*; Klafter, J., Lim, S., Metzler, R., Eds.; World Scientific: Singapore, 2011; pp. 207–223.
19. Herrmann, R. *Fractional Calculus: An Introduction for Physicists*, 2nd ed.; World Scientific Publishing: Singapore, 2014.
20. Mainardi, F. *Fractional Calculus and Waves in Linear Viscoelasticity. An Introduction to Mathematical Models*; Imperial College Press: London, UK, 2010.
21. Ray, S.S. *Fractional Calculus with Applications for Nuclear Reactor Dynamics*; CRC Press Taylor & Francis Group: Boca Raton, FL, USA; London, UK; New York, NY, USA, 2016.
22. Anastassiou, G.A.; Argyros, I.K. *Intelligent Numerical Methods: Applications to Fractional Calculus*; Springer International Publishing: Cham, Switzerland, 2016.
23. Bratu, P. Viscous nonlinearity for interval energy dissipation. *Int. J. Acoust. Vib.* **2000**, *4*, 321–326.
24. Marin, M.; Baleanu, D.; Vlase, S. Effect of microtemperatures for micropolar thermoelastic bodies. *Struct. Eng. Mech.* **2017**, *61*, 381–387. [CrossRef]
25. Iancu, V.; Vasile, O.; Gillich, G.R. Modelling and Characterization of Hybrid Rubber-Based Earthquake Isolation Systems. *Mater. Plast.* **2012**, *4*, 237–241.
26. Nastac, S. Theoretical and experimental researches regarding the dynamic behavior of the passive vibration isolation of systems. In *Research Trends in Mechanics*; Dinel, P., Chiroiu, V., Toma, I., Eds.; Romanian Academy Publishing House: Bucharest, Romania, 2008; Volume II, pp. 234–262.
27. Leopa, A.; Nastac, S.; Debeleac, C. Researches on damage identification in passive vibro-isolation devices. *Shock Vib.* **2012**, *19*, 803–809. [CrossRef]
28. Nastac, S. On Nonlinear Computational Assessments of Passive Elastomeric Elements for Vibration Isolation. *RJAV* **2014**, *2*, 130–135.
29. Nastac, S.; Debeleac, C. On Shape and Material Nonlinearities Influences about the Internal Thermal Dissipation for Elastomer-Based Vibration Isolators. *PAMM* **2014**, *1*, 751–752. [CrossRef]

30. Nastac, S. On Fractional Order Dynamics of Elastomers-based Vibration Insulation Devices. *Acta Electrotech.* **2019**, *60*, 169–182.
31. Xiao, D.; Fang, F.; Buchan, A.G.; Pain, C.C.; Navon, I.M.; Du, J.; Hu, G. Non-linear model reduction for the Navier–Stokes equations using residual DEIM method. *J. Comput. Phys.* **2014**, *263*, 1–18. [CrossRef]

© 2019 by the authors. Licensee MDPI, Basel, Switzerland. This article is an open access article distributed under the terms and conditions of the Creative Commons Attribution (CC BY) license (http://creativecommons.org/licenses/by/4.0/).

Article

Thermoelasticity of Initially Stressed Bodies with Voids: A Domain of Influence

Marin Marin [1,*], Mohamed I. A. Othman [2], Sorin Vlase [3] and Lavinia Codarcea-Munteanu [1]

1. Department of Mathematics and Computer Science, Transilvania University of Brasov, 500093 Brasov, Romania; codarcealavinia@unitbv.ro
2. Department of Mathematics, Faculty of Science, Zagazig University, P.O. Box 44519 Zagazig, Egypt; m_i_a_othman@yahoo.com
3. Department of Mechanical Engineering, Transilvania University of Brasov, 500096 Brasov, Romania; svlase@unitbv.ro
* Correspondence: m.marin@unitbv.ro

Received: 9 April 2019 ; Accepted: 18 April 2019; Published: 19 April 2019

Abstract: In our study, we will extend the domain of influence in order to cover the thermoelasticity of initially stressed bodies with voids. In what follows, we prove that, for a finite time $t > 0$, the displacement field u_i, the dipolar displacement field φ_{jk}, the temperature θ and the change in volume fraction ϕ generate no disturbance outside a bounded domain B.

Keywords: initially stressed bodies; dipolar structure; voids; volume fraction; domain of influence

1. Introduction

We must outline that our bodies are included in the thermoelasticity of bodies with voids. As we know, Nunziato and Cowin, in the paper [1], have initiated the approach of the bodies with vacuous pores (or voids). In this theory, the authors introduce an additional degree of freedom in order to develop the mechanical behavior of a body in which the skeletal material is elastic and interstices are voids of material. There are many modern applications of this theory, of which we just mention the manufactured porous materials and the geological materials like soils and rocks. In the linear case, this theory of bodies with pores was approached by Cowin and Nunziato in the study [2]. In this paper, the authors demonstrated a result of the uniqueness regarding the solution for the mixed problem and obtained a result of weak stability for the respective solutions. See also [3]. The equations of the thermoelasticity of bodies with pores were obtained by Iesan in work [4]. Other results regarding the contributions of voids of material in the theory of micropolar body with pores can be found in the papers [5,6]. Lately, the number of papers devoted to various aspects of microstructure has greatly increased (see [7–23], so that our work can be considered a continuation in this respect. In our present study, we extend the previous results in order to cover the thermoelasticity of initially stressed material with voids. Thus, after we put down the main equations, the initial conditions and the boundary data of the mixed value problem in this context, we define the so-called domain of influence, denoted by B_t, which corresponds to the data at time t and it is associated with the mixed problem. As in previous studies (see for instance [5,24,25]), we will use a specific method in order to prove a theorem regarding the domain of influence.

Our basic result asserts that the solutions of the mixed initial-boundary value problem, in the context of the theory of thermoelasticity of bodies with voids, decay to zero outside B_t, for any finite variable $t > 0$.

2. Basic Equations

In our paper, we will consider some elastic bodies. We suppose that such anisotropic body is situated in regular domain B, include in R^3, that is, the three-dimensional Euclidian space. The border of this domain is a smooth surface, denoted by ∂B. The closure of B is usual denoted by $\bar{B}$.

A fix system of Cartesian axes Ox_i, $(i = 1, 2, 3)$ is used and the Cartesian notation is used for vectors and tensors. For the time derivative of a function, it is used a superposed dot. For the partial derivatives of a function with respect its spatial variables, we will use a comma which is followed by a subscript. partial derivatives with respect to the spatial coordinates. In the case of the repeated indices, the Einstein summation rule is used.

In addition, if there is no possibility of confusion, then dependence of function with regards to its spatial or time variables will be omitted. The evolution of body with dipolar structure will be described with the help of the following specific variables:

$$u_i(x,t), \; \varphi_{ij}(x,t), \; (x,t) \in B \times [0, t_0). \tag{1}$$

Here, we denoted by u_i the components of the displacement vector field and by φ_{ij} the components of the dipolar displacement tensor field.

Using the above variables $u_i(x,t)$, $\varphi_{ij}(x,t)$, we will introduce the components of the tensors of strain, namely ε_{ij}, k_{ij} and γ_i, as follows:

$$2\varepsilon_{ij} = u_{j,i} + u_{i,j}, \; \gamma_{ij} = u_{j,i} - \varphi_{ij}, \; \chi_{ijk} = \varphi_{ij,k}. \tag{2}$$

Being in the context of linear theory, it is natural to consider that internal energy density is a quadratic form with the following expression:

$$\begin{aligned}
\varrho_0 e =& \frac{1}{2} A_{ijmn} \varepsilon_{ij} \varepsilon_{mn} + G_{ijmn} \varepsilon_{ij} \gamma_{mn} + F_{ijmnr} \varepsilon_{ij} \chi_{mnr} \\
&+ \frac{1}{2} B_{ijmn} \gamma_{ij} \gamma_{mn} + D_{ijmnr} \gamma_{ij} \chi_{mnr} + \frac{1}{2} C_{ijkmnr} \chi_{ijk} \chi_{mnr} + \\
&+ u_{j,k} P_{ki} \varepsilon_{ij} - \varphi_{jk} Q_{ki} \gamma_{ij} + u_{j,r} N_{irk} \chi_{ijk} + \\
&+ d_{ijm} \varepsilon_{ij} \phi_{,m} + e_{ijm} \gamma_{ij} \phi_{,m} + f_{ijkm} \chi_{ijk} \phi_{,m} + \\
&+ a_{ij} \varepsilon_{ij} \phi + b_{ij} \gamma_{ij} \phi + c_{ijk} \chi_{ijk} \phi + \frac{1}{2} A_{ij} \phi_{,i} \phi_{,j} - \\
&+ d_i \phi \phi_{,i} + a_i \theta \phi_{,i} + \xi \phi^2 - m \theta \phi + \frac{1}{2} a \theta^2 + \\
&- \alpha_{ij} \varepsilon_{ij} \theta - \beta_{ij} \gamma_{ij} \theta - \delta_{ijk} \chi_{ijk} \theta + \frac{1}{2} k_{ij} \theta_{,i} \theta_{,j}.
\end{aligned} \tag{3}$$

We will use a procedure similar to that used by Nunziato and Cowin in [3]. Thus, taking into account that

$$\tau_{ij} = \frac{\partial e}{\partial \varepsilon_{ij}}, \; \eta_{ij} = \frac{\partial e}{\partial \gamma_{ij}}, \; \mu_{ijk} = \frac{\partial e}{\partial \chi_{ijk}},$$
$$h_i = \frac{\partial e}{\partial \phi_{,i}}, \; g = -\frac{\partial e}{\partial \phi}, \; \eta = \frac{\partial e}{\partial \theta}, \; q_i = \frac{\partial e}{\partial \theta_{,i}}, \tag{4}$$

we obtain the following constitutive equations

$$\tau_{ij} = u_{j,k} P_{ki} + A_{ijmn} \varepsilon_{mn} + G_{ijmn} \gamma_{mn} + F_{mnrij} \chi_{mnr} +$$
$$+ a_{ij} \phi + D_{ijk} \phi_{,k} - \alpha_{ij} \theta,$$
$$\eta_{ij} = -\varphi_{jk} Q_{ik} + \varphi_{jk,r} N_{rik} + G_{ijmn} \varepsilon_{mn} + B_{ijmn} \gamma_{mn} +$$
$$+ D_{ijmnr} \chi_{mnr} + b_{ij} \phi + e_{ijk} \phi_{,k} - \beta_{ij} \theta,$$
$$\mu_{ijk} = u_{j,r} N_{rik} + F_{ijkmn} \varepsilon_{mn} + D_{mnijk} \gamma_{mn} + C_{ijkmnr} \chi_{mnr} +$$
$$+ c_{ijk} \phi + f_{ijkm} \phi_{,m} - \delta_{ijk} \theta, \qquad (5)$$
$$h_i = d_{mni} \varepsilon_{mn} + e_{mni} \gamma_{mn} + f_{mnri} \chi_{mnr} + d_i \phi + A_{ij} \phi_{,j} - a_i \theta,$$
$$g = -a_{ij} \varepsilon_{ij} - b_{ij} \gamma_{ij} - c_{ijk} \chi_{ijk} - \xi \phi - d_i \phi_{,i} + m \theta,$$
$$\eta = \alpha_{ij} \varepsilon_{ij} + \beta_{ij} \gamma_{ij} + \delta_{ijk} \chi_{ijk} - m \phi + a_i \phi_{,i} + a \theta,$$
$$q_i = k_{ij} \theta_{,j}.$$

The main equations that govern the thermoelasticity of initially stressed materials with pores are (see [5]):
- the motion equations:

$$\left(\tau_{ij} + \eta_{ij}\right)_{,j} + \varrho f_i = \varrho \ddot{u}_i,$$
$$\mu_{ijk,i} + \eta_{jk} + u_{j,i} Q_{ik} + \varphi_{ki} Q_{ji} - \qquad (6)$$
$$- \varphi_{kr,i} N_{ijr} + \varrho g_{jk} = I_{kr} \ddot{\varphi}_{jr};$$

- the equation for the equilibrium of the forces:

$$h_{i,i} + g + \varrho l = \varrho \kappa \ddot{\phi}, \qquad (7)$$

- the equation of energy:

$$\varrho T_0 \dot{\eta} = q_{i,i} + \varrho r. \qquad (8)$$

We complete the above equations with:
- the kinetic relations

$$\varepsilon_{ij} = \frac{1}{2} \left(u_{j,i} + u_{i,j}\right), \; \gamma_{ij} = u_{j,i} - \varphi_{ij},$$
$$\chi_{ijk} = \varphi_{jk,i}, \; \theta = T - T_0, \; \phi = \varphi - \varphi_0. \qquad (9)$$

The meaning of the notations that we used in Equations (3)–(8) is as follows: ϱ—the density of mass, which is a constant; η—the entropy per unit mass; T_0—the temperature of the material in its undeformed state; I_{ij}—the tensor of microinertia; κ—the variable inertia the equilibrated forces; u_i—the vector of the vector of moving; φ_{jk}—the tensor for dipolar moving; φ—the function for the volume fraction that in the undeformed state has the value φ_0; ϕ—a measure for change in volume, regarding the undeformed state; θ—the variation of the temperature, regarding the temperature T_0 from the reference state; ε_{ij}, γ_{ij}, χ_{ijk}—the strain tensors; τ_{ij}, η_{ij}, μ_{ijk}—the the stress tensors; h_i—the components of the vector for equilibrated stress; q_i—the vector for the flux of the heat; f_i—the body forces; g_{jk}—the dipolar charges; r—the measure of the supply heat; g—a measure for balancing of intrinsic force; L—a measure for balancing of extrinsic force; $A_{ijmn}, B_{ijmn}, ..., k_{ij}$—the functions what characterize the elastic properties of the material. These satisfy the following symmetry relations:

$$A_{ijmn} = A_{mnij} = A_{jimn}, \; B_{ijmn} = B_{mnij}, \; a_{ij} = a_{ji},$$
$$d_{ijk} = d_{jik}, \; g_{ij} = g_{ji}, \; C_{ijkmnr} = C_{mnrijk}, \; F_{ijknm} = F_{ijknm}, \quad (10)$$
$$G_{ijmn} = G_{ijnm}, \; P_{ij} = P_{ji}, \; k_{ij} = k_{ji}.$$

In the above relations (1) and (3), the quantities P_{ij}, Q_{ij} and N_{ijk} are prescribed functions which satisfy the following equations:

$$\left(P_{ij} + Q_{ij}\right)_{,j} = 0, \; N_{ijk,i} + Q_{jk} = 0.$$

Based on the inequality of the entropy production, we can deduce that

$$k_{ij}\theta_{,i}\theta_{,j} \geq 0. \quad (11)$$

In order to complete our mixed initial-boundary value problem, we add to the basic Equations (2)–(7), the following prescribed initial data

$$u_i(x,0) = u_i^0(x), \; \dot{u}_i(x,0) = u_i^1(x), \; \varphi_{jk}(x,0) = \varphi_{jk}^0(x), \; \dot{\varphi}_{jk}(x,0) = \varphi_{jk}^1(x),$$
$$\theta(x,0) = \theta^0(x), \; \phi(x,0) = \phi^0(x), \; \dot{\phi}(x,0) = \phi^1(x), \; x \in \bar{B}. \quad (12)$$

We also consider the given conditions to the limit

$$u_i = \bar{u}_i \text{ on } \partial B_1 \times [0, t_0), \; t_i \equiv \left(\tau_{ij} + \eta_{ij}\right) n_j = \bar{t}_i \text{ on } \partial B_1^c \times [0, t_0),$$
$$\varphi_{jk} = \bar{\varphi}_{jk} \text{ on } \partial B_2 \times [0, t_0), \; m_{jk} \equiv \mu_{ijk} n_i = \bar{m}_{jk} \text{ on } \partial B_2^c \times [0, t_0), \quad (13)$$
$$\phi = \bar{\phi} \text{ on } \partial B_3 \times [0, t_0), \; h \equiv h_i n_i = \bar{h} \text{ on } \partial B_3^c \times [0, t_0),$$
$$\theta = \bar{\theta} \text{ on } \partial B_4 \times [0, t_0), \; q \equiv q_i n_i = \bar{q} \text{ on } \partial B_4^c \times [0, t_0).$$

$n = (n_i)$ is the unit normal to the surface ∂B, outward oriented. In addition, we denoted by $\partial B_1, \partial B_2, \partial B_3$ and ∂B_4 the subsets of ∂B, considered together with their corresponding complements $\partial B_1^c, \partial B_2^c, \partial B_3^c$ and ∂B_4^c. The time t_0 can be infinite. The functions $u_i^0, u_i^1, \varphi_{jk}^0, \varphi_{jk}^1, \theta^0, \phi^0, \phi^1, \bar{u}_i, \bar{t}_i, \bar{\varphi}_{jk}, \bar{\mu}_{jk}, \bar{\phi}, \bar{\theta}, \bar{q}, \bar{h}$ are prescribed and regular in all points where are defined.

By introducing the geometric Equation (7) and the constitutive Equation (6) into Equations (2), (3) and (5), we obtain the following system of equations:

$$\varrho \ddot{u}_i = \left[u_{j,k} P_{ki} - \varphi_{jk,r} N_{rik} + \left(A_{ijmn} + G_{ijmn}\right) \varepsilon_{mn} + \left(G_{mnij} + B_{ijmn}\right) \gamma_{mn} + \right.$$
$$\left. + \left(F_{mnrij} + D_{ijmnr}\right) \chi_{mnr} + \left(a_{ij} + b_{ij}\right) \phi + \left(d_{ijk} + e_{ijk}\right) \phi_{,k} - \left(\alpha_{ij} + \beta_{ij}\right) \theta\right]_{,j} + \varrho f_i,$$

$$I_{kr} \ddot{\varphi}_{jr} = \left(u_{j,r} N_{irk} + F_{ijkmn} \varepsilon_{mn} + D_{mnijk} \gamma_{mn} + C_{ijkmnr} \chi_{mnr} + c_{ijk} \phi + f_{ijkm} \phi_{,m} - \delta_{ijk} \theta\right)_{,i} \quad (14)$$
$$- \varphi_{ji} Q_{ki} + \varphi_{ji,r} N_{rki} + G_{jkmn} \varepsilon_{mn} + B_{jkmn} \gamma_{mn} + D_{jkmnr} \chi_{mnr} + b_{jk} \phi + e_{jki} \phi_{,i} - \beta_{jk} \theta + \varrho g_{jk},$$

$$\varrho \kappa \ddot{\phi} = \left(d_{mni} \varepsilon_{mn} + e_{mni} \gamma_{mn} + f_{mnri} \chi_{mnr} + d_i \phi + A_{ij} \phi_{,j} - a_i \theta\right)_{,i} + \varrho l -$$
$$- a_{ij} \varepsilon_{ij} - b_{ij} \gamma_{ij} - c_{ijk} \chi_{ijk} - \xi \phi - d_i \phi_{,i} + m \theta,$$

$$a \dot{\theta} = \frac{1}{\varrho T_0} (k_{ij} \theta_{,j})_{,i} + \frac{1}{T_0} r - \alpha_{ij} \dot{\varepsilon}_{ij} - \beta_{ij} \dot{\gamma}_{ij} - \delta_{ijk} \dot{\chi}_{ijk} - m \dot{\phi} - a_i \dot{\phi}_{,i}. \quad (15)$$

We can define the solution of our mixed initial boundary value problem in the context of the theory of thermoelasticity of initially stressed bodies with voids in the cylinder $\Omega_0 = B \times [0, t_0)$ as being the array $(u_i, \varphi_{jk}, \theta, \phi)$ that verify all equations of the system (14)–(15) for any $(x,t) \in \Omega_0$, and satisfy the initial data (12) and also the conditions to the limit (13).

3. Main Result

First of all, we start by introducing the definition of the concept of a domain of influence. Then, we will prove an inequality which will underpin the influence theorem. This inequality is a counterpart of that demonstrated in the study [5]. The main result of our work is a theorem regarding the existence of a domain of influence in the context of thermoelasticity of porous materials.

In order to obtain our results, we need to impose the following hypotheses on the properties of the bodies:

(i) $\varrho > 0$, $\kappa > 0$, $I_{ij} > 0$, $T_0 > 0$, $a > 0$;

(ii)
$$A_{ijmn}x_{ij}x_{mn} + 2G_{ijmn}x_{ij}y_{mn} + B_{ijmn}y_{ij}y_{mn} + 2F_{mnrij}x_{ij}z_{mnr} +$$
$$+ 2D_{ijmnr}y_{ij}z_{mnr} + C_{ijkmnr}z_{ijk}z_{mnr} + P_{ki}x_{jk}x_{ji} - 2Q_{ik}x_{ji}y_{jk} +$$
$$+ N_{rik}x_{ji}z_{jkr} + 2a_{ij}x_{ij}\omega + 2b_{ij}y_{ij}\omega + 2c_{ijk}z_{ijk}\omega + 2d_{ijk}x_{ij}\omega_k +$$
$$+ 2e_{ijk}y_{ij}\omega_k + 2f_{ijkm}z_{ij}\omega_m + 2d_i\omega_i\omega + \xi\omega^2 + A_{ij}\omega_i\omega_j \geq$$
$$\geq \alpha(x_{ij}x_{ij} + y_{ij}y_{ij} + z_{ijk}z_{ijk} + \omega_i\omega_i + \omega^2), \tag{16}$$

for all $x_{ij} = x_{ji}$, y_{ji}, z_{ijk}, ω_i, ω;

(iii) $k_{ij}\eta_i\eta_j \geq \gamma\eta_i\eta_i$, for all η_i.

These hypotheses are not considered as very restrictive, as they are commonly imposed in mechanics of continuum media. As an example, the hypothesis *iii* is deduced from the corollary (9) and this can be obtained from the entropy production inequality.

By analogy with the step function of Heaviside, we will consider a smooth non-decreasing function $\mathcal{U}_e(z)$ as follows:

$$\mathcal{U}_e(z) = \begin{cases} 0, & \text{if } z \in (-\infty, 0], \\ 1, & \text{if } z \in [\alpha, \infty), \end{cases}$$

for a sufficiently small $e > 0$.

We now fix two constants $R > 0$ and $t > 0$ and use $d = |x - x_0|$, in order to define, with the help of the above function $\mathcal{U}_e$, the following useful function

$$V : B \times [0, t] \to R, \quad V(x, s) = \mathcal{U}_e\left(\frac{R-d}{v} + t - s\right). \tag{17}$$

x_0 is an arbitrary point fixed in B. Here, $v > 0$ is a constant which have the amplitude of speed, which will be determined later.

Using a sphere $S(x_0, \mathcal{R})$ of the form

$$S(x_0, \mathcal{R}) = \{x \in R^3 : |x - x_0| < R\}, \tag{18}$$

we define the set $\mathcal{A}$ by

$$\mathcal{A} = \bigcup_{\alpha \in [0,t]} S[x_0, R + v(t - \alpha)].$$

It is easy to see that $V(x,s)$ is a regular function in all points of $B \times [0,t]$, which decay to zero outside set $\mathcal{A}$.

We now prove an inequality which is useful in what follows.

Proposition 1. *If the ordered array $(u_i, \varphi_{ij}, \phi, \theta)$ satisfies the system of Equations (14) and (15) and verifies Equations (12) and (13), then the following inequality takes place:*

$$\begin{aligned}
&\Big[\varrho \dot{u}_i \dot{u}_i + I_{kr}\dot{\varphi}_{jr}\dot{\varphi}_{jk} + \varrho\kappa\dot{\phi}^2 + a\dot{\theta}^2 + A_{ijmn}\varepsilon_{ij}\varepsilon_{mn} + \\
&+ 2G_{ijmn}\varepsilon_{ij}\gamma_{mn} + B_{ijmn}\gamma_{ij}\gamma_{mn} + 2F_{mnrij}\varepsilon_{ij}\chi_{mnr} + \\
&+ 2D_{ijmnr}\gamma_{ij}\chi_{mnr} + C_{ijkmnr}\chi_{ijk}\chi_{mnr} + 2a_{ij}\varepsilon_{ij}\phi + \\
&+ 2b_{ij}\gamma_{ij}\phi + 2c_{ijk}\chi_{ijk}\phi + 2d_{ijk}\varepsilon_{ij}\phi_{,k} + 2e_{ijk}\gamma_{ij}\phi_{,k} + \\
&+ 2f_{ijkm}\chi_{ijk}\phi_{,k} + 2d_i\phi\phi_{,i} + 2A_{ij}\phi_{,i}\phi_{,j} + \xi\phi^2\Big] \geq \\
&\geq \Big[\varrho \dot{u}_i \dot{u}_i + I_{kr}\dot{\varphi}_{jr}\dot{\varphi}_{jk} + \varrho\kappa\dot{\phi}^2 + a\dot{\theta}^2 + \\
&+ \varepsilon_{ij}\varepsilon_{ij} + \gamma_{ij}\gamma_{ij} + \chi_{ijk}\chi_{ijk} + \phi^2 + \phi_{,i}\phi_{,i}\Big]
\end{aligned} \quad (19)$$

for all $(x,s) \in B \times [0,t]$.

Proof. This result can be immediately deduced by taking into account the above assumptions *i* and *ii*. □

Let us define a function $P(x,s)$ by

$$\begin{aligned}
P = \frac{1}{2}\Big[&\varrho \dot{u}_i \dot{u}_i + I_{kr}\dot{\varphi}_{jr}\dot{\varphi}_{jk} + \varrho\kappa\dot{\phi}^2 + a\dot{\theta}^2 + A_{ijmn}\varepsilon_{ij}\varepsilon_{mn} + \\
&+ 2G_{ijmn}\varepsilon_{ij}\gamma_{mn} + B_{ijmn}\gamma_{ij}\gamma_{mn} + 2F_{mnrij}\varepsilon_{ij}\chi_{mnr} + \\
&+ 2D_{ijmnr}\gamma_{ij}\chi_{mnr} + C_{ijkmnr}\chi_{ijk}\chi_{mnr} + P_{ki}u_{j,k}u_{j,i} - \\
&- 2Q_{ik}u_{j,i}\varphi_{jk} + 2N_{rik}u_{j,i}\varphi_{jk,r} + 2a_{ij}\varepsilon_{ij}\phi + 2b_{ij}\gamma_{ij}\phi + \\
&+ 2c_{ijk}\chi_{ijk}\phi + 2d_{ijk}\varepsilon_{ij}\phi_{,k} + 2e_{ijk}\gamma_{ij}\phi_{,k} + \\
&+ 2f_{ijkm}\chi_{ijk}\phi_{,k} + 2d_i\phi\phi_{,i} + A_{ij}\phi_{,i}\phi_{,j} + \xi\phi^2\Big].
\end{aligned} \quad (20)$$

From definition (20), we can deduce that P, as a function of (t,s) is, in fact, the potential density energy.

In the following, we also use the function $K(x,s)$ defined by

$$\begin{aligned}
K = \frac{1}{2}\Big[&\varrho \dot{u}_i \dot{u}_i + I_{kr}\dot{\varphi}_{jr}\dot{\varphi}_{jk} + \varrho\kappa\dot{\phi}^2 + a\dot{\theta}^2 + \\
&+ \varepsilon_{ij}\varepsilon_{ij} + \gamma_{ij}\gamma_{ij} + \chi_{ijk}\chi_{ijk} + \phi^2 + \phi_{,i}\phi_{,i}\Big].
\end{aligned} \quad (21)$$

Clearly, this function is kinetic energy.

If we take into account the hypotheses *i* and *ii*, from Equations (20) and (21), we are led to the conclusion

$$P(x,\tau) \geq K(x,\tau), \quad \forall (x,\tau) \in B \times [0,t]. \quad (22)$$

In the following theorem, we will prove an inequality, which is helpful to obtain our main result.

Theorem 1. If $(u_i, \varphi_{ij}, \theta, \phi)$ is a solution of the system of Equations (14) and (15) that verifies (12) and (13), then the next inequality is satisfied for any (x, τ) in the cylinder $B \times [0, t]$:

$$\int_{D[x_0,R]} P(x,t) dv + \frac{1}{T_0} \int_0^t \int_{D[x_0,R+c(t-s)]} k_{ij}\theta_{,i}\theta_{,j} dv \leq \int_{D[x_0,R+ct]} P(x,0) dv +$$
$$+ \int_0^t \int_{D[x_0,R+c(t-s)]} \varrho \left[f_i \dot{u}_i + g_{jk}\dot{\varphi}_{jk} + l\dot{\phi} + \frac{1}{T_0} r\theta \right] dvds +$$
$$+ \int_0^t \int_{\partial D[x_0,R+c(t-s)]} \left[\bar{t}_i \dot{u}_i + \bar{\mu}_{jk}\dot{\varphi}_{jk} + \bar{h}\dot{\phi} + \frac{1}{T_0} \bar{q}\theta \right] dSds, \tag{23}$$

for any $R > 0, t > 0$ and $x_0 \in B$.

Here, $D(x_0, R) = \{\alpha \in B : |\alpha - x_0| < R\}$, $\partial D(x_0, R) = \{\alpha \in \partial B : |\alpha - x_0| < R\}$.

Proof. If we multiply the both members of Equation (14)$_1$ by $V\dot{u}_i$, we obtain

$$\frac{1}{2} V \frac{d}{dt}(\varrho \dot{u}_i \dot{u}_i) = \varrho V f_i \dot{u}_i + \left[V\left(\tau_{ij} + \eta_{ij}\right)\dot{u}_i\right]_{,j} - V_{,j}\left(\tau_{ij} + \eta_{ij}\right)\dot{u}_i -$$
$$- V \left(A_{ijmn}\varepsilon_{mn} + G_{ijmn}\gamma_{mn} + F_{ijmnr}\chi_{mn} + a_{ij}\phi + d_{ijk}\phi_{,k} - \alpha_{ij}\theta\right) \dot{u}_{i,j}. \tag{24}$$

Analogously, we multiply both sides of Equation (14)$_2$ by $V\dot{\varphi}_{jk}$, so that we get the equality

$$\frac{1}{2} V \frac{d}{dt}(I_{kr}\dot{\varphi}_{jr}\dot{\varphi}_{jk}) = \varrho V m_{jk}\dot{\varphi}_{jk} + (V\mu_{ijk}\dot{\varphi}_{jk})_{,i} - V_{,i}\mu_{ijk}\dot{\varphi}_{jk} -$$
$$- V \left(F_{ijkmn}\varepsilon_{mn} + D_{mnijk}\gamma_{mn} + C_{ijkmnr}\chi_{mnr} + c_{ijk}\phi + f_{ijkm}\phi_{,m} - \delta_{ijk}\right) \dot{\varphi}_{jk,i}. \tag{25}$$

Furthermore, multiplying both sides of (14)$_3$ by $V\dot{\phi}$, we obtain the identity

$$\frac{1}{2} V \frac{d}{dt}(\varrho\kappa\dot{\phi}^2) = \varrho V l\dot{\phi} + (Vh_i\dot{\phi})_{,i} - V_{,i}h_i\dot{\phi} -$$
$$- V(A_{ij}\phi_{,j}\dot{\phi}_{,i} + d_{mni}\varepsilon_{mn}\dot{\phi}_{,i} + e_{mni}\gamma_{mn}\dot{\phi}_{,i} + f_{mnri}\chi_{mnr}\dot{\phi}_{,i} + d_i\phi\dot{\phi}_{,i} - a_i\theta\dot{\phi}_{,i}) - \tag{26}$$
$$- V(a_{ij}\varepsilon_{ij}\dot{\phi} + b_{ij}\gamma_{ij}\dot{\phi} + c_{ijk}\chi_{ijk}\dot{\phi} + \xi\phi\dot{\phi} + d_i\phi_{,i}\dot{\phi} - m\theta\dot{\phi}).$$

At last, by multiplying both sides of (14)$_4$ by $V\theta$, we will obtain

$$\frac{1}{2} G \frac{d}{dt}(a\theta^2) = \frac{1}{T_0} Gr\theta + \frac{1}{\varrho T_0}\left[(G\theta q_i)_{,i} - G_{,i}\theta q_i\right] -$$
$$- \frac{1}{\varrho T_0} Gk_{ij}\theta_{,i}\theta_{,j} - G\left(\alpha_{ij}\theta\dot{\varepsilon}_{ij} + \beta_{ij}\theta\dot{\gamma}_{ij} + \delta_{ijk}\theta\dot{\chi}_{ij} + m\theta\dot{\phi} + a_i\theta\dot{\phi}_{,i}\right). \tag{27}$$

Now, by summing up, term by term, Equations (24)–(27), it is easy to deduce the identity

$$
\begin{aligned}
\frac{1}{2}V\frac{d}{dt}&(\varrho\dot{u}_i\dot{u}_i + I_{kr}\dot{\varphi}_{jr}\dot{\varphi}_{jk} + \varrho\kappa\dot{\phi}^2 + a\dot{\theta}^2) = \\
&= \varrho V f_i \dot{u}_i + \varrho V g_{jk}\dot{\varphi}_{jk} + \varrho V l \dot{\phi} + \frac{1}{T_0}V r \theta + \\
&+ V\left[(\tau_{ij} + \eta_{ij})\dot{u}_j + \mu_{ijk}\dot{\varphi}_{jk} + h_i\dot{\phi} + \frac{1}{\varrho T_0}\theta q_i\right]_{,i} - \\
&- V\left[A_{ijmn}\varepsilon_{mn}\dot{\varepsilon}_{ij} + G_{ijmn}\left(\varepsilon_{mn}\dot{\gamma}_{ij} + \dot{\varepsilon}_{mn}\gamma_{ij}\right) +\right. \\
&+ B_{ijmn}\gamma_{mn}\dot{\gamma}_{ij} + F_{mnrij}\left(\varepsilon_{ij}\dot{\chi}_{mnr} + \dot{\varepsilon}_{ij}\chi_{mnr}\right) + \\
&+ D_{ijmnr}\left(\gamma_{ij}\dot{\chi}_{mnr} + \dot{\gamma}_{ij}\chi_{mnr}\right) + C_{ijkmnr}\chi_{ijk}\dot{\chi}_{mnr} + \\
&+ a_{ij}\left(\dot{\varepsilon}_{ij}\phi + \varepsilon_{ij}\dot{\phi}\right) + b_{ij}\left(\dot{\gamma}_{ij}\phi + \gamma_{ij}\dot{\phi}\right) + c_{ijk}\left(\dot{\chi}_{ijk}\phi + \chi_{ijk}\dot{\phi}\right) + \\
&+ d_{ijk}\left(\varepsilon_{ij}\dot{\phi}_{,k} + \dot{\varepsilon}_{ij}\phi_{,k}\right) + e_{ijk}\left(\gamma_{ij}\dot{\phi}_{,k} + \dot{\gamma}_{ij}\phi_{,k}\right) + \\
&+ f_{ijkm}\left(\chi_{ijk}\dot{\phi}_{,m} + \dot{\chi}_{ijk}\phi_{,m}\right) + d_i\left(\dot{\phi}\phi_{,i} + \phi\dot{\phi}_{,i}\right) + \\
&\left.+ A_{ij}\phi_{,i}\dot{\phi}_{,j} + \xi\phi\dot{\phi}\right] - V_{,j}\left(\tau_{ij} + \eta_{ij}\right)\dot{u}_i - V_{,i}\mu_{ijk}\dot{\varphi}_{jk} - \\
&- V_{,i}h_i\dot{\phi} - \frac{1}{\varrho T_0}V_{,i}q_i\theta - \frac{1}{\varrho T_0}V k_{ij}\theta_{,i}\theta_{,j}.
\end{aligned}
\tag{28}
$$

It is not difficult to notice that relation (28) can be rewritten in the following equivalent form:

$$
\begin{aligned}
\frac{1}{2}G\frac{d}{dt}&\left(\varrho\dot{u}_i\dot{u}_i + I_{kr}\dot{\varphi}_{jr}\dot{\varphi}_{jk} + \varrho\kappa\dot{\phi}^2 + a\dot{\theta}^2 + \right. \\
&+ A_{ijmn}\varepsilon_{mn}\varepsilon_{ij} + 2G_{ijmn}\gamma_{mn}\varepsilon_{ij} + B_{ijmn}\gamma_{mn}\gamma_{ij} + \\
&+ 2F_{mnrij}\varepsilon_{ij}\chi_{mnr} + 2D_{ijmnr}\gamma_{ij}\chi_{mnr} + C_{ijkmnr}\chi_{ijk}\chi_{mnr} + \\
&+ 2a_{ij}\varepsilon_{ij}\phi + 2b_{ij}\gamma_{ij}\phi + 2c_{ij}\chi_{ij}\phi + 2d_{ijk}\varepsilon_{ij}\phi_{,k} + \\
&+ 2e_{ijk}\gamma_{ij}\phi_{,k} + 2f_{ijkm}\chi_{ijk}\phi_{,m} + 2d_i\phi\phi_{,i} + A_{ij}\phi_{,i}\phi_{,j} + \\
&\left.+ 2a_i\theta\phi_{,i} - 2m\theta\phi + a\theta^2 + \xi\phi^2\right) = \\
&= \varrho V\left(f_i\dot{u}_i + g_{jk}\dot{\varphi}_{jk} + l\dot{\phi} + \frac{1}{T_0}r\theta\right) + \\
&+ V\left[(\tau_{ij} + \eta_{ij})\dot{u}_i + \mu_{ijk}\dot{\varphi}_{jk} + h_j\dot{\phi} + \frac{1}{\varrho T_0}\theta q_j\right]_{,j} - \\
&- V_{,j}\left(\tau_{ij} + \eta_{ij}\right)\dot{u}_i - V_{,i}\mu_{ijk}\dot{\varphi}_{jk} - V_{,i}h_i\dot{\phi} - V_{,i}\frac{1}{\varrho T_0}\theta q_i - \frac{1}{\varrho T_0}k_{ij}\theta_{,i}\theta_{,j}.
\end{aligned}
\tag{29}
$$

Taking into account the expression of the potential energy P from (30), identity (29) can be restated in the form

$$
\begin{aligned}
\frac{1}{2}V\dot{P} &+ \frac{1}{\varrho T_0}k_{ij}\theta_{,i}\theta_{,j} = \\
&= V\left(\varrho f_i\dot{u}_i + \varrho g_{jk}\dot{\varphi}_{jk} + \varrho l\dot{\phi} + \frac{1}{T_0}\varrho r\theta\right) + \\
&+ V\left[(\tau_{ij} + \eta_{ij})\dot{u}_j + \mu_{ijk}\dot{\varphi}_{jk} + h_i\dot{\phi} + \frac{1}{\varrho T_0}\theta q_i\right]_{,i} - \\
&- V_{,i}\left[(\tau_{ij} + \eta_{ij})\dot{u}_j + \mu_{ijk}\dot{\varphi}_{jk} + h_i\dot{\phi} + \frac{1}{\varrho T_0}\theta q_i\right].
\end{aligned}
\tag{30}
$$

By integrating, over $B \times [0,t]$, both sides of identity (30) so that, by using the divergence theorem and the conditions to the limit (13), we are led to the following equality:

$$\int_B VP(x,t)dv + \frac{1}{\varrho T_0}\int_0^t \int_B Vk_{ij}\theta_{,i}\theta_{,j}dvds = \int_B VP(x,0)dv + \\
+ \int_0^t \int_{\partial B} V\left(\bar{t}_i\dot{u}_i + \bar{\mu}_{jk}\dot{\varphi}_{jk} + \bar{h}\dot{\phi} + \frac{1}{\varrho T_0}\bar{q}\theta\right)dvds + \\
+ \int_0^t \int_B \varrho V\left(f_i\dot{u}_i + g_{jk}\dot{\varphi}_{jk} + l\dot{\phi} + \frac{1}{T_0}r\theta\right)dvds + \\
+ \int_0^t \int_B \dot{V}P(x,s)dvds - \int_0^t \int_B V_{,i}\left[(\tau_{ij}+\eta_{ij})\dot{u}_j + \mu_{ijk}\dot{\varphi}_{jk} + h_i\dot{\phi} + \frac{1}{\varrho T_0}q_i\theta\right]dvds. \quad (31)$$

Now, we consider the definition (17) of the function V in order to the identity:

$$\left|-V_{,j}(\tau_{ij}+\eta_{ij})\dot{u}_i - V_{,i}\mu_{ijk}\dot{\varphi}_{jk} - V_{,i}h_i\dot{\phi} - \frac{1}{\varrho T_0}V_{,i}q_i\theta\right| = \\
= \left|\frac{1}{c}\mathcal{U}_e'\frac{x_j}{\mathbf{r}}(\tau_{ij}+\eta_{ij})\dot{u}_i + \frac{1}{c}\mathcal{U}_e'\frac{x_i}{\mathbf{r}}\mu_{ijk}\dot{\varphi}_{jk} + \frac{1}{c}\mathcal{U}_e'\frac{x_i}{\mathbf{r}}h_i\dot{\phi} + \frac{1}{c\varrho T_0}\mathcal{U}_e'\frac{x_i}{\mathbf{r}}q_i\theta\right| = \\
= \left|\frac{1}{c}\mathcal{U}_e'\frac{1}{\mathbf{r}}\left[(A_{ijmn}\varepsilon_{mn}x_j + G_{ijmn}\gamma_{mn}x_j + F_{ijmnr}\chi_{mnr}x_j + \\
+ (a_{ij}+b_{ij})\phi x_j + (d_{ijk}+e_{ijk})\phi_{,k}x_j - (\alpha_{ij}+\beta_{ij})\theta x_j\right]\dot{u}_i + \\
+ \left(F_{jkmnr}\varepsilon_{mn}x_r + D_{jkmnr}\gamma_{mn}x_r + C_{ijkmnr}\chi_{mnr}x_i\right)\dot{\varphi}_{jk} + \\
+ (D_{mni}\varepsilon_{mn}x_i + E_{mni}\gamma_{mn}x_i + A_{ij}\phi_{,j}x_i + d_i\phi x_i - a_i\theta x_i)\dot{\phi} + \frac{1}{\varrho T_0}k_{ij}\theta_{,j}\theta x_i\right]\right|, \quad (32)$$

where we used the notation:

$$\mathcal{U}_e' = \frac{d\mathcal{U}_e}{d\mathbf{r}}.$$

For the terms on the right-hand side of identity (32), we will use the arithmetic-geometric mean inequality in the form

$$ab \leq \frac{1}{2}\left(\frac{a^2}{m^2} + b^2 m^2\right) \quad (33)$$

such that if we choose some suitable parameters m, we can find v (from the definition of $\mathcal{U}_e$) to satisfy the inequality

$$\left|-V_{,j}(\tau_{ij}+\eta_{ij})\dot{u}_i - V_{,j}\mu_{ijk}\dot{\varphi}_{jk} - V_{,i}h_i\dot{\phi} - \frac{1}{T_0}V_{,i}q_i\theta\right| \leq \mathcal{U}_e' K(x,s). \quad (34)$$

In addition, we obtain the inequality

$$\int_0^t \int_B \dot{V}P(x,s)dvds - \\
- \int_0^t \int_B \left(V_{,j}(\tau_{ij}+\eta_{ij})\dot{u}_i + V_{,j}\mu_{ijk}\dot{\varphi}_{jk} + V_{,i}h_i\dot{\phi} + \frac{1}{T_0}V_{,i}q_i\theta\right)dvds \leq \\
\leq \int_0^t \int_B \mathcal{U}_e'(x,\tau)[K(x,\tau) - P(x,\tau)]dvd\tau \leq 0. \quad (35)$$

Considering inequality (35), we are led to the conclusion that identity (31) receives the following form:

$$\int_B VP(x,t)dV + \frac{1}{T_0}\int_0^t\int_B Vk_{ij}\theta_{,i}\theta_{,j}dvds \le \int_B VP(x,0)dv +$$
$$+ \int_0^t\int_B \varrho V\left(f_i\dot{u}_i + g_{jk}\dot{\varphi}_{jk} + l\dot{\phi} + \frac{1}{\varrho^2 T_0}r\theta\right)dvds +$$
$$+ \int_0^t\int_{\partial B} V\left(\bar{t}_i\dot{u}_i + \bar{\mu}_{jk}\dot{\varphi}_{jk} + \bar{h}\dot{\phi} + \frac{1}{\varrho T_0}\bar{q}\theta\right)dvds. \qquad (36)$$

Finally, if we will pass to the limit in relation (36), as $e \to 0$, then we deduce that the (boundedly) limit of the function V is the indicator function, also called the characteristic function, for the set $\mathcal{A}$, defined after (18). As an immediate consequence, we are led to the inequality (23), such that Theorem 1 is concluded. □

The previous estimations obtained in Theorem 1, Proposition 1 will be used to obtain the main theorem of our present work, which is a generalization of the domain of influence result.

Let us denote by $B(t)$ the set that contains the points x from $\bar{B}$ such that:

(1) for $x \in B$, $u_i^0 \ne 0$ or $u_i^1 \ne 0$ or $\varphi_{jk}^0 \ne 0$ or $\varphi_{jk}^1 \ne 0$ or $\phi^0 \ne 0$ or $\phi^1 \ne 0$ or $\theta^0 \ne 0$ or
$\exists \alpha \in [0,t]$ so that $f_i(x,\alpha) \ne 0$ or $m_i(x,\alpha) \ne 0$ or $l(x,\alpha) \ne 0$ or $r(x,\alpha) \ne 0$;
(2) for $x \in \partial B_1$, $\exists \alpha \in [0,t]$ so that $\bar{u}_i(x,\alpha) \ne 0$;
(3) for $x \in \partial B_1^c$, $\exists \alpha \in [0,t]$ so that $\bar{t}_i(x,\alpha) \ne 0$;
(4) for $x \in \partial B_2$, $\exists \alpha \in [0,t]$ so that $\bar{\varphi}_{jk}(x,\alpha) \ne 0$;
(5) for $x \in \partial B_2^c$, $\exists \alpha \in [0,t]$ so that $\bar{\mu}_{jk}(x,\alpha) \ne 0$;
(6) for $x \in \partial B_3$, $\exists \alpha \in [0,t]$ so that $\bar{\phi}(x,\alpha) \ne 0$;
(7) for $x \in \partial B_3^c$, $\exists \alpha \in [0,t]$ so that $\bar{h}(x,\alpha) \ne 0$;
(8) for $x \in \partial B_4$, $\exists \alpha \in [0,t]$ so that $\bar{\theta}(x,\alpha) \ne 0$;
(9) for $x \in \partial B_4^c$, $\exists \alpha \in [0,t]$ so that $\bar{q}(x,\alpha) \ne 0$.

At the instant t, the domain of influence of the data, B_t, is a set defined by

$$B_t = \{x_0 \in \bar{B} : B(t) \cap \bar{S}(x_0,vt) \ne \Phi\}, \qquad (37)$$

where Φ is the notation for the empty set and the sphere $S(x_0,vt)$ is defined in Equation (18). Now, we can prove the main result of our study.

Theorem 2. *If the array $(u_i, \varphi_{ij}, \theta, \phi)$ verifies all equations of the system of Equations (14) and (15) and satisfies the conditions (12) and (13), we obtain a characterization of the solution as follows:*

$$u_i = 0, \ \varphi_{ij} = 0, \ \theta = 0 \ \text{and} \ \phi = 0, \ \text{for} \ (x,\tau) \in \{\bar{B} \setminus B_t\} \times [0,t]. \qquad (38)$$

Proof. We will use inequality (23) considered for an arbitrary x_0, $x_0 \in \bar{B} \setminus B_t$ and $\tau \in [0,t]$, by taking the values $t = \tau$ and $R = v(t-\tau)$. Then, we obtain

$$\int_{D[x_0,v(t-\tau)]} P(x,\tau)dV + \frac{1}{T_0}\int_0^\tau \int_{D[x_0,v(t-s)]} k_{ij}\theta_{,i}\theta_{,j}dVds \le$$
$$\le \int_{D[x_0,vt)]} P(0,x)dv + \int_0^\tau \int_{D[x_0,v(t-s)]} \varrho\left(f_i\dot{u}_i + g_{jk}\dot{\varphi}_{jk} + l\dot{\phi} + \frac{1}{T_0}r\theta\right)dVds +$$
$$+ \int_0^\tau \int_{\partial D[x_0,v(t-s)]} \varrho\left(\bar{t}_i\dot{u}_i + \bar{\mu}_{jk}\dot{\varphi}_{jk} + \bar{h}\dot{\phi} + \frac{1}{T_0}\bar{q}\theta\right)dSds. \qquad (39)$$

However, $x_0 \in \bar{B} \setminus B_t$, so we can deduce that $x \in D(x_0, vt)$ and therefore $x \notin B(t)$. Thus, we are led to the conclusion that

$$\int_{D[x_0,vt]} P(0,x)dv = 0. \tag{40}$$

Taking into account that, because of $D[x_0, v(t-s)] \subseteq D(x_0, vt)$, we get

$$\int_0^\tau \int_{D[x_0,v(t-s)]} \varrho\left(f_i \dot{u}_i + g_{jk}\dot{\varphi}_{jk} + l\dot{\phi} + \frac{1}{T_0}r\theta\right) dV ds = 0 \tag{41}$$

and

$$\int_0^\tau \int_{D[x_0,v(t-s)]} \left(\bar{t}_i \dot{u}_i + \bar{\mu}_{jk}\dot{\varphi}_{jk} + \bar{h}\dot{\phi} + \frac{1}{T_0}\bar{q}\theta\right) dV ds = 0. \tag{42}$$

Now, considering the hypothesis *iii* and taking into account the relations (40)–(42), we deduce

$$\int_{D[x_0,v(t-s)]} P(s,x)dv \leq 0. \tag{43}$$

The inequality (43) together with the inequality (22) lead to the conclusion

$$\int_{D[x_0,v(t-s)]} K(x,s)dV \leq 0 \tag{44}$$

so that, considering the definition (21) of the function K, we obtain

$$\dot{u}_i(x_0,s) = 0, \ \dot{\varphi}_{jk}(x_0,s) = 0, \ \theta(x_0,s) = 0, \ \phi(x_0,s) = 0$$

for all $(x_0,s) \in \{\bar{B} \setminus B_t\} \times [0,t]$.

At last, because $u_i(x_0,0) = 0, \ \varphi_{jk}(x_0,0) = 0$ for all $x_0 \in \bar{B} \setminus B_t$, we get

$$u_i(x_0,s) = 0, \ \varphi_{jk}(x_0,s) = 0, \ \theta(x_0,s) = 0, \ \phi(x_0,s) = 0$$

for all $(x_0,s) \in \{\bar{B} \setminus B_t\} \times [0,t]$ so that the proof of Theorem 2 is complete. □

4. Conclusions

We want to emphasize that our main result from the present study is a generalization of the result regarding the domain of influence theorem from classical elasticity and this extension is made in a more complex context, one of the theory of thermoelastic body with dipolar structure and with voids. Thus, we have proven that the result regarding the domain of influence is still valid even if we are out of the framework of classical elasticity.

Namely, we need to emphasize that the validity of the domain of influence result was not affected by the fact that we considered the effect of thermal treatment, the effect of the dipolar structure and the effect of voids.

Author Contributions: All authors contributed equally to the writing of this paper. All authors read and approved the final form of the manuscript.

Funding: This research received no external funding

Acknowledgments: We want to thank the reviewers who have read the manuscript carefully and have proposed pertinent corrections that have led to an improvement in our manuscript.

Conflicts of Interest: The authors declare no conflict of interest.

References

1. Nunziato, J.W.; Cowin, S.C. A nonlinear theory of materials with voids. *Arch. Rat. Mech. Anal.* **1979**, *72*, 175–201. [CrossRef]
2. Cowin, S.C Nunziato, J.W. Linear elastic materials with voids. *J. Elast.* **1983**, *13*, 125–147. [CrossRef]
3. Goodman, M.A.; Cowin, S.C. A continuum theory of granular materia. *Arch. Rat. Mech. Anal.* **1971**, *44*, 249–266. [CrossRef]
4. Iesan, D. A theory of thermoelastic material with voids. *Acta Mech.* **1986**, *60*, 67–89. [CrossRef]
5. Marin, M.; Marinescu, C. Thermoelasticity of initially stressed bodies, asymptotic equipartition of energies. *Int. J. Eng. Sci.* **1998**, *36*, 73–86. [CrossRef]
6. Marin, M. Cesaro means in thermoelasticity of dipolar bodies. *Acta Mech.* **1997**, *122*, 155–168. [CrossRef]
7. Eringen, A.C. Theory of micromorphic materials with memory. *Int. J. Eng. Sci.* **1972**, *10* 623–641. [CrossRef]
8. Eringen, A.C. Theory of thermo-microstretch elastic solids. *Int. J. Eng. Sci.* **1990**, *28*, 1291–1301. [CrossRef]
9. Mindlin, R.D. Micro-structure in linear elasticity. *Arch. Ration. Mech. Anal.* **1964**, *16*, 51–78. [CrossRef]
10. Green, A.E. Rivlin, R.S. Multipolar continuum mechanics. *Arch. Ration. Mech. Anal.* **1964** *17*, 113–147. [CrossRef]
11. Fried, E.; Gurtin, M.E. Thermomechanics of the interface between a body and its environment. *Continuum Mech. Thermodyn.* **2007**, *19*, 253–271. [CrossRef]
12. Abbas, I.A. A GN model based upon two-temperature generalized thermoelastic theory in an unbounded medium with a spherical cavity. *Appl Math Comput.* **2014**, *245*, 108–115. [CrossRef]
13. Abbas, I.A. Eigenvalue approach for an unbounded medium with a spherical cavity based upon two-temperature generalized thermoelastic theory. *J Mech. Sci. Technol.* **2014**, *28*, 4193–4198. [CrossRef]
14. Abbas, I.A. Abo-Dahab, S.M. On the numerical solution of thermal shock problem for generalized magneto-thermoelasticity for an infinitely long annular cylinder with variable thermal conductivity. *J. Comput. Theor. Nanosci.* **2014**, *11*, 607–618. [CrossRef]
15. Othman, M.I.A. State Space Approach to Generalized Thermoelasticity Plane Waves with Two Relaxation Times under the Dependence of the Modulus of Elasticity on Reference Temperature. *Can. J. Phys.* **2003**, *81*, 1403–1418 [CrossRef]
16. Sharma, J.N.; Othman, M.I.A. Effect of Rotation on Generalized Thermo-viscoelastic Rayleigh-Lamb Waves. *Int. J. Solids Struct.* **2007**, *44*, 4243–4255. [CrossRef]
17. Othman, M.I.A. Hasona, W.M. Abd-Elaziz, E.M. Effect of Rotation on Micropolar Generalized Thermoelasticity with Two-Temperatures using a Dual-Phase-Lag Model. *Can. J. Phys.* **2014**, *92*, 149–158. [CrossRef]
18. Marin, M.; Öchsner, A. The effect of a dipolar structure on the Holder stability in Green-Naghdi thermoelasticity. *Contin. Mech. Thermodyn.* **2017**, *29*, 1365–1374. [CrossRef]
19. Hassan, M.; Marin, M.; Alsharif, A.; Ellahi, R. Convective heat transfer flow of nanofluid in a porous medium over wavy surface. *Phys. Lett. A* **2018**, *382*, 2749–2753. [CrossRef]
20. Marin, M.; Nicaise, S. Existence and stability results for thermoelastic dipolar bodies with double porosity. *Contin. Mech. Thermodyn.* **2016**, *28*, 1645–1657. [CrossRef]
21. Marin, M. Lagrange identity method for microstretch thermoelastic materials. *J. Math. Anal. Appl.* **2010**, *363*, 275–286. [CrossRef]
22. Modrea, A.; Vlase, S.; Calin, R.; Peterlicean, A. The influence of dimensional and structural shifts of the elastic constant values in cylinder fiber composites. *J. Optoelectron. Adv. Mater.* **2013**, *15*, 278–283.
23. Niculita, C.; Vlase, S.; Bencze, A.; Mihalcica, M. Optimum stacking in a multi-ply laminate used for the skin of adaptive wings. *J. Optoelectron. Adv. Mater.* **2011**, *5*, 1233–1236.
24. Carbonaro, B.; Russo, R. Energy inequalities in classical elastodynamics. *J. Elast.* **1984**, *14*, 163–174. [CrossRef]
25. Chandrasekharaiah, D.S. An uniqueness theorem in the theory of elastic. *J. Elast.* **1987**, *18*, 173–179. [CrossRef]

© 2019 by the authors. Licensee MDPI, Basel, Switzerland. This article is an open access article distributed under the terms and conditions of the Creative Commons Attribution (CC BY) license (http://creativecommons.org/licenses/by/4.0/).

MDPI
St. Alban-Anlage 66
4052 Basel
Switzerland
Tel. +41 61 683 77 34
Fax +41 61 302 89 18
www.mdpi.com

Symmetry Editorial Office
E-mail: symmetry@mdpi.com
www.mdpi.com/journal/symmetry

www.ingramcontent.com/pod-product-compliance
Lightning Source LLC
LaVergne TN
LVHW070419100526
838202LV00014B/1487